AF342006

Kelvin–Helmholtz Instability in Solar Atmospheric Jets

Kelvin–Helmholtz Instability in Solar Atmospheric Jets

Ivan Zhelyazkov
Sofia University, Bulgaria

Ramesh Chandra
Kumaun University, India

World Scientific

NEW JERSEY · LONDON · SINGAPORE · BEIJING · SHANGHAI · HONG KONG · TAIPEI · CHENNAI · TOKYO

Published by

World Scientific Publishing Co. Pte. Ltd.

5 Toh Tuck Link, Singapore 596224

USA office: 27 Warren Street, Suite 401-402, Hackensack, NJ 07601

UK office: 57 Shelton Street, Covent Garden, London WC2H 9HE

British Library Cataloguing-in-Publication Data
A catalogue record for this book is available from the British Library.

ISBN 978-981-122-374-7 (hardcover)
ISBN 978-981-122-375-4 (ebook for institutions)
ISBN 978-981-122-376-1 (ebook for individuals)

For any available supplementary material, please visit
https://www.worldscientific.com/worldscibooks/10.1142/11920#t=suppl

Typeset by Stallion Press
Email: enquiries@stallionpress.com

Preface

The study of the Sun provides us an opportunity to understand the physics of other stars. One of the lively problems in solar physics is heating of the solar corona. Currently there is a competition between two mechanisms for its heating, i.e., dissipation of energy by waves and small-scale frequent coronal magnetic reconnection. However, some studies indicate this may be a joint effect of these two possible mechanisms. Kelvin–Helmholtz instability (KHI) of propagating magnetohydrodynamic modes in solar flowing structures is considered to play an important role in the solar atmosphere. It can trigger the onset of wave turbulence leading to an effective plasma heating and particle acceleration. Kelvin–Helmholtz instability is a multifaceted phenomenon and the purpose of this book is to illuminate its (instability) manifestation in various solar jets like spicules, dark mottles, surges, macrospicules, Extreme Ultraviolet (EUV) and X-ray jets, as well as rotating, tornado-like, jets, solar wind, and coronal mass ejections. The modeling of KHI is performed in the framework of ideal magnetohydrodynamics (and also in Hall magnetohydrodynamics for the solar wind). The book consists of 12 chapters and is intended primarily for advanced undergraduate and postgraduate students, early carrier researchers. Our one-dimensional approach cannot compete, with respect to comprehensiveness of jets' topology, with the sophisticated 2.5- and 3D numerical codes, but nonetheless yields reasonable instability characteristics in good agreement with observations. Moreover, our approach has a key advantage: every plot in this book can be recovered by the reader because, after all, the problem comes down to finding the solutions to the derived in closed form wave dispersion relations in complex variables. That task, in general, is not easy one, but nevertheless solvable. We would like to encourage the reader to make modeling of KHI in any jet-like eruption studied in the current literature, or better, to take its basic parameters directly from the data set of the Atmospheric Imaging Assembly on board *Solar Dynamics Observatory (SDO)* or from the *Interface Region Imaging Spectrograph (IRIS)*.

Acknowlegdments

We will be grateful to readers to notify us of typographical or other errors they find: izh@phys.uni-sofia.bg or rchandra.ntl@gmail.com.

We are indebted to Professors Teimuraz Zaqarashvili, Leon Ofman, and Abhishek Srivastava for their constructive criticism and useful comments during the preparation of many articles on this subject. We wish also to thank NASA for using snapshots of various solar atmospheric events as illustrations in the book, and Dr. Snezhana Yordanova for plotting a couple of figures.

I. Zhelyazkov, R. Chandra

Contents

Chapter 1

The Sun: General Introduction

The Sun is our nearest star of average size. Its position in Hertzsprung–Russel (HR) diagram is in main sequence having G2V spectral classification. The HR diagram is proposed independently by Ejnar Hertzsprung of Denmark and Henry Norris Russell in 1912 using observational comparisons and acts like a periodic table in chemistry. This diagram tells about the distribution of stars and their physical parameters such as: temperature, magnitude, spectral class and luminosity. The apparent magnitude of the Sun is -26.74, while its absolute magnitude is 4.83 (the apparent magnitude of star when it is situated at a distance of ten parsec).

Due to its proximity to Earth, solar eruptions directly affect us. Solar eruptions are due to magnetic reconnection and the eruptions can also drive the shocks. These two phenomena can generate solar energetic particles like: protons, electrons, etc. These particles can affect or damage our satellites. When the energy of these particles is in the range of GeV, they are termed as Ground-Level Enhancements (GLEs), and it directly affects Earth's surface. In addition to this, the solar geomagnetic storm can also disturb our Earth magnetic field. In a nutshell, the study of solar activity and its prediction is very crucial for humanity. Its closest distance from Earth in comparison to other stars provides an opportunity to observe its surface features with better spatial resolution. Therefore, it serves as a natural laboratory to understand the physics of other distinctly situated stars. Together with this, it also provides an excellent laboratory to test the magnetohydrodynamics (MHD) simulations. The basic physical parameters of the Sun are given in Table 1.1.

With this little introduction, now we will discuss here the structure of the Sun. Primarily, we can divide it into two major parts namely: internal and external. The internal atmosphere consists of the core, radiative zone, and the convective zone. The external atmosphere includes the photosphere, chromosphere, and the corona

1

Table 1.1. Basic physical parameters of the Sun.

Age	4.6×10^9 years
Radius	6.95×10^5 km
Mass	2×10^{30} kg
Escape velocity at surface	$620 \, \mathrm{km\,s^{-1}}$
Core temperature	1.5×10^7 K
Surface temperature	6600 K
Absolute magnitude	$+4.83$
Apparent magnitude	-26.74
Spectral class	G2V
Average distance from earth	1 AU (1.496×10^8 km)

(Stix, 2002; Aschwanden, 2004; Priest, 2014). The cartoon of the different layers containing internal and external structure is displayed in Fig. 1.1. The details about these two structures are presented in the next sections and subsections.

1.1 Internal structure

The internal structure of the Sun cannot be directly observed. Theoretical models are used to investigate it. However, to test the theory, the observational evidences are needed. Therefore, for the observations, people use a special technique. This technique is known as Helioseismology. Helioseismology is the method which is based on the oscillations. These oscillations are principally caused by sound waves.

The core, radiation zone and the convective zone, three internal part of the solar atmosphere, are summarized as follows.

1.1.1 *Core*

Core is the innermost part of the Sun. It is expanded up to the fourth fraction of the solar radius. Solar radius is equal to 696 Mm and it is defined as the distance between the center of the core and the visible surface of the Sun, i.e., photosphere. The core of the Sun has high temperature and density. The average values of these are 1.5×10^7 K and $150 \, \mathrm{g\,cm^{-3}}$, respectively. These physical conditions at the solar surface are favorable conditions for the nuclear reactions. As a result of this, the nuclear reactions start in the core of the Sun and power the Sun. The evidence of the nuclear reaction at the solar core is the observational detections of neutrinos.

There are two possible mechanisms of the nuclear energy generation in the Sun. They are named as proton–proton (p–p) and CNO (carbon–nitrogen–oxygen) cycles. The p–p is a chain of reactions leading to the formation of helium nuclei

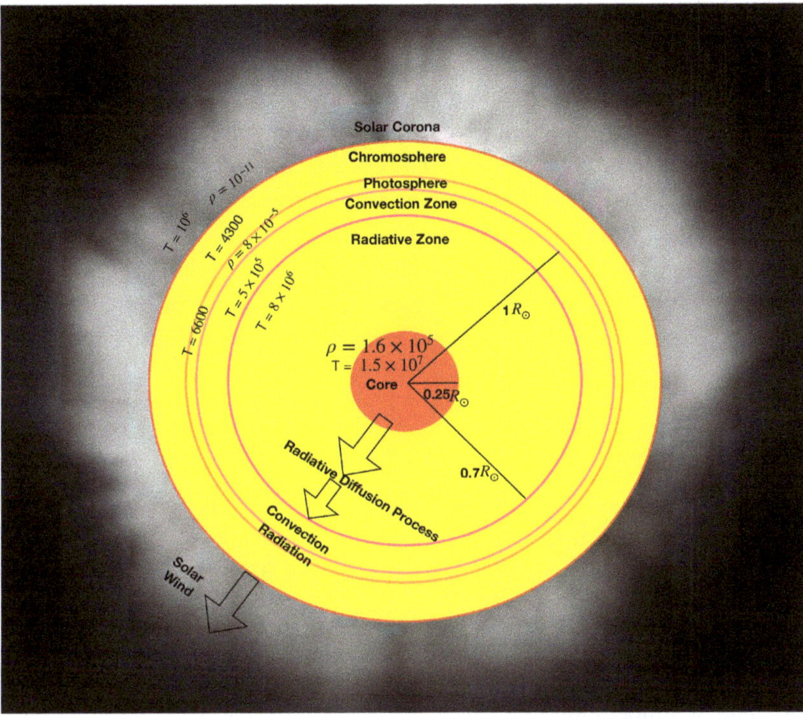

Fig. 1.1. Schematic diagram indicating different layers of the solar atmosphere. The density, and the temperature in the diagram are in the units of $\mathrm{kg\,m^{-3}}$ and kelvins, respectively.

from the conversion of hydrogen. In the CNO cycle, four protons fuse, using carbon, nitrogen, and oxygen isotopes as catalysts, to produce one alpha particle, two positrons and two electron neutrinos. In the p–p cycle, the nuclear burning starts with the collision of two protons, which produce a positron, a deuterium and a neutrino:

$$\mathrm{H}^1 + \mathrm{H}^1 \longrightarrow \mathrm{e}^+ + \mathrm{D}^2 + \nu.$$

In the next step, a proton collides with the deuteron and forms a He^3 nucleus with a gamma ray:

$$\mathrm{H}^1 + \mathrm{D}^2 \longrightarrow \mathrm{He}^3 + \gamma.$$

In the final step, two He^3 nucleus produce a He^4 nucleus and liberate two protons:

$$\mathrm{He}^3 + \mathrm{He}^3 \longrightarrow \mathrm{He}^4 + \mathrm{H}^1 + \mathrm{H}^1.$$

1.1.2 *Radiative zone*

After the energy generation in the core of the Sun, it starts to transport through the radiation process. The radiation zone ranges from 0.25 to 0.70 R_{Sun}. In the radiative zone, the solar material is hot and dense enough that thermal radiation transfers the heat of the core outward. The nuclear reaction produces a tremendous amount of gamma radiation. These photons of energy are emitted, absorbed, and then emitted again by various particles in the radiative zone. The path that photons take is called the "random walk". Instead of going in a straight beam of light, they travel in a zigzag direction, eventually reaching the surface of the Sun. Due to this, a single photon can take a 200 000 years to make the journey through the radiative zone of the Sun.

1.1.3 *Convective zone*

After the radiative zone, when the temperature gradient becomes very steep, convective energy transfer mechanism becomes "more efficient". In the convection process, the gas in the solar interior moves up and down. In this phenomenon, the hot gas blobs move up and cold gas blobs move down. The criteria for the convection is given by the famous Schwarzschild stability condition proposed by Schwarzschild in 1906. The Schwarzschild criterion is defined as follows: The lesser value of temperature gradient than the critical value of

$$\left|\frac{\mathrm{d}T}{\mathrm{d}r}\right| < \left(1 - \frac{1}{\gamma}\right)\frac{T}{p}\left|\frac{\mathrm{d}p}{\mathrm{d}r}\right|$$

makes the atmosphere unstable for convection. Here, T is temperature in the atmosphere, $\frac{\mathrm{d}p}{\mathrm{d}r}$ is the pressure gradient, γ is ratio of specific heats at constant pressure and volume. The bottom of the convective zone is located at about 0.7 R_{Sun} from the center of the Sun. Whereas the top of the convection zone is at the base of the photosphere. The results of the convection phenomenon can be observed as the observations of convective cells. These cells are known as granulations. They appear all over the photosphere and can be observed clearly in G-band observations.

1.2 External structure

After the convection zone, the upper layers can be directly observed by telescopes equipped with backend Charge Coupled Camera (CCD). These layers collective named as the external structure. These different external structures are described in the next three subsections.

1.2.1 *Photosphere*

Above the convective zone, there is a visible surface. This visible surface of solar atmosphere is named as solar photosphere. The existence of photosphere is due to the drop in the opacity of gas in this region. The gas is called opaque when the propagating photons can only travel small distance before they deflect. The optical depth along a path between t_0 and t is defined as

$$\tau_\nu = \int_{t_0}^{t} \alpha_\nu(t)\mathrm{d}t, \tag{1.1}$$

where α_ν is defined as absorption coefficient. The medium is said to be optically thick or opaque, when the value of α_ν is much greater than unity.

The most prominent features that observe on solar photosphere are the sunspots. Sunspots appear as dark spots on the surface of the Sun. They are cooler than the surrounding atmosphere. Their temperature is about 4000 K (compared to 5700 K for the surrounding photosphere). Their diameter can be up to 50 Mm. They typically last for several days. However, the large ones can live for several weeks. Sunspots are magnetic regions on the Sun with strong magnetic field (up to 4000 G). Sunspots are usually observed in groups with two sets of spots. One set will have positive while the other set will have negative magnetic field. The field is strongest in the darker parts of the sunspots—the umbra. The field is weaker and more horizontal in the lighter part—the penumbra.

In addition to sunspots, the features observed on the photosphere are faculae, granules and supergranules.

The solar photosphere with various features are presented in Fig. 1.2.

1.2.2 *Chromosphere*

As we move outwards from the solar photosphere, a thin layer of solar atmosphere is observed. The thickness of this layer is ≈ 2500 km and it is named as solar chromosphere. The density of chromosphere is low. It is about 10^{-4} times of the photospheric density. Due to its lower density in comparison to photosphere, it is not visible. It is only visible when we observe the solar eclipse. At the base of the chromosphere, the temperature is about 4500 K, and as we move upwards the temperature starts to increase and it becomes $10\,000$ K at the top of the solar chromosphere. The chromosphere is describe as a irregular or jagged, as it does not have a smooth boundary. The chromospheric structures are greatly observed by monochromatic imaging in strong chromospheric lines such as: Hα and CaII and H and K. Different features that are visible in the chromosphere are displayed in Fig. 1.3. The chromospheric features observed on the chromosphere are given

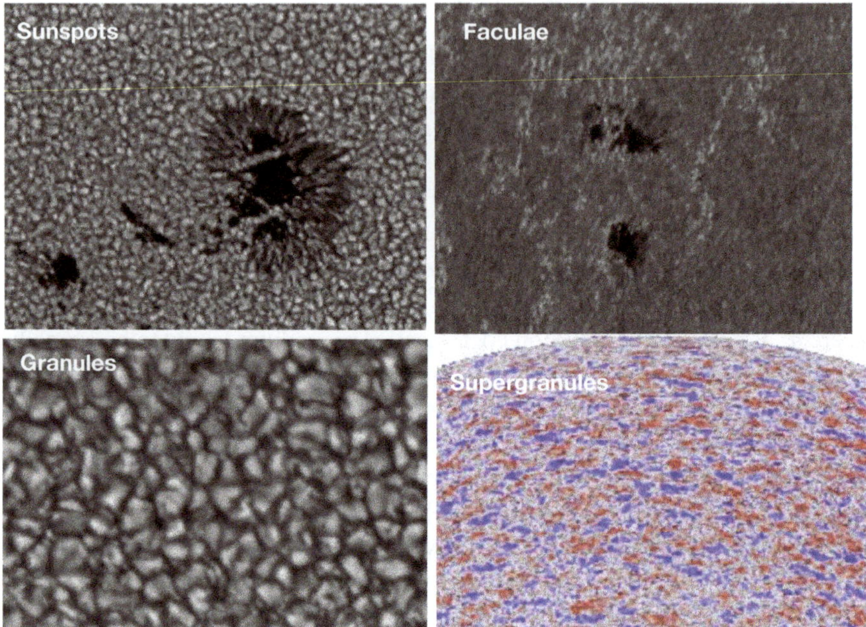

Fig. 1.2. Different quiet (granulations and supergranulations) and active features (sunspots) visible on the solar photosphere.

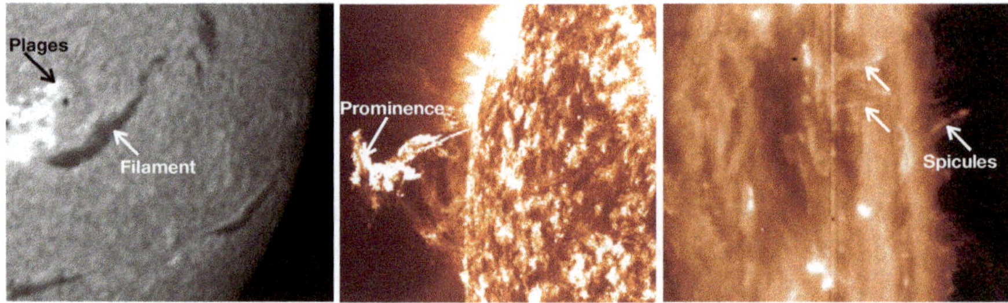

Fig. 1.3. Different chromospheric features: Solar filament (prominence), plages and spicules.

as follows:

1. Sunspots: As discussed in photospheric subsection, they are the regions of high magnetic field and also observed in the solar chromosphere.
2. Filaments/Prominences: Filaments (known as prominences, when appear on the solar limb) are intriguing structures on the solar atmosphere. The

observations suggest that these features are dense and cool chromospheric material suspended in the million degree hot corona at height up to 100 Mm above the chromosphere. They are found along the polarity inversion line.

3. Spicules: The elongated, tilted structures known as spicules are most evident and visible at the solar limb. The height of the largest spicules is about 10–15 arcsec above the limb. There thickness is about 1–2 arcsec.

4. Plages: Plages are originated from French word and easily observed near sunspots at any position on the disk.

1.2.3 *Corona*

After the chromosphere, the solar corona starts. The chromosphere is connected to corona through the transition region (TR). In this region, the temperature increased from 10^4 to 10^6 K. The solar corona is highly inhomogeneous. The solar corona extends from the visible disk of the Sun outward, eventually enveloping the Earth. The temperature in the corona reached up to million degree Kelvin while the density is very small ($\sim 10^{-11}$ kg m^{-3}). Due to its low density, it is visible only during the solar eclipse. It primarily emits in the X-ray and the ultra-violet radiation of the electromagnetic spectrum. Figure 1.4 presents the image of solar corona during solar eclipse in X-ray wavebands.

The corona has small magnetic diffusivity (~ 1 m^2 s^{-1}) and large length scale (loop height $\sim 10^7$ m). Because of this property, the coronal plasma satisfy the condition of flux-freezing. Due to the flux-freezing property, the magnetic field plays an important role for the coronal physics. An important parameter, the plasma β, is low in the corona. The plasma β parameter is defined as the ratio of the gas pressure to the magnetic pressure. After the computations, plasma β comes less than unity, which suggests that the magnetic pressure dominates over the gas pressure (Gary, 2001).

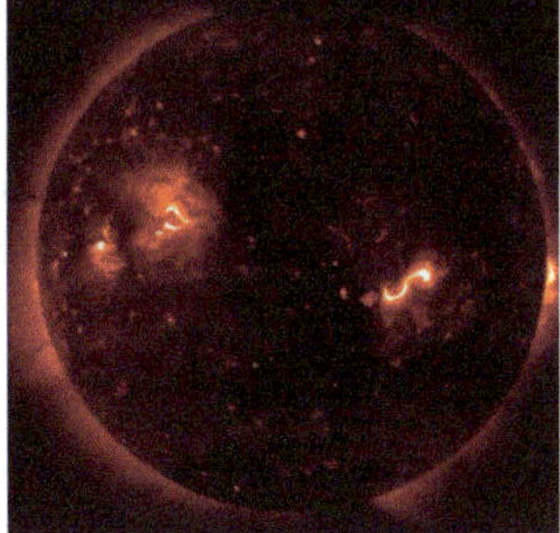

Fig. 1.4. Solar corona observed during solar eclipse and in X-ray. (*Courtesy: Hinode home page*).

Moreover, the magnetic filed in the corona is very small (varies from few gauss to 40 gauss). Therefore, its direct measurement is not possible with the help of Zeeman effect or other methods. Hence, the coronal magnetic field is measured by the extrapolation of the photospheric or chromospheric magnetic fields. Moreover, NASA latest Solar Probe mission will provide us *in-situ* measurement of the coronal magnetic field.

Solar coronal heating is one of the important unsolved problem in the area of solar physics. Different mechanisms have been proposed to explain it. The main mechanisms for this phenomena are the magnetic reconnection and the wave mechanism. According to magnetic reconnection, in the tedious corona, several micro- and nano-flares are ongoing due to small magnetic reconnection and the energy released during this process is responsible for this heating. Another popular mechanism is the wave mechanism. According to this, in the solar atmosphere, several wave phenomena are observed and when these waves propagate from photosphere to corona, they dissipate their energy and this released energy can be responsible for the heating. However, none of the mechanisms is responsible alone to explain the solar coronal heating problem. Therefore, now people think the combination of these two possible mechanisms can be responsible for the corona heating. To solve this problem, we further need the high spatial and spectral resolution observations of solar corona and other solar layers.

1.3   Quiet and active Sun

The Sun can be divided into two parts, namely quiet and active. The Sun without activity on the surface is called the quiet Sun. The quiet solar features do not change in short interval of time and survive for a longer time. The examples of the quiet features are granulations (giant bubbles at top of convection zone and visible on the photosphere), super granulations, spicules (cool high density material in chromosphere), quiet prominences/filaments (the cool material suspended in the corona), coronal holes (regions of open magnetic flux and the source of solar wind). However, the active Sun is defined as the properties changing in a short interval of time. The main active solar features are: sunspots, flares, active filaments (or prominences), coronal mass ejections (CMEs), etc. In the following subsections, we shall discuss these features in more detail.

1.3.1 *Prominences/Filaments*

Prominences (known as filaments when observed as a dark feature against the bright solar disk) are relatively cool and dense material, starting from chromosphere and reaching up to the height of several hundreds of kilometers in the hotter solar corona

(Hirayama, 1985; Tandberg-Hassen, 1995; Mackay *et al.*, 2010). Various classification of filaments/prominences have been made based on morphology, location and the magnetic properties (Tandberg-Hassen, 1995; Parenti, 2014). According to their observed properties, filaments can be broadly divided into two categories, i.e., active and quiet filaments. The active filaments are located in the active regions, whereas the quiet filaments are found away from the active region and they are found in all latitudes of the Sun. The lifetime of active filaments is small (few hours to days), whereas the quiet filaments can stay for longer time (weeks to months) on the solar surface. As mentioned above, the quiet filaments have more lifetime in comparison to active filaments.

Occasionally, the filaments/prominences become unstable, which subsequently leads to eruption. The causes of the instabilities of filaments can be emerging of magnetic fluxes, photospheric motion of magnetic polarities, and flux cancellation close to the filament locations. However, the cause of the instabilities is still not clear. In most of the cases, the initial rise phase is quiet slow (speed ~ 1–$15\,\mathrm{km\,s^{-1}}$) and it can last a few hours, while this phase can be much shorter (10 min) for active region structures (Sterling *et al.*, 2004; Williams *et al.*, 2005). The initial phase is followed by an accelerated eruption phase in which the filament may reach a speed of about 100–$1000\,\mathrm{km\,s^{-1}}$, together with the rest of the expelled coronal material to form the coronal mass ejections (CMEs) (Schrijver *et al.*, 2008; Gosain *et al.*, 2009). In the final stage, the structure expands at a nearly constant speed or may decelerate.

1.3.2 *Solar flares*

Solar flare is a sudden flash of light by release of enormous amount of energy in the order of 10^{26}–10^{32} ergs in a very short time of 100 to 1000 seconds. They emit over a wide range of radiation extending from radio, visible, ultraviolet (UV), X-rays, γ-rays. A single solar flare can create an explosion equivalent to several billions of hydrogen bombs each of 100 megaton TNT destructive power, exploded simultaneously.

Solar flares mostly occur in active regions, i.e., in the vicinity of complex sunspot groups. Large-scale highly sheared magnetic fields in active regions store huge amounts of free magnetic energy. This energy is unleashed above the photosphere, either in chromosphere or in corona. According to magnetic topology, flares are of two kinds: confined and eruptive events. The confined or impulsive flares are generally observed as abrupt brightening in the low-lying coronal loops, during which the energy release takes place in a compact region without eruptive activity. The second category comprises the long duration events (LDEs), which are eruptive in nature.

Solar flares are classified as A, B, C, M, and X depending on their peak 1–8 Å. An X-ray flux near Earth, as measured on the *GOES* spacecraft. Each flare class has a peak flux ten times greater than the preceding one. The typical values of flux for A, B, C, M, and X are 10^{-8}–10^{-7}, 10^{-7}–10^{-6}, 10^{-6}–10^{-5}, 10^{-5}–10^{-4}, and $> 10^{-4}$ W m^{-2}, respectively. Within each class, there is a scale from 1 to 9, such that the flux in an X3 flare is 50% greater than an X2 flare.

Magnetic reconnection has been recognized as the most fundamental process responsible for the solar flares. This process is responsible for the rapid conversion of stored magnetic energy into thermal and kinetic energy of plasma and particles during solar eruptive events (Priest and Forbes, 2002).

However, the physical mechanisms responsible for energy build-up and release processes in solar flares are not yet completely understood. Multi-wavelength and multi-viewpoint observations are very crucial to probe the underlying physical processes occurring at different layers and regions at and above the photosphere during a solar flare.

An example of M-class solar flare is presented in Fig. 1.5.

1.3.3 *Solar jets*

Solar jets are defined as the small scale plasma ejections from the solar surface to coronal heights (see, for example, Roy, 1973a; Schmieder *et al.*, 1988; Shibata *et al.*, 1992; Uddin *et al.*, 2012; Sterling *et al.*, 2015, and references cited therein). Based on the morphological description of coronal jets using the *Hinode*/XRT data, Moore *et al.* (2010) divided the solar jets into two categories, named as "standard" and "blowout" jets. Standard jets have a narrow spire with a relative dim base, whereas the blowout jets reveal an initial phase quite similar to standard jets, but with a bright base and a narrow spire. This is afterwards followed by a violent flux rope eruption and then consequent broadening of the spire (Moore *et al.*, 2013).

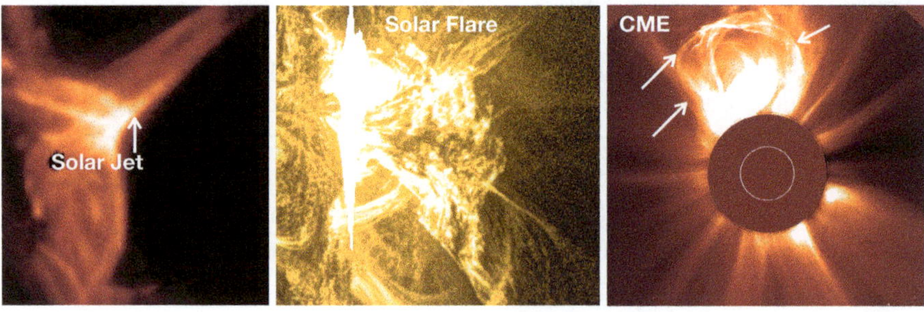

Fig. 1.5. Solar eruptive features: solar flares, jet, and CME.

The origin, classification and basic physical parameters of solar jets are describe in detail in Chapter 2.

1.3.4 *Coronal mass ejections*

When the solar eruption leaves solar corona, it is named as coronal mass ejection (CME). CMEs are huge bubbles of gas threaded with magnetic field lines that are ejected from the Sun over the course of several hours. A typical CME has a mass around 10^{11}–10^{12} kg and traverses outward from the Sun with a speed and traverses outward from the Sun with a speed between few hundreds and few thousand $km\,s^{-1}$. CMEs are observed by the coronagraph instruments. Coronagraph is an instrument, where the artificial eclipse is produced. It measure mainly photospheric photons scattered by free electrons in the coronal plasma (Thomson scattering), yielding the integrated density along the line-of-sight, providing a white-light image.

The CMEs were discovered by coronagraph on board *7th Orbiting Solar Observatory* (*OSO* 7) mission in the beginning of 1970s. Later on till present the Large Angle and Spectrometric Coronagraph (LASCO) on board the *Solar and Heliospheric Observatory* (*SOHO*) (Brueckner *et al.*, 1995) is providing the observations of CMEs continuously. We have also multi-view point observations of CMEs from the Sun Earth Connection Coronal and Heliospheric Investigation (SECCHI) on board the *Solar Terrestrial Relations Observatory* (*STEREO*) that includes the inner coronagraph COR1, the outer coronagraph COR2, the two heliospheric imagers covering the entire space (Howard *et al.*, 2008). An example of CME observed by LASCO coronagraph is displayed in Fig. 1.5.

1.4 Solar cycle

It is observed that the solar active features shows periodic behavior of 11-year period. This 11-year periodic behavior is termed as the name *"Solar Cycle"*. The solar cycle was discovered in 1843 by Schwabe (1843) using the 17 years of observations of average number of sunspots. The state of solar cycle is measured by counting the number of sunspots (f) and sunspot groups (g) visible on the solar disk by using an empirical relation given by Wolf (1848), which is as follows:

$$R = k(10g + f), \tag{1.2}$$

where f is the number of individual spots, g is the number of sunspot groups and k is a factor that varies with location and instrumentation.

During the solar cycle, the magnetic polarity of the global solar magnetic field is reversed. Every 22 year the original magnetic configuration is restored. This

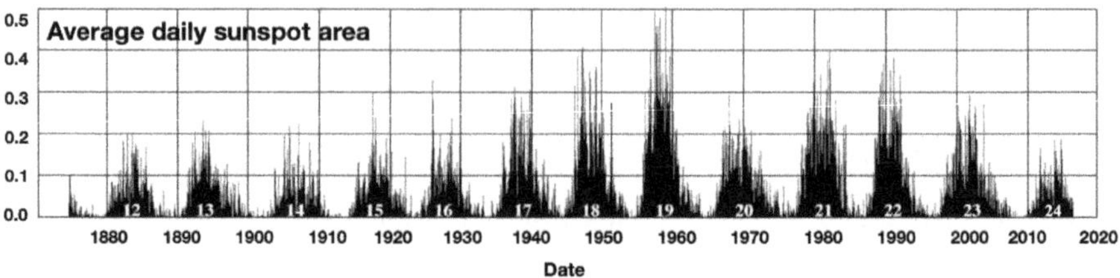

Fig. 1.6. Solar Cycle: Evolution of sunspot number with time.

22 year cycle is called a *Hale cycle*. The variation of sunspot number demonstrating the Solar Cycle is displayed in Fig. 1.6.

It is widely accepted that the filament eruption, solar flares and CMEs are different observational manifestations of same physical phenomena and these are commonly known as "Solar Eruptions" (Forbes, 2000). These all three phenomena will occur or not depend on the pre-eruption magnetic field configuration of the solar surface.

The filament eruption can be associated with solar flares, CMEs, etc. However, there are cases where the filament eruption is not associated with solar flares. The cases where the filament cannot leave the solar surface is called failed eruption (Török and Kliem, 2005).

1.5 Solar eruption mechanisms

Because the aim of this book in not to discuss the solar eruptions in details, we wish to introduce an available eruptive model, which is presented briefly below.

To explain the filament eruption and associated phenomenon, the so called *CSHKP* model was proposed by Carmichael (1964), Sturrock (1966), Hirayama (1974), and Kopp and Pneuman (1976). Afterwards, it is refined in several studies. Very recently, the *CSHKP* model is extended to 3D by Aulanier *et al.* (2012, 2013) and Janvier *et al.* (2013). This model explained the several observational features of the eruption such as: flare ribbon formation and separation, moving of flare kernels. However, there 2 and 3D models are not able to address the initiation mechanisms of the eruption.

For the triggering of solar eruptions, different models have been proposed. These models include the *tether-cutting* (Moore *et al.*, 2001), *magnetic breakout* (Antiochos, 1998; Antiochos *et al.*, 1999), and *kink instability* (Fan and Gibson, 2004; Török and Kliem, 2005). In the *tether-cutting* model, the reconnection occurs below the erupting filament. For the magnetic *breakout* model, the reconnection occurs high in the corona at magnetic null-point. This model requires a multipolar magnetic flux system. The *kink instability* occurs, when the twist reaches at a critical value.

However, it is believed that all the above-mentioned models bring the flux rope into unstable state and efficient mechanism for the solar eruption is the loss of equilibrium, i.e., *torus instability* (Forbes and Isenberg, 1991; Kliem and Török, 2006; Aulanier *et al.*, 2010; Schmieder *et al.*, 2013).

Chapter 2

Solar Jets: Origin, Classification and Basic Physical Parameters

Solar jets are collimated hot plasma flows along vertical or oblique magnetic field lines. They are observed in the solar atmosphere from the photosphere to the outer corona and represent important manifestations of ubiquitous solar transients, which may be the source of significant mass and energy input to the upper solar atmosphere and the solar wind (Shibata *et al.*, 1992). Observational studies over the years have suggested that most solar jets are formed due to the magnetic reconnection between emerging fluxes and their ambient open magnetic field lines (Panesar *et al.*, 2016). However, sometimes magnetic flux cancellation is also important to the generation of solar jets (Joshi *et al.*, 2017). Recently, there was reported that large-scale coronal waves can also lead to coronal jets by disturbing the coronal magnetic field at the boundary of coronal holes (Narang *et al.*, 2016). Since the occurrence rate of jets or jet-like activities is very high in the solar atmosphere, solar physicists believe in that solar jets could be a possible candidate source for heating the coronal plasma and accelerating the fast solar wind (Howson *et al.*, 2017). There are three main classification methods for solar jets. Firstly, solar jets can be divided into spicules (Shetye *et al.*, 2016; Martínez-Sykora *et al.*, 2017) and surges (Hα), extreme ultraviolet jets (EUV) (Kubo *et al.*, 2016), and X-ray jets (X-ray) (Shibata *et al.*, 1994) according to different observing wavelengths. Secondly, according to their different driving mechanisms and morphology, solar jets are classified into the so-called *anemone jet* and *two-sided loop jet*. Thirdly, on the base of a statistical analysis it is established that there is a dichotomy of coronal jets according to their different eruption characteristics, which implies that one can talk about "standard jets" and "blowout jets" (Moore *et al.*, 2010). The physical parameters of various solar jets, notably electron number densities, temperatures, jets' widths, heights, lifetimes, and magnetic fields vary in very wide ranges. The aim of this chapter is to introduce the reader to the accepted terminology, the methods of jets' observations, and most importantly to their (jets) physical parameters.

Solar jet are observed in various locations on solar surface. These locations are quiet regions, active regions and coronal holes. They are detected in all over the solar cycles. However, more active region jets are during the peak phase of solar cycle, due to appearance of more solar active regions. First they are observed in X-rays and later on registered in EUV (Wang *et al.*, 1998; Alexander and Fletcher, 1999; Innes *et al.*, 2011; Sterling *et al.*, 2015; Chandra *et al.*, 2015; Joshi *et al.*, 2017) wavebands. The cooler part of jets are observed in Hα and termed as solar surges. An example of jet observed in various EUV wavelengths is displayed in Fig. 2.1.

As far as the origin of solar jets, it is believed and evident that the magnetic reconnection is their trigger mechanism (Heyvaerts *et al.*, 1997). Many models discussed about the the existence of a null point in the coronal magnetic field configuration that gives rise to the jet. In the coronal null, there is a possibility of build-up of thin, strong current sheets where reconnection can occur in an explosive manner (Antiochos, 1990). Such a region can formed whenever a bipole of one polarity is embedded within a larger-scale domain of the opposite polarity. This

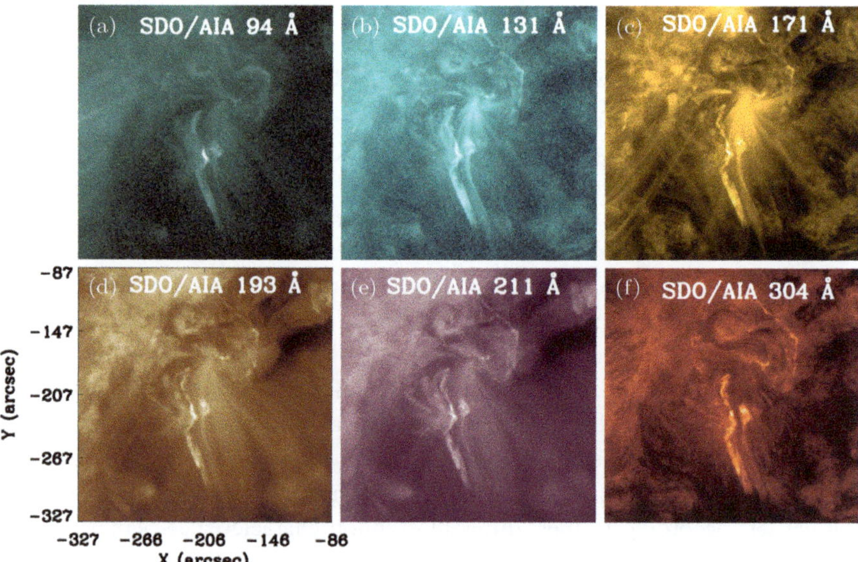

Fig. 2.1. An example of standard jet on 16 April 2014, observed at different AIA wavelengths. *Credit to: AIA/SDO NASA.*

background flux consists of magnetic field lines either open or a larger-scale loop. The evidence of null topology before the jet has been inferred from the "Eiffel tower" shape in X-ray or EUV wavelengths (Shibata *et al.*, 1992), circular flare ribbons observed in the chromosphere at the base of some jets (Wang and Liu, 2012). This is also confirmed from magnetic field extrapolations (Moreno-Insertis *et al.*, 2008; Schmieder *et al.*, 2013).

Three main scenarios have been observed for the trigger of coronal jets namely magnetic flux emergence, magnetic flux cancellation and the loss of equilibrium. In the magnetic flux emergence process, the newly emerged magnetic flux interact with ambient magnetic flux and as a result of this interaction, the jet phenomena is observed. Using the X-ray observations of *Yohkoh* satellite, Shibata *et al.* (1992) found that the small loop system was appeared close to the ambient large loop system at the jet's locations. Their study suggested that the reconnection starts between these two flux loop system. This physical mechanism is displayed in Fig. 2.2.

Magnetic flux cancellation has also been observed at the jet eruption site (Adams *et al.*, 2014; Young and Muglach, 2014a). Further, Sterling *et al.* (2015) proposed (specially in case of coronal hole jets) that the solar jets can be triggered by the small-scale filament eruptions called as mini-filament. In their data sample, they have used the data of 20 solar jets from AIA telescope on board *SDO* satellite. They proposed that the mini-filament erupts due to the magnetic flux cancellation near the magnetic neutral line and below the mini-filament. Later on mini-filament eruption as a trigger mechanism for the solar jets has been confirmed in more observations (see, for example, Panesar *et al.*, 2016, 2017). The physical scenario proposed by Sterling *et al.* (2015) is depicted in Fig. 2.3. In the loss of equilibrium, the stressed, non-potential, closed flux beneath the null point begins to reconnect with the ambient, quasi-potential flux exterior to the fan surface.

Fig. 2.2. Possible physical picture proposed by Shibata *et al.* (1992). The left, middle and the right panel shows the pre-jet, jet and post-jet phases (adapted from Shibata *et al.*, 1992).

Based on this observations, to explain the trigger mechanism of solar coronal jets, first dynamical model in two dimensions (2D) was proposed by Yokoyama and Shibata (1995, 1996) and Nishizuka *et al.* (2008). In this 2D model, the magnetic reconnection was driven by magnetic flux emergence. In their simulation two configurations were considered, one is of anemone type and another is of two-sided type. These simulations can reproduce the cool as well as the hot jets.

3D numerical models for the coronal jets were done by Moreno-Insertis *et al.* (2008). They studied in details the coronal consequences of the emergence of magnetized plasma from below the photosphere. One can identify the bright and hot plasma apparent in the observations at the base of the jet with the null point and current sheet structures resulting in those simulations (see the scheme in Fig. 11, right panel): the collision of the emerging magnetized plasma with the pre-existing coronal magnetic system leads, when the mutual orientation of the magnetic field is sufficiently different, to the formation of an elongated current sheet harboring a null point and to reconnection. As a next step in the pattern identification, the hot plasma loops apparent in the southern vault in the AIA 193 Å image and the temperature panels of their Fig. 2.6 should correspond to the hot post-reconnection loop system in the numerical models (as apparent in Figs. 2.3 and 2.4 of the paper by (Moreno-Insertis *et al.*, 2008), or along the paper by Moreno-Insertis and Galsgaard (2013).

Pariat *et al.* (2009) did 3D numerical simulations for the jets imposing the horizontal photospheric motion of polarities. In the above said studies, an open magnetic field was present in the background corona. In addition to that the high-lying coronal magnetic field was observed instead of open field (see, for example, Uddin *et al.*, 2012). Both in 2D and 3D models, the solar jet is a direct consequence of the magnetic reconnection, which takes place in a current sheet at the interface between the perturbed and pre-existing magnetic domains. Further

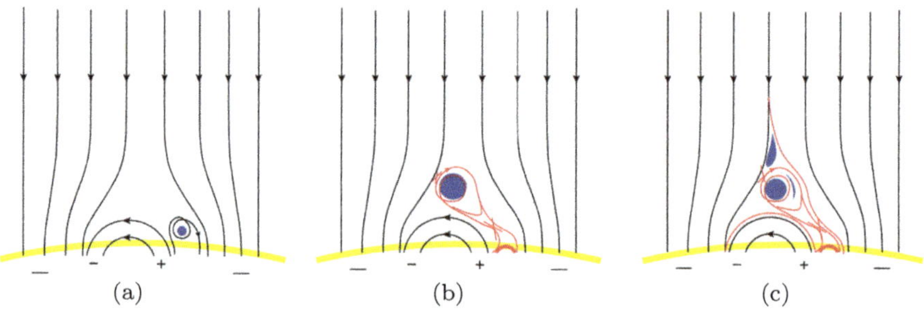

Fig. 2.3. Physical mechanism proposed by Sterling *et al.* (2015) to explain the jets triggered by mini-filament eruption (adapted from Sterling *et al.*, 2015).

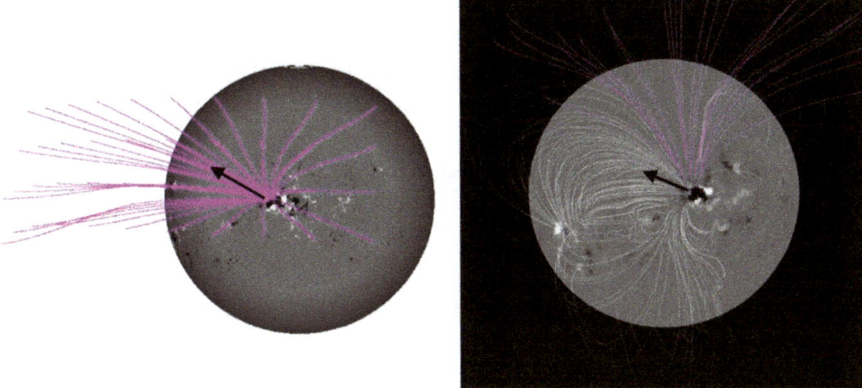

Fig. 2.4. An example of magnetic open and high-laying field lines along the jet ejection direction (Chandra *et al.*, 2017 and Uddin *et al.*, 2012).

Pariat *et al.* (2015, 2016) proposed a model based on a fully ionized single-fluid plasma having no atmospheric stratification for the eruption of straight and helical solar jets. They explained that a straight jet is a result of slow reconnection between the closed flux of an embedded bipole region and the open coronal hole flux surrounded the region. In their model, the slow reconnection is driven by the continuous stressing of the closed flux by photospheric disturbances or due to continuous flux emergence. Whereas, the helical jets are due to an explosive/ robust burst of the interchange reconnection. This explosive reconnection is a result of kink instability in a closed field region. To reach the stage of the kink instability, magnetic stress needs to approach beyond a certain level. Recently Wyper *et al.* (2017, 2018) proposed a model for the coronal hole jets in which a filament channel is formed beneath the spine-fan topology of a three-dimensional (3D) null point surrounded by vertical ambient magnetic field. Magnetic breakout reconnection at the null leads to the eruption of the filament channel, producing a helical jet. Wyper *et al.* (2017) also proposed a universal model for the solar eruptions. According to them the phenomena of filament eruption and the jet's have same physical origin. Therefore, a single mechanism may be responsible for both of them.

The magnetic field topology includes, a coronal magnetic null point, the build up of currents in its fan-separatrix surface and the sudden release of energy in heating followed by jet ejection. A topology with null point is supported by many observations (Török *et al.*, 2009; Raouafi *et al.*, 2016). However, other magnetic topologies computed from coronal extrapolation models have also been found associated with jet observations. Quasi-separatrix layers (QSLs) were identified with no magnetic null points (Mandrini *et al.*, 1996; Guo *et al.*, 2013). Recently Chandra *et al.* (2017) analyzed the jets during 21 to 24 October 2003 from the active region NOAA AR 10484. They found the foot point of jet location was well correlated with the bald patches (BPs) topology. BPs are defined as the portions where the magnetic field lines are curved upwards and the horizontal component of the magnetic field crosses the inversion line from the negative to the positive polarity (i.e., in the opposite way as compared to normal portions of the polarity inversion line (PIL) (Titov *et al.*, 1993).

In several observations, it is reported that the small closed-loop system and the open-loop system is found at the jet's locations and the material of jet is erupted along the open field lines. The particle generated during the jet process finally generate the type III radio bursts. Another possibility of magnetic configuration at the jet's location is very high-laying coronal loops instead of open field lines. This phenomena is presented in the study of Uddin *et al.* (2012). The magnetic field topology of open and very high-laying coronal field lines using the PFSS model is shown in Fig. 2.4.

2.1　Classification

Based on the morphological description of coronal jets using the *Hinode*/XRT data, Moore *et al.* (2010) divided the solar jets into two categories, named as "standard" and "blowout" jets. Standard jets have a narrow spire with a relative dim base, whereas the blowout jets reveal an initial phase quite similar to standard jets, but with a bright base and a narrow spire. This is afterwards followed by a violent flux rope eruption and then consequent broadening of the spire (Sterling *et al.*, 2010; Liu *et al.*, 2011; Moore *et al.*, 2013). Later on this violent flux rope eruption can be observed as Coronal Mass Ejections (CMEs). According to Hundhausen *et al.* (1984), the CME can be defined as "an observable change in coronal structure that (1) occurs on a timescale between a few minutes and several hours and (2) involves the appearance of a new, discrete, bright white-light feature in the coronagraph field of view." Examples of standard and blowout jets are presented in Figs. 2.1 and 2.5.

According to various observations of jet's Moore *et al.* (2010) suggested that the standard jets can be explained by the model proposed by Shibata *et al.* (1992) and Shibata (2000). Moreover, it is suggested that the blowout jets are also produced like a standard jet's. However, during the reconnection process of standard jets, sometimes the emerging bipole is triggered unstable and erupts outward. This eruption blows out the bipole field and the surrounding field, carrying outward the cool material and turned into blowout jets. In the case of blowout jets, the emerging flux have more free energy then in the case of standard jets. According

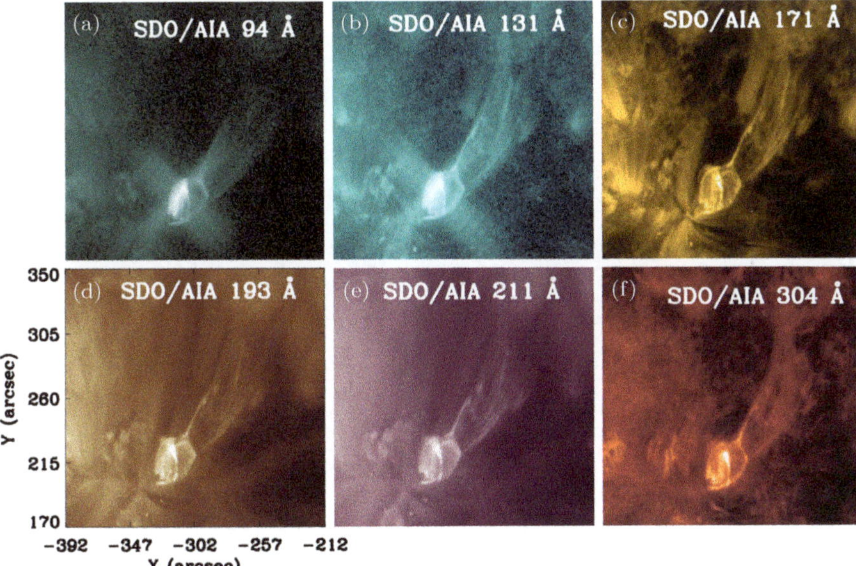

Fig. 2.5. Evolution of blow-out jet of 28 April 2013 in EUV/UV wavelengths observed by AIA instruments. *Credit to: AIA*/SDO *NASA.*

to Moore *et al.* (2010, 2013), the proposed process during which the blowout jet's is emerging is schematically presented in Fig. 2.6.

2.2 Basic physical parameters

Since from the first observations of jet's in X-rays, several studies have been done using different time to time ground and space-borne instruments. In this section, we are discussing the different physical parameters like: speeds, density, temperature, lengths, heights, widths, etc., of solar jets.

First study of solar jets using X-ray observations was carried out by Shibata *et al.* (1992), they found the speed range from 30 to 300 km s^{-1}. Immediately after this Shimojo *et al.*, 1996 did statistical study of 100 X-ray jets observed by *Yohkoh* during 1991 November 1 to 1992 April 30. Their study found the apparent projected velocity of these jets ranges from 10 to 1000 km s^{-1} with an average value of $\approx$200 km s^{-1}. Later on the joint analysis of EUV and white light coronal hole jets were carried out by Wang *et al.* (1998) using the *SOHO* satellite data. They have calculated the speed of these jets and the average values were $\approx$250 km s^{-1}. Further using the *LASCO* observations Wang and Sheeley, Jr. (2002) found the average speed of the jets during the maximum phase of the

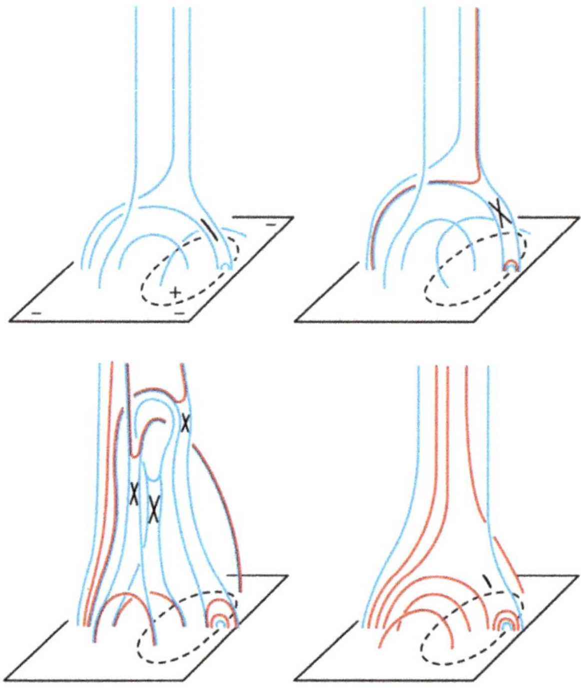

Fig. 2.6. Topology, eruption, and reconnection of the magnetic field proposed by Moore *et al.* (2010) scenario for blowout solar jets (adapted from Moore *et al.*, 2010).

solar cycle was estimated as $600\,\text{km s}^{-1}$. After the launch of *TRACE* satellite, the detailed study of the coronal jets using *TRACE* EUV observations were made by Alexander and Fletcher (1999) and they found the speed varies between 90 and $310\,\text{km s}^{-1}$.

Using the STEREO multi-view point observations, Nisticò *et al.* (2009) did a statistical study of coronal hole jets. They found the speeds are in the range of 250–$400\,\text{km s}^{-1}$ and of 100–$400\,\text{km s}^{-1}$ for EUV and WL jets, respectively. From the statistical studies of Nisticò *et al.* (2009) and Paraschiv *et al.* (2010), the average speeds of EUVI and COR1 jets are both around 300–$400\,\text{km s}^{-1}$. Now with the high spatial and temporal resolution observations of *Solar Dynamics Observatory* (*SDO*), the speed of the EUV jets varies from 100 to $500\,\text{km s}^{-1}$.

Density and temperature of solar jets are very important parameters for the purpose of their theoretical modeling. The spectroscopic observations provides an important tools for their computation. *Hinode*/EIS instruments contributes very crucial information about these parameters. According to Young and Muglach

(2014a), the commutated values of density and temperature for the coronal hole jets are $(0.9\text{--}2.8) \times 10^8\,\mathrm{cm}^{-3}$ and $1.4\text{--}1.6\,\mathrm{MK}$, respectively. In addition to this, for the active region jets the temperature values varies from 2 to $3\,\mathrm{MK}$. The spectroscopic observation also provides the information about the Doppler velocities and it is evident the presence of blue as well as red shifts in the jets eruptions. The observation of both (red and blue) Doppler shifts provides the confirmation of twisted structure of jets.

Due to the high temporal and spatial observations of *SDO* observations in different EUV and UV filters, the number of quiet as well as active region jets was studied and it is reported that their average height varies from 10 to $100\,\mathrm{Mm}$. The widths of the jets range from one to tens of megameters.

Chapter 3

Magnetohydrodynamic Waves and Instabilities

3.1 Magnetohydrodynamics basic equations

Magnetohydrodynamics (MHD) is a fluid model that describes the macroscopic magnetic properties of electrically conducting fluid such as plasma or liquid metals. The fundamental concept behind MHD is that magnetic fields can induce currents in a moving conductive fluid, which in turn alters the magnetic field itself. As the MHD theory describes dynamics of the electrically conducting fluids, therefore, it is the fusion of fluid dynamics and electromagnetism. The behavior of the non-viscous fluid is governed by hydrodynamic equations, while Maxwell's equations along with Ohm's law provide the complete description of the electromagnetism. In the MHD framework, the behavior of the plasma is governed by hydrodynamic equations, Maxwell's equations along with Ohm's law, and ideal gas law. MHD generally describes large scale, slow dynamics of plasmas. One can apply MHD when

(a) characteristic time of the process is much larger than the ion gyroperiod and mean free path time:

(b) characteristic length scale, L, is much larger than ion gyroradius and mean free path length:

$$L \gg 300 \left(\frac{T}{10^6 \, \text{K}} \right)^2 \left(\frac{n}{10^{17} \, \text{m}^{-3}} \right)^{-1} \text{km}$$

- in chromosphere with $T = 10^4 \, \text{K}$, $n = 10^{20} \, \text{m}^{-3}$, $L = 3 \, \text{cm}$,
- in corona at $T = 10^6 \, \text{K}$, $n = 10^{16} \, \text{m}^{-3}$, $L = 30 \, \text{km}$;

25

(c) plasma fluid velocities are not relativistic ($v \ll c$), where c is the speed of light.

A self-consistent set of MHD equations connects the plasma mass density ρ, the plasma velocity $\boldsymbol{v}$, the gas or kinetic pressure p and the magnetic field $\boldsymbol{B}$. We note that in a strict derivation of MHD, one should neglect the motion of electrons and consider only heavy ions. Hence, in deriving the basic set of MHD equation we should merge electrodynamics and hydrodynamics equations.

We start the derivation of MHD equations from Ampère's law

$$\nabla \times \boldsymbol{B} = \mu \boldsymbol{j},$$

(where μ is the magnetic permeability of the vacuum and $\boldsymbol{j}$ is the current density), in which the displacement current $(1/c^2)\partial \boldsymbol{E}/\partial t$ is neglected, which in non-relativistic limit, $v \ll c$ (where v is a characteristic speed of the process, and c the speed of light) is much smaller than the $\nabla \times \boldsymbol{B}$ term—the ratio $|\partial \boldsymbol{E}/\partial t|/|\nabla \times \boldsymbol{B}| \sim v^2/c^2$.

The following equations are the Faraday law

$$\nabla \times \boldsymbol{E} = -\frac{\partial \boldsymbol{B}}{\partial t}$$

and Ohm's law in its simplest form

$$\boldsymbol{j} = \sigma (\boldsymbol{E} + \boldsymbol{v} \times \boldsymbol{B}).$$

Here, σ is the plasma conductivity and $\boldsymbol{E} + \boldsymbol{v} \times \boldsymbol{B}$ is the electric field in reference frame of moving plasma. Note that in the Ohm's law the electron inertia and the Hall term are neglected.

The equation of motion or momentum equation is

$$\rho \frac{d\boldsymbol{v}}{dt} = -\nabla p + \boldsymbol{j} \times \boldsymbol{B} + \rho \boldsymbol{g} + \boldsymbol{F}_{\mathrm{visc}}.$$

The right-hand side of this equation is the total force (per unit volume): the gradient of pressure p, Lorentz force, gravity and viscosity.

- Viscosity is anisotropic in magnetized plasma and often is neglected.
- Gravity is usually negligible in high solar atmosphere.

Using the Ampère's law, we note that

$$\boldsymbol{j} \times \boldsymbol{B} = \frac{1}{\mu}(\nabla \times \boldsymbol{B}) \times \boldsymbol{B}.$$

The left-hand side of momentum equation is the mass (per unit volume) multiplied by the acceleration seen by a moving fluid element—this is given by a convective (Lagrangian) derivative

$$\frac{\mathrm{d}\boldsymbol{v}}{\mathrm{d}t} = \frac{\partial \boldsymbol{v}}{\partial t} + \boldsymbol{v} \cdot \nabla \boldsymbol{v}.$$

Another basic hydrodynamic equations are the continuity equation and the equation of state. The continuity equation is an equation of mass conservation (rate of change of mass in a small volume is equal to mass flux into the volume—no sources/sinks of plasma):

$$\frac{\partial \rho}{\partial t} + \nabla \cdot (\rho \boldsymbol{v}) = 0.$$

For incompressible flows, this equation reduces to $\nabla \cdot \boldsymbol{v} = 0$.

The equation of state is usually taken as the perfect gas law

$$p = \frac{k_{\mathrm{B}}}{m} \rho T,$$

where k_{B} is the Boltzmann constant and m is the mean particle mass. Note that for a fully-ionized hydrogen plasma (protons and electrons)

$$m \approx \frac{1}{2} m_{\mathrm{p}}.$$

Finally, we need an energy equation. Fairly generally this can be written as

$$\frac{\rho^{\gamma}}{\gamma - 1} \left(\frac{\partial}{\partial t} + \boldsymbol{v} \cdot \nabla \right) \frac{p}{\rho^{\gamma}} = -\nabla \cdot (\boldsymbol{\kappa} \cdot \nabla T) - \rho^2 Q(T) + \frac{j^2}{\sigma} + H,$$

where γ is the adiabatic index—usually one uses $\gamma = 5/3$.

The left-hand side is the rate of change of internal energy (per unit volume).

The right-hand side is the sources/links of energy: conduction (where the thermal conductivity tensor $\boldsymbol{\kappa}$ is related to the collision frequency and is also non-isotropic—different across field/along the magnetic field); optically-thin radiation, characterized by function Q; Ohmic heating, and other sources of heat.

Often we can use the simpler adiabatic equation

$$\frac{\mathrm{d}}{\mathrm{d}t} \left(\frac{p}{\rho^{\gamma}} \right) = 0.$$

In addition, Maxwell's equations give us a constraint on the magnetic field: magnetic monopoles do not exist:

$$\nabla \cdot \boldsymbol{B} = 0.$$

Thus, the basic MHD equations are as follows:

$$\frac{\partial \boldsymbol{B}}{\partial t} = \nabla \times (\boldsymbol{v} \times \boldsymbol{B}) + \eta \nabla^2 \boldsymbol{B} \quad \text{with } \eta = \frac{1}{\mu \sigma}, \tag{3.1}$$

$$\rho \left(\frac{\partial \boldsymbol{v}}{\partial t} + \boldsymbol{v} \cdot \nabla \boldsymbol{v} \right) \equiv \frac{\mathrm{d}\boldsymbol{v}}{\mathrm{d}t} = -\nabla p + \frac{1}{\mu}(\nabla \times \boldsymbol{B}) \times \boldsymbol{B}, \tag{3.2}$$

$$\frac{\partial \rho}{\partial t} + \nabla \cdot (\rho \boldsymbol{v}) = 0, \tag{3.3}$$

$$p = \frac{k_{\mathrm{B}}}{m} \rho T, \tag{3.4}$$

$$\left(\frac{\partial \rho}{\partial t} + \boldsymbol{v} \cdot \nabla \right) \frac{p}{\rho^{\gamma}} = 0, \tag{3.5}$$

and the constraint

$$\nabla \cdot \boldsymbol{B} = 0. \tag{3.6}$$

In above equations, μ is the magnetic permeability of the free space and η is the *magnetic diffusivity*. Thus, we have nine highly nonlinear, coupled equations in nine unknowns (components of $\boldsymbol{B}$, $\boldsymbol{v}$; p, ρ, T).

In MHD, the fluid velocity $\boldsymbol{v}$ and the magnetic field $\boldsymbol{B}$ are *primary variables* while the current density $\boldsymbol{j} = \nabla \times \boldsymbol{B}/\mu$ and the electric field $\boldsymbol{E} = -\boldsymbol{v} \times \boldsymbol{B} + \boldsymbol{J}/\sigma$ are *secondary variables*—they can be derived as required from Ampère's law and Ohm's law.

Note that

- plasma flows generate magnetic field (induction equation),
- magnetic field creates plasma flows (momentum equation).

We note also that more general MHD versions are available, for example,

— multi-fluid (Hall-MHD)—treats two-fluid effects, add neutral fluid, etc.,
— relativistic MHD (used in high-energy astrophysics),
— radiative MHD (coupled with radiative transfer).

A question which immediately rises is what is the relative importance of plasma motions and Ohmic resistivity, $1/\sigma$, in induction equation? Consider the ratio of the two terms on right-hand side of Eq. (3.1):

$$\frac{|\nabla \times \boldsymbol{B}|}{|\eta \nabla^2 \boldsymbol{B}|} \approx \frac{vB/L}{\eta B/L^2} = \frac{Lv}{\eta},$$

where L and v are typical length-scale and velocity. This ratio is the *magnetic Reynolds number*, Re_{m}—a dimensionless number similar to Reynolds number in fluid dynamics

$$Re_{\mathrm{m}} = \frac{Lv}{\eta} = \mu \sigma L v. \tag{3.7}$$

Magnetic Reynolds number tells us the relative importance of the resistivity and plasma flows in determining the evolution of $\boldsymbol{B}$. We can consider two cases:

- $Re_{\mathrm{m}} \gg 1$: flows dominate, resistivity is negligible,
- $Re_{\mathrm{m}} \ll 1$: highly resistive plasma.

Astrophysical plasmas are very good conductors, also L is very large in astrophysical plasmas, which implies that Re_{m} is usually very large.

Consider the two limits separately.

The limit $Re_{\mathrm{m}} \ll 1$ (high resistivity)

Induction equation becomes diffusion equation

$$\frac{\partial \boldsymbol{B}}{\partial t} = \eta \nabla^2 \boldsymbol{B}, \quad \text{where} \quad \frac{B}{t_{\mathrm{diffusion}}} \approx \frac{L^2}{\eta}.$$

Field gradients/currents diffuse away due to Ohmic resistivity, $1/\sigma$, on a diffusion timescale

$$t_{\mathrm{diffusion}} = \frac{L^2}{\eta}.$$

The diffusion is usually very slow process, e.g., in solar corona, taking $L = 1000\,\mathrm{km}$ and $T = 10^6\,\mathrm{K}$

$$t_{\mathrm{diffusion}} \approx 30\,000\,\mathrm{years}.$$

The limit $Re_{\mathrm{m}} \gg 1$ (high conductivity)

The limit $Re_{\mathrm{m}} \gg 1$ (high conductivity) is known as ideal MHD. It is relevant for astrophysics. The induction equation becomes

$$\frac{\partial \boldsymbol{B}}{\partial t} = \nabla \times (\boldsymbol{v} \times \boldsymbol{B}).$$

Analogous to the Kelvin–Helmholtz theorem in ideal hydrodynamics, which states that vorticity lines move with the inviscid fluid, the ideal induction equation yields the Alfvén's theorem of the frozen field lines (Alfvén, 1942):

> *In an ideal perfectly conductive plasma, the total amount of magnetic flux passing through any closed circuit moving with the local fluid velocity is constant in time.* In other words, ideal MHD equations imply the conservation of magnetic flux and connectivity of the field lines.

Further on we will focus on plasmas in which all dissipative processes (finite viscosity, electrical resistivity, thermal conduction, or radiative cooling) will be neglected. Such media are termed ideal and the closed set of ideal MHD equations is as follows:

$$\frac{\partial \rho}{\partial t} + \nabla \cdot (\rho \boldsymbol{v}) = 0 \quad \text{Continuity equation,}$$

$$\rho \frac{d\boldsymbol{v}}{dt} = -\nabla p - \frac{1}{\mu} \boldsymbol{B} \times (\nabla \times \boldsymbol{B}) \quad \text{Momentum equation,}$$

$$\frac{d}{dt}\left(\frac{p}{\rho^{\gamma}}\right) = 0 \quad \text{Energy equation,}$$

$$\frac{\partial \boldsymbol{B}}{\partial t} = \nabla \times (\boldsymbol{v} \times \boldsymbol{B}) \quad \text{Induction equation.}$$

In addition, the magnetic field is subject to the constraint

$$\nabla \cdot \boldsymbol{B} = 0.$$

3.2   Magnetohydrodynamic equilibrium

The conditions for static equilibrium are

$$\boldsymbol{v} = 0, \quad \frac{\partial}{\partial t} = 0. \tag{3.8}$$

These conditions identically satisfy the continuity, energy and induction equations. From momentum equation, we obtain

$$-\nabla p - \frac{1}{\mu}\boldsymbol{B} \times (\nabla \times \boldsymbol{B}) = 0, \tag{3.9}$$

which is called the equation of magnetostatics. This equation must be supplemented with the constraint (3.6). Equation (3.9) can be rewritten in the form

$$-\nabla\left(p + \frac{B^2}{2\mu}\right) + \frac{1}{\mu}(\boldsymbol{B} \cdot \nabla)\boldsymbol{B} = 0. \tag{3.10}$$

The first term represents the gradient of the total pressure, which is the sum of the kinetic (thermodynamic) pressure p and the magnetic pressure $B^2/(2\mu)$. The second term is magnetic tension. The force is directed anti-parallel to the radius of the magnetic field line curvature. The magnetic curvature force can be thought of as a tension that tries to straighten curved field lines.

The ratio of the gas/kinetic and magnetic pressure defines an important plasma parameter, called plasma β:

$$\beta \equiv \frac{\text{gas pressure}}{\text{magnetic pressure}} = \frac{p}{B^2/2\mu}. \tag{3.11}$$

Plasma β can be estimated by the formula,

$$\beta = 3.5 \times 10^{-21} n\, T\, B^{-2},$$

where the particle number density n is in m^{-3}, temperature T in K, and magnetic field B in G. For example,

- In the solar corona, $T = 10^6\,\mathrm{K}$, $n = 10^{14}\,\mathrm{m}^{-3}$, $B = 10\,\mathrm{G}$, and $\beta = 3.5 \times 10^{-3}$.
- In photospheric magnetic flux tubes, $T = 6 \times 10^3\,\mathrm{K}$, $n = 10^{23}\,\mathrm{m}^{-3}$, $B = 1000\,\mathrm{G}$, and $\beta = 2$.
- In the solar wind at 1 AU, $T = 2 \times 10^5\,\mathrm{K}$, $n = 10^7\,\mathrm{m}^{-3}$, $B = 6 \times 10^{-5}\,\mathrm{G}$, and $\beta = 2$.

3.3 Magnetic reconnection

Magnetic reconnection is a physical process in highly conducting plasmas in which the magnetic topology is rearranged and magnetic energy is converted to kinetic energy, thermal energy and particle acceleration. Magnetic reconnection occurs on timescales intermediate between slow resistive diffusion of the magnetic field and fast Alfvénic timescales. The qualitative description of the reconnection process is such that magnetic field lines from different magnetic domains (defined by the field line connectivity) are spliced to one another, changing their patterns of connectivity with respect to the sources. It is a violation of an approximate conservation law in plasma physics, called the Alfvén's theorem, and can concentrate mechanical or magnetic energy in both space and time. Solar flares, the largest explosions in the solar system, may involve the reconnection of large systems of magnetic flux on the Sun, releasing, in minutes, energy that has been stored in the magnetic field over a period of hours to days.

In an electrically conductive plasma, magnetic field lines are grouped into "domains"—bundles of field lines that connect from a particular place to another particular place, and that are topologically distinct from other field lines nearby. This topology is approximately preserved even when the magnetic field itself is strongly distorted by the presence of variable currents or motion of magnetic sources, because effects that might otherwise change the magnetic topology instead induce eddy currents in the plasma; the eddy currents have the effect of canceling out the topological change. In two dimensions, the most common type of magnetic reconnection is separator reconnection, in which four separate magnetic domains exchange magnetic field lines. Domains in a magnetic plasma are separated by separatrix surfaces: curved surfaces in space that divide different bundles of flux. Field lines on one side of the separatrix all terminate at a particular magnetic pole, while field lines on the other side all terminate at a different pole of similar sign. Since each field line generally begins at a north magnetic pole and ends at a south magnetic pole, the most general way of dividing simple flux systems involves four domains separated by two separatrices: one separatrix surface divides the flux into two bundles, each of which shares a south pole, and the other separatrix surface divides the flux into two bundles, each of which shares a north pole. The intersection of the separatrices forms a separator, a single line that is at the boundary of the four separate domains. In separator reconnection, field lines enter the separator from two of the domains, and are spliced one to the other, exiting the separator in the other two domains (see Fig. 3.1).

According to simple resistive MHD theory, reconnection happens because the plasma's electrical resistivity near the boundary layer opposes the currents necessary to sustain the change in the magnetic field. The need for such a current can be seen from one of Maxwell's equations, namely Ampère's law

$$\nabla \times \boldsymbol{B} = \mu \boldsymbol{j}.$$

The resistivity of the current layer allows magnetic flux from either side to diffuse through the current layer, canceling out flux from the other side of the boundary. When this happens, the plasma is pulled out by magnetic tension along the direction of the magnetic field lines. The resulting drop in pressure pulls more plasma and magnetic flux into the central region, yielding a self-sustaining process.

A current problem in plasma physics is that observed reconnection happens much faster than predicted by MHD in high Lundquist number plasmas: solar flares, for example, proceed 13–14 orders of magnitude faster than a naive calculation would suggest, and several orders of magnitude faster than current theoretical models that include turbulence and kinetic effects. Note that the Lundquist number (denoted by S) is a dimensionless ratio which compares the timescale of an Alfvén wave crossing to the timescale of resistive diffusion. It is a special case of the magnetic Reynolds number when the Alfvén velocity is the typical velocity

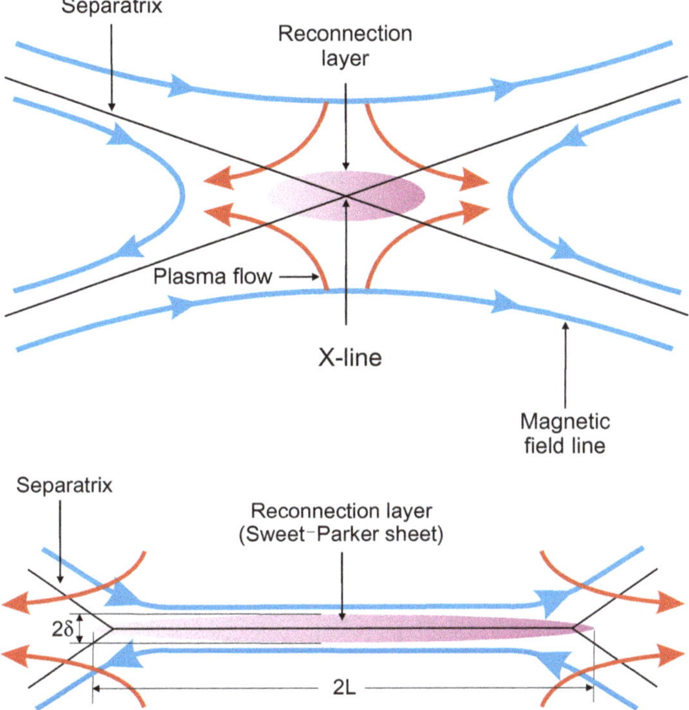

Fig. 3.1. Cross-section through four magnetic domains undergoing separator magnetic reconnection. Two separatrices divide space into four magnetic domains with a separator at the center of the figure. Field lines (and associated plasma) flow inward from above and below the separator, reconnect, and spring outward horizontally. (*Top panel*) Petschek magnetic reconnection. (*Bottom panel*) Sweet–Parker magnetic reconnection.

scale of the system, and is given by

$$S = \frac{L v_{\mathrm{A}}}{\eta},$$
(3.12)

where L is the typical length scale of the system, η is the magnetic diffusivity and v_{A} is the Alfvén velocity. In the solar corona S lies in the range of 10^8–10^{12}.

The first theoretical 2D model of magnetic reconnection was proposed by Sweet (1958) and Parker (1957). (Actually, at an IAU Symposium in 1956, Peter Sweet pointed out that by pushing two plasmas with oppositely directed magnetic fields together, resistive diffusion is able to occur on a length scale much shorter than a typical equilibrium length scale. Eugene Parker was in attendance at this conference and developed scaling relations for this model during his return travel.)

The Sweet–Parker model describes time-independent magnetic reconnection in the resistive MHD framework when the reconnecting magnetic fields are antiparallel (oppositely directed) and effects related to viscosity and compressibility are unimportant. The ideal Ohm's law then yields the relation

$$E_y = v_{\text{in}} B_{\text{in}},$$

where E_y is the out-of-plane electric field, v_{in} is the characteristic inflow velocity, and B_{in} is the characteristic upstream magnetic field strength. Ampère's law, $\boldsymbol{j} = (\nabla \times \boldsymbol{B})/\mu$, gives the relation

$$j_y = \frac{B_{\text{in}}}{\mu \delta},$$

where δ is the current sheet half-thickness. This relation uses that the magnetic field reverses over a distance of $\sim 2\delta$. By matching the ideal electric field outside of the layer with the resistive electric field $\boldsymbol{E} = (1/\sigma)\boldsymbol{j}$ inside the layer (using Ohm's law), we find that

$$v_{\text{in}} = \frac{E_y}{B_{\text{in}}} \sim \frac{1}{\mu \sigma \delta} = \frac{\eta}{\delta},$$

where η is the magnetic diffusivity. When the inflow density is comparable to the outflow density, conservation of mass yields the relationship

$$v_{\text{in}} L \sim v_{\text{out}} \delta,$$

where L is the half-length of the current sheet and v_{out} is the outflow velocity. The left- and right-hand sides of the above relation represent the mass flux into the layer and out of the layer, respectively. Equating the upstream magnetic pressure with the downstream dynamic pressure gives

$$\frac{B_{\text{in}}^2}{2\mu} \sim \frac{\rho v_{\text{out}}^2}{2},$$

where ρ is the mass density of the plasma. Solving for the outflow velocity then yields

$$v_{\text{out}} \sim \frac{B_{\text{in}}}{\sqrt{\mu \rho}} \equiv v_{\text{A}},$$

where v_{A} is the Alfvén velocity. With the above relations, the dimensionless reconnection rate R can then be written in two forms, the first in terms of η, δ, and v_{A}

using the result earlier derived from Ohm's law, the second in terms of δ and L from the conservation of mass as

$$R = \frac{v_{\text{in}}}{v_{\text{out}}} \sim \frac{\eta}{v_A \delta} \sim \frac{\delta}{L}.$$

Since the dimensionless Lundquist number S, given by Eq. (3.12), is $S \equiv L v_A / \eta$, the two different expressions of R are multiplied by each other and then square-rooted, giving a simple relation between the reconnection rate R and the Lundquist number S

$$R \sim \sqrt{\frac{\eta}{v_A L}} = \frac{1}{S^{1/2}}.$$

Sweet–Parker reconnection allows for reconnection rates much faster than global diffusion, but is not able to explain the fast reconnection rates observed in solar flares and the Earth's magnetosphere. Additionally, Sweet–Parker reconnection neglects three-dimensional effects, collisionless physics, time-dependent effects, viscosity, compressibility, and downstream pressure. Results from the Magnetic Reconnection Experiment (MRX) of collisional reconnection show agreement with a generalized Sweet–Parker model which incorporates compressibility, down-stream pressure, and anomalous resistivity (Ji *et al.*, 1999).

One of the reasons why Sweet–Parker reconnection is slow is that the aspect ratio of the reconnection layer is very large in high Lundquist number plasmas. The inflow velocity, and thus the reconnection rate, must then be very small. In 1964, Petschek (1964) first pointed out that a much larger reconnection rate would be possible if the diffusion region were much shorter. In his model, a tiny diffusion region is bound by two pairs of slow shocks, the outflow region plasma is strongly accelerated and heated by the slow shocks. The maximum reconnection rate in this model is about

$$\frac{v_{\text{in}}}{v_A} \cong \frac{\pi}{8 \ln S}.$$

This expression allows fast reconnection almost independent of the Lundquist number. A typical value of the Lundquist number for solar corona is $S = 10^8$. Then the Sweet–Parker reconnection rate is $S_{\text{S–P}} = 10^{-4}$ while that of Petschek is $S_{\text{P}} = 2 \times 10^{-2}$. A more detailed comparison between the two reconnection models as well as an improvement of Petschek's theory the reader can find in Kulsrud (2001). A generalization of Petschek's solution is discussed by Forbes (2001) who also is giving alternative approaches of both (Sweet–Parker and Petschek) models.

3.4 Magnetohydrodynamic waves

Any disturbance of equilibrium magnetic field or pressure leads to the excitation of oscillation and waves. In the general case, the theory of MHD waves is nonlinear due to the nonlinearity of governing momentum and induction equations. For small perturbations, however, one can linearize the basic equations and subsequently obtain the dependence of the wave angular frequency ω on the wave vector $\boldsymbol{k}$, $\omega = f(\boldsymbol{k})$, which is termed *dispersion relation*.

We will investigate the MHD equations for small amplitude, plane wave solutions. Into the MHD equations, we substitute:

$$\boldsymbol{B} = \boldsymbol{B}_0 + \boldsymbol{B}_1(\boldsymbol{r}, t),$$

$$\boldsymbol{v} = \boldsymbol{0} + \boldsymbol{v}_1(\boldsymbol{r}, t),$$

$$p = p_0 + p_1(\boldsymbol{r}, t),$$

$$\rho = \rho_0 + \rho_1(\boldsymbol{r}, t),$$

where quantities with subscript "1" represent small, linear perturbations from the values (subscript "0") obtained from an equilibrium solution of the MHD equations. Higher order perturbations are ignored. The equilibrium is assumed to be uniform, with $\rho_0 = $ const, $\boldsymbol{B}_0 = $ const, and $p_0 = $ const. Then we linearize by substituting into the MHD equations, cancel the terms which appear in the equilibrium solutions to obtain

$$\frac{\partial \rho_1}{\partial t} + \rho_0 \nabla \cdot \boldsymbol{v}_1 = 0, \tag{3.13}$$

$$\rho_0 \frac{\partial \boldsymbol{v}_1}{\partial t} = -\nabla \left\{ p_1 + \frac{\boldsymbol{B}_0 \cdot \boldsymbol{B}_1}{\mu} \right\} + \frac{\boldsymbol{B}_0}{\mu} \cdot \nabla \boldsymbol{B}_1, \tag{3.14}$$

$$\frac{\partial p_1}{\partial t} - \frac{\gamma p_0}{\rho_0} \frac{\partial \rho_1}{\partial t} = 0, \tag{3.15}$$

$$\frac{\partial \boldsymbol{B}_1}{\partial t} = \nabla \times (\boldsymbol{v}_1 \times \boldsymbol{B}_0) = 0. \tag{3.16}$$

We note that the right-hand side of Eq. (3.14) is derived by linearization of the magnetostatic Eq. (3.10). Moreover, by substituting $\partial \rho_1 / \partial t = -\rho_0 \nabla \cdot \boldsymbol{v}_1$ from continuity equation into the energy Eq. (3.15) we obtain

$$\frac{\partial p_1}{\partial t} = -\gamma p_0 \nabla \cdot \boldsymbol{v}_1. \tag{3.17}$$

Finally, the induction equation (3.16) can be rewritten in the form:

$$\frac{\partial \boldsymbol{B}_1}{\partial t} = -\boldsymbol{B}_0 \nabla \cdot \boldsymbol{v}_1 + \boldsymbol{B}_0 \cdot \nabla \boldsymbol{v}_1. \tag{3.18}$$

The first-order density ρ_1 does not appear in the above equations, so that the linearized form of the continuity equation is unnecessary; it serves only to determine ρ_1 after the equations have been solved. Similarly, the perturbations $\boldsymbol{E}_1$ and $\boldsymbol{j}_1$ can be found from the linearized forms of Ampère's law and of Ohm's law in the limit of infinite conductivity. This set of subsidiary equations is

$$\frac{\partial \rho_1}{\partial t} = -\rho_0 \nabla \cdot \boldsymbol{v}_1, \tag{3.19}$$

$$\boldsymbol{E}_1 = -\boldsymbol{v}_1 \times \boldsymbol{B}_0, \tag{3.20}$$

$$\boldsymbol{j}_1 = \frac{1}{\mu} \nabla \times \boldsymbol{B}_1. \tag{3.21}$$

If Eq. (3.14) is differentiated with respect to time and Eqs. (3.17) and (3.18) are used to eliminate p_1 and $\boldsymbol{B}_1$, we get an equation of second order:

$$\begin{aligned}
\rho_0 \frac{\partial^2 \boldsymbol{v}_1}{\partial t^2} = &\left\{ \nabla \left\{ \left(\gamma\, p_0 + \frac{B_0^2}{\mu} \right) \nabla \cdot \boldsymbol{v}_1 \right\} - \nabla \left\{ \frac{\boldsymbol{B}_0}{\mu} \cdot (\boldsymbol{B}_0 \cdot \nabla) \boldsymbol{v}_1 \right\} \right. \\
&\left. - \frac{1}{\mu}(\boldsymbol{B}_0 \cdot \nabla)\left\{ \boldsymbol{B}_0 \nabla \cdot \boldsymbol{v}_1 - (\boldsymbol{B}_0 \cdot \nabla)\boldsymbol{v}_1 \right\} \right\}.
\end{aligned} \tag{3.22}$$

This equation makes no assumptions about the uniformity of the medium. The zeroth-order quantities are independent of time but may be functions of position. If, in addition, we assume that the zeroth-order quantities do not vary in space, we get a wave equation for a uniform medium:

$$\begin{aligned}
\rho_0 \frac{\partial^2 \boldsymbol{v}_1}{\partial t^2} = &(v_A^2 + c_s^2)\nabla(\nabla \cdot \boldsymbol{v}_1) - \nabla(\boldsymbol{v}_A \cdot \nabla)(\boldsymbol{v}_A \cdot \boldsymbol{v}_1) \\
&- \boldsymbol{v}_A(\boldsymbol{v}_A \cdot \nabla)(\nabla \cdot \boldsymbol{v}_1) + (\boldsymbol{v}_A \cdot \nabla)(\boldsymbol{v}_A \cdot \nabla)\boldsymbol{v}_1,
\end{aligned} \tag{3.23}$$

where

$$\boldsymbol{v}_A = \frac{\boldsymbol{B}_0}{\sqrt{\mu \rho_0}} \quad \text{and} \quad c_s = \sqrt{\frac{\gamma\, p_0}{\rho_0}} \tag{3.24}$$

have the dimensions of velocity. The quantity $\boldsymbol{v}_A$ is called the Alfvén velocity and c_s is termed the speed of sound.

We assume a harmonic wave solution to Eq. (3.23) such that the wave propagates along the direction of the wavevector $\boldsymbol{k}$ whose magnitude is equal to the wavenumber $k = 2\pi/\lambda$, where λ is the wavelength. A complex plane harmonic wave then has time and space behavior given by

$$A \exp\{i(\boldsymbol{k} \cdot \boldsymbol{r} - \omega t)\}, \tag{3.25}$$

where A is a complex amplitude and the actual physical quantity is the real part of this representation. This plane waves advances with a velocity ω/k in the $\hat{\boldsymbol{k}}$-direction. We define the phase velocity

$$\boldsymbol{v}_{\mathrm{ph}} = \frac{\omega}{k}\hat{\boldsymbol{k}}. \tag{3.26}$$

The operators ∇ and $\partial/\partial t$ can then be replaced by $i\boldsymbol{k}$ and $-i\omega$ so that Eq. (3.23) becomes

$$\{\omega^2 - (\boldsymbol{k} \cdot \boldsymbol{v}_{\mathrm{A}})^2\}\boldsymbol{v}_1 - \boldsymbol{k}(v_{\mathrm{A}}^2 + c_{\mathrm{s}}^2)\boldsymbol{k} \cdot \boldsymbol{v}_1 + (\boldsymbol{k} \cdot \boldsymbol{v}_{\mathrm{A}})\{\boldsymbol{k}(\boldsymbol{v}_{\mathrm{A}} \cdot \boldsymbol{v}_1) + \boldsymbol{v}_{\mathrm{A}}(\boldsymbol{k} \cdot \boldsymbol{v}_1)\}$$
$$= 0. \tag{3.27}$$

Let us assume that the equilibrium magnetic field $\boldsymbol{B}_0$ is in the z-direction and the wavevector $\boldsymbol{k}$ is in the xz-plane (see Fig. 3.2). Then, the above equation can be written in the following form:

$$\begin{pmatrix} \omega^2 - k_z^2 v_{\mathrm{A}}^2 - k_x^2\left(v_{\mathrm{A}}^2 + c_{\mathrm{s}}^2\right) & 0 & -k_x k_z c_{\mathrm{s}}^2 \\ 0 & \omega^2 - k_z^2 v_{\mathrm{A}}^2 & 0 \\ -k_x k_z c_{\mathrm{s}}^2 & 0 & \omega^2 - k_z^2 c_{\mathrm{s}}^2 \end{pmatrix} \begin{pmatrix} v_{1x} \\ v_{1y} \\ v_{1z} \end{pmatrix} = 0. \tag{3.28}$$

It can be shown that v_{1y} is determined independently from v_{1x} and v_{1z}. This equation only has a non-trivial solution if the determinant of its coefficients is zero. This condition gives a relationship between the angular frequency ω and the

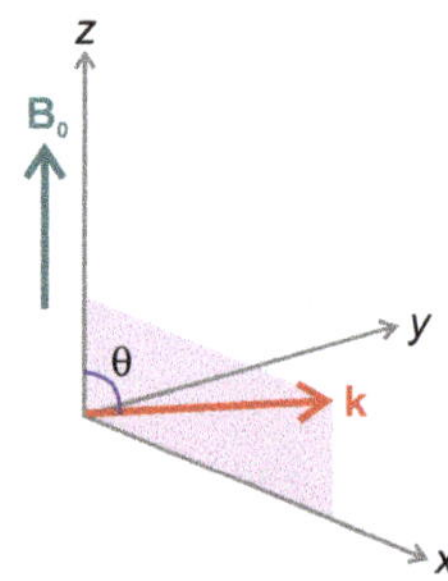

Fig. 3.2. Equilibrium magnetic field $\boldsymbol{B}_0$ and the wavevector $\boldsymbol{k}$ of a propagating Alfvén wave.

wavevector k and, thus, determines the wave speed. Such a relationship, as we already said, is called a dispersion relation.

The determinant of Eq. (3.28) has two factors and, thus, there are two dispersion relations, one determining a wave in which the velocity perturbation is in the y-direction and the other one in which it is in the xz-plane. These are as follows:

$$\omega^2 - k_z^2 v_A^2 = 0, \tag{3.29}$$

$$\omega^4 - k^2 (v_A^2 + c_s^2)\omega^2 + k^2 k_z^2 v_A^2 c_s^2 = 0. \tag{3.30}$$

Let us examine the dispersion relation (3.29). It clearly has an unusual property in that it is independent of the component of k normal to the magnetic field. The implication is that k_z and k_y may be arbitrarily fixed and only k_z and ω are related. Thus, the phase velocity, which has components ω/k_x, ω/k_y and ω/k_z, can have any value provided that its component along the z-direction is equal to v_A, the Alfvén speed. The other dispersion relation is quadratic in ω^2. It therefore describes two waves with different phase velocities.

The phase velocity of the wave whose dispersion relation is given by Eq. (3.29) is

$$v_{\text{ph}} = v_A \cos\theta, \tag{3.31}$$

where θ is the angle between k and B_0 (see Fig. 3.2). The wave with phase velocity given by (3.31) is called the *transverse* or *shear Alfvén wave* because, as previously noted, the velocity perturbation is in the y-direction which is perpendicular to the plane containing the wavevector k and the magnetic field B_0. It is highly anisotropic. The component of the phase velocity along the magnetic field is always equal to the Alfvén speed. Alfvén waves can be linearly polarized, elliptically polarized, or circularly polarized. For the Alfvén waves, magnetic tension force of the field lines is the only restoring force, therefore, these waves are purely incompressible waves, $\nabla \cdot v_1 = 0$.

The group velocity $v_{\text{gr}} = \partial\omega/\partial k$ is the rate at which the overall shape of the waves' amplitudes propagates through space. The group velocity of Alfvén waves is always parallel to the magnetic field, $v_{\text{gr}} \| B_0$, by contrast to the phase velocity which can be oblique to the field. In general, $v_{\text{ph}} \nparallel v_{\text{gr}}$. The absolute value of the group speed equals the Alfvén speed, v_A.

The dispersion relation (3.30) gives rise to two different characteristic waves with phase velocities given by

$$v_{\text{ph}}^2 = \frac{1}{2}\{(v_A^2 + c_s^2) \pm \sqrt{(v_A^2 + c_s^2)^2 - 4v_A^2 c_s^2 \cos^2\theta}\}. \tag{3.32}$$

These relations depend on both the Alfvén speed and the sound speed. They are, therefore, called the *magnetosonic waves*. Examination of the expressions for the phase velocities shows that one of these waves always has a phase speed which is greater than that of the other. They are, therefore, appropriately named the *fast* and *slow magnetosonic waves*. The fast mode is fast because the pressure and tension are nearly in phase. The slow mode is slow because the tension and pressure are nearly out of phase.

Limiting cases: We briefly consider some special cases of the magnetosonic waves.

Case (a): Propagation parallel to the field. When $\theta = 0$, the expression for the phase velocity of the magnetosonic waves (3.32) becomes

$$v_{\mathrm{ph}} = v_{\mathrm{A}} \quad \mathrm{or} \quad c_{\mathrm{s}}.$$

One of the waves is propagated along the field with the Alfvén speed and the other with the sound speed. Which is the fast and which is the slow wave that depends on the magnitudes of the two characteristic speeds.

Case (b): Propagation perpendicular to the field. When $\theta = 90°$, the fast wave has a velocity

$$v_{\mathrm{ph}} = \sqrt{v_{\mathrm{A}}^2 + c_{\mathrm{s}}^2}$$

and the slow wave has zero velocity.

Case (c): $v_{\mathrm{A}} \gg c_{\mathrm{s}}$. The square root in (3.32) may be expanded to first order in $c_{\mathrm{s}}/v_{\mathrm{A}}$. The phase velocities of the two waves are given by

$$v_{\mathrm{ph}} = v_{\mathrm{A}} \quad \mathrm{or} \quad c_{\mathrm{s}} \cos\theta.$$

The fast wave now has a phase speed which is equal to the Alfvén speed, no matter what the direction of propagation. It is called the isotropic Alfvén wave. The slow wave is a sound wave, propagated with a much smaller velocity but constrained to the magnetic-field direction. It has similar properties to the transverse Alfvén wave. This is because the plasma cannot have a compressional velocity perpendicular to the magnetic field: the plasma is frozen to the field lines and does not have sufficient energy to distort them.

Case (d): $c_{\mathrm{s}} \gg v_{\mathrm{A}}$. In a similar fashion,

$$v_{\mathrm{ph}} = v_{\mathrm{s}} \quad \mathrm{or} \quad v_{\mathrm{A}} \cos\theta.$$

The fast wave is now an isotropic sound wave in a gas, unaffected by the presence of the weak magnetic field. The slow wave now has the same phase velocity as the transverse Alfvén wave. It can only be propagated parallel to the magnetic field.

An overall view of the phase velocity dependencies of the three MHD waves on angle θ can be seen from Friedrichs diagrams pictured in Fig. 3.3.

3.4.1 *MHD modes in magnetic flux tubes*

In the beginning of Sec. 3.3, we have given brief outline about MHD waves in the ideal MHD context and in isotropic magneto-plasma system. Present subsection is focused on various MHD modes in the localized magnetic flux tubes. Such magnetic flux tubes are the basic building blocks of the solar atmosphere and they serve as an ideal MHD waveguide also. Magnetic flux tubes are the localized magnetic field regions, which fan out in the solar atmosphere as pressure balanced structures. Here, we will discuss the magnetic cylindrical geometry by neglecting the twist in the magnetic field, which closely resembles various structures in the solar atmosphere (e.g., coronal loops, polar plumes, spicules, jets, etc.). The basic understanding of the MHD modes excited in the magnetic cylinder is extremely important to understand the wave dynamics in the various solar atmospheric structures. Under the discussion of MHD modes in this geometry, we have assumed a cylindrical magnetic flux tube with radius a which is filled with plasma (see Fig. 3.4). The surrounding environment of the flux tube is also the plasma along with magnetic field, however, the properties of these two regions are different from each other. Density, pressure and magnetic field of the flux tube are denoted by ρ_i, p_i and $\boldsymbol{B}_i$, respectively, while ρ_e, p_e and $\boldsymbol{B}_e$ are the same quantities for the outside environment. In the equilibrium condition, the magnetostatic

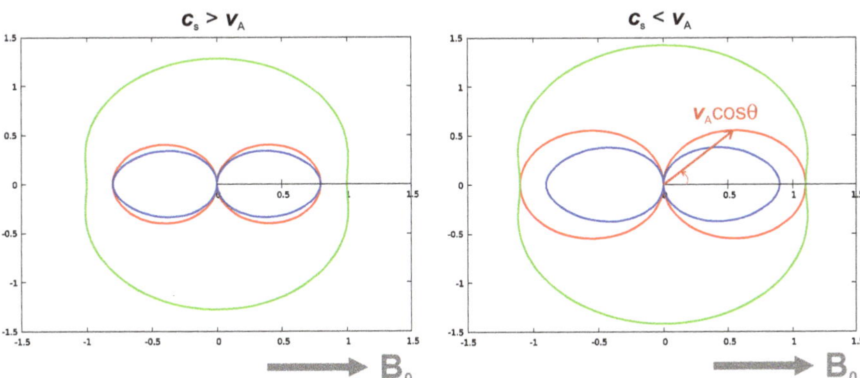

Fig. 3.3. Dependencies of phase velocities on the angle between the magnetic field $\boldsymbol{B}_0$, directed on the horizontal axes, and the wavevector $\boldsymbol{k}$. The red colored closed curves are for the Alfvén wave, the green ones for the fast magnetosonic wave, and the blue for the slow magnetosonic wave.

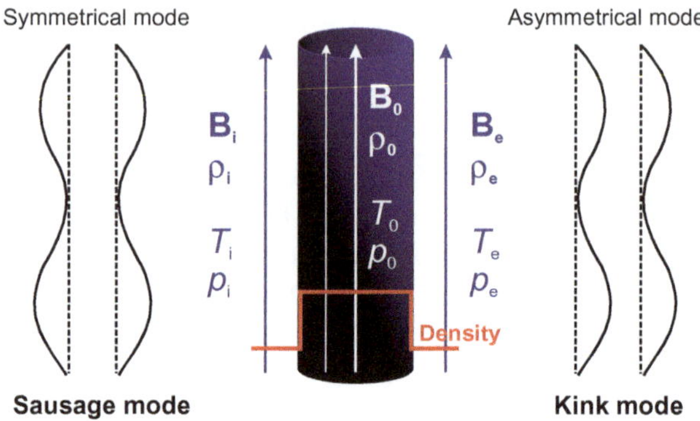

Fig. 3.4. Magnetic field topology and plasma density profile in a cylindrical magnetic flux tube.

equation (3.9)

$$\nabla p_0 + \frac{1}{\mu} \boldsymbol{B}_0 \times (\nabla \times \boldsymbol{B}_0 = \nabla \left(p_0 + \frac{B_0^2}{2\mu} \right) - \frac{1}{\mu} \boldsymbol{B}_0 \cdot \nabla \boldsymbol{B}_0 = 0,$$

with $\boldsymbol{B}_0 = (0, B_{0\phi}(r), B_{0z}(r))$ and $p_0 = p_0(r)$ has the form:

$$\frac{\mathrm{d}}{\mathrm{d}r} \left(p_0(r) + \frac{B_{0\phi}^2 + B_{0z}^2}{2\mu} \right) + \frac{B_{0\phi}^2}{\mu r} = 0.$$

For straight magnetic fields, that is, $B_{0\phi}(r) = 0$ and $B_{0z}(r) = B_0$, this implies that the total internal pressure (thermal plus magnetic) should be equal to the total outer pressure

$$p_\mathrm{i} + \frac{B_\mathrm{i}^2}{2\mu} = p_\mathrm{e} + \frac{B_\mathrm{e}^2}{2\mu}. \tag{3.33}$$

Sound speed, c_s, and Alfvén speed, v_A, are the characteristics of the flux tube. These sound and Alfvén speeds are associated with a slow speed, c_T which is called the tube (or cusp) speed (Roberts and Webb, 1978)

$$c_\mathrm{T} = \frac{c_\mathrm{s} v_\mathrm{A}}{\sqrt{c_\mathrm{s}^2 + v_\mathrm{A}^2}}. \tag{3.34}$$

On the other hand, the densities of these two regions are related through sound and Alfvén speeds in the following manner:

$$\frac{\rho_\mathrm{e}}{\rho_\mathrm{i}} = \frac{c_\mathrm{si}^2 + \frac{1}{2}\gamma\, v_\mathrm{Ai}^2}{c_\mathrm{se}^2 + \frac{1}{2}\gamma\, v_\mathrm{Ae}^2}. \tag{3.35}$$

Here, c_{si}, c_{se}, v_{Ai}, v_{Ae} and γ are flux tube's sound speed, environment's sound speed, flux tube's Alfvén, speed environment's Alfvén speed, and ratio of specific heats, respectively. Using linearization technique, one can derive the dispersion relation of the fast and slow magnetosonic waves from the MHD equations. Such a derivation will be presented in the next chapter. Here, we shall only mention that there is two types of modes, namely *sausage* (symmetrical) mode and *kink* (antisymmetrical) mode (see Fig. 3.4). One should mention also that pure slow acoustic modes can exist in the flux tubes along with pure incompressible torsional Alfvén waves which are resulted in the flux tube due to the perturbations of the azimuthal components of the velocity and the magnetic field (see Fig. 3.5).

In magnetically structured entities like slabs or flux tubes, one observes jump conditions across the so-called *discontinuity* (the interface between the two media). There exist two types of discontinuities, namely *contact* and *tangential* discontinuities. These discontinuities are transition layers across which there is no particle transport. Thus, in the frame moving with the discontinuity, $v_{n1} = v_{n2}$ (the subscript n refers to the normal component of a vector with respect the discontinuity, and the subscripts 1 and 2 refer to the two states of the plasma on each side of the discontinuity).

Contact discontinuities are discontinuities for which the thermal pressure, the magnetic field and the velocity are continuous. Only the mass density and temperature change (see the left panel of Fig. 3.6).

Tangential discontinuities are discontinuities for which the total pressure (sum of the thermal and magnetic pressures) is conserved. The normal component of the magnetic field is identically zero. The density, thermal pressure and tangential

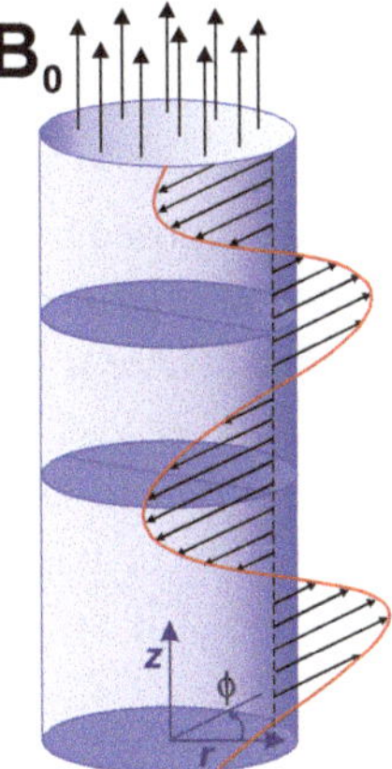

Fig. 3.5. Torsional ($m = 0$) Alfvén wave in a cylindrical flux tube.

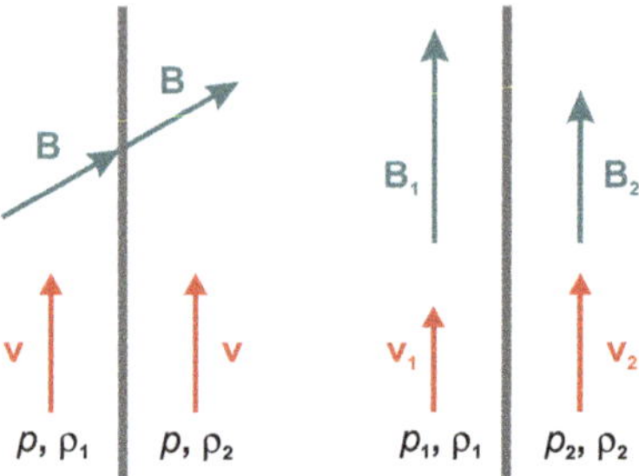

Fig. 3.6. (*Left panel*) Contact discontinuity. (*Right panel*) Tangential discontinuity.

component of the magnetic field vector can be discontinuous across the layer (see the right panel of Fig. 3.6).

Discontinuities play an important role in solving any problem in the real astrophysical structures. The finding solution to the MHD equations is not enough to solve the problem. It is necessary to impose boundary conditions in the places where discontinuities exists. It is the type of these discontinuities that determine the kind of boundary conditions.

3.5   Magnetohydrodynamic instabilities

Plasma equilibria could be unstable: this happens when a small perturbation grows in time or in space. Violent instabilities of plasma equilibria occurring in the configuration space, causing a major reconstruction of the plasma configuration, are known as "macro-instabilities" or "MHD-instabilities." They may be responsible for triggering impulsive energy releases in astrophysical and laboratory plasmas.

3.5.1   *Rayleigh–Taylor instability*

The Rayleigh–Taylor instability, first proposed by Lord Rayleigh (Rayleigh, 1882) and Taylor (Taylor, 1950), plays a key role in many astrophysical systems. For a contact discontinuity that is formed where a heavy fluid is supported above a light fluid against gravity, this boundary is unstable to perturbations that grow by converting gravitational potential energy into kinetic energy creating rising and falling fingers. If $\rho_0^{(+)}$ is the density of the heavier fluid and $\rho_0^{(-)}$ of the lighter one (see Fig. 3.7) and we account for gravity, if the magnetic field is vertical, $\boldsymbol{B}_0 \| \hat{z}$, then the wave dispersion relation is

$$\omega^2 = -gk\frac{\rho_0^{(+)} - \rho_0^{(-)}}{\rho_0^{(+)} + \rho_0^{(-)}} = -gkA, \tag{3.36}$$

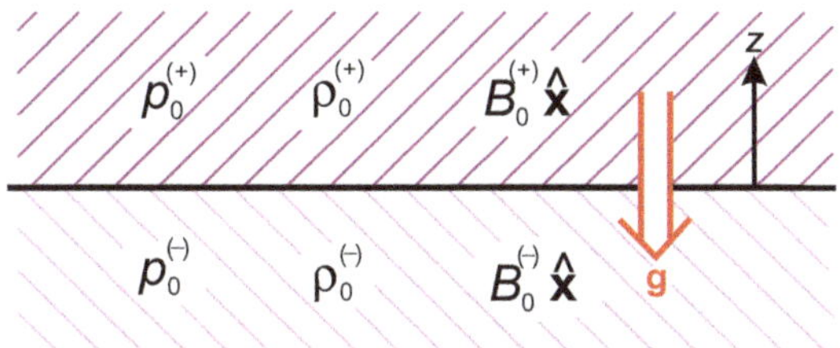

Fig. 3.7. Typical Rayleigh–Taylor instability configuration.

where

$$A = \frac{\rho_0^{(+)} - \rho_0^{(-)}}{\rho_0^{(+)} + \rho_0^{(-)}}$$

is the so-called *Atwood number*. If $A > 0$, that is, $\rho_0^{(+)} > \rho_0^{(-)}$, the wave is unstable ($\omega^2$ is negative) and its growth rate is $\mathrm{Im}(\omega) = gAk$.

If the magnetic field is horizontal, $\boldsymbol{B}_0 \| \hat{x}$, the wave dispersion relation is

$$\omega^2 = -gk\frac{\rho_0^{(+)} - \rho_0^{(-)}}{\rho_0^{(+)} + \rho_0^{(-)}} + \frac{2(\boldsymbol{k} \cdot \boldsymbol{B}_0)^2}{\mu(\rho_0^{(+)} + \rho_0^{(-)})}. \tag{3.37}$$

Here the first term on the right-hand side is exactly the same as the hydrodynamic version of the instability as given in Eq. (3.36). The second term is that of a surface Alfvén wave and signifies how a magnetic field of strength B_0 in an arbitrary horizontal direction can suppress the growth of the magnetic Rayleigh–Taylor instability through magnetic tension. As the magnetic field has a direction, the influence of this force is anisotropic. As seen from Eq. (3.37), there will be no effect if $\boldsymbol{k} \perp \boldsymbol{B}_0$. If, however, $k = k_x$, the instability occurs only if $k < k_{\mathrm{crit}}$, where

$$k_{\mathrm{crit}} = \frac{g\mu(\rho_0^{(+)} - \rho_0^{(-)})}{2B_0^2}$$

is the *instability threshold*.

Note that if $g = 0$ and $\rho_0^{(+)} = \rho_0^{(-)}$, Eq. (3.37) describes Alfvén waves, $\omega^2 = v_{\mathrm{A}}^2 k^2 \cos^2\theta = v_{\mathrm{A}}^2 k_x^2$, running in x-direction.

3.5.2 *Kelvin–Helmholtz instability*

The Kelvin–Helmholtz instability (KHI) was first documented by Hermann von Helmholtz (Helmholtz, 1868) and Lord Kelvin (Kelvin, 1871). The KHI occurs when there are two adjacent fluids moving with a velocity shear between them

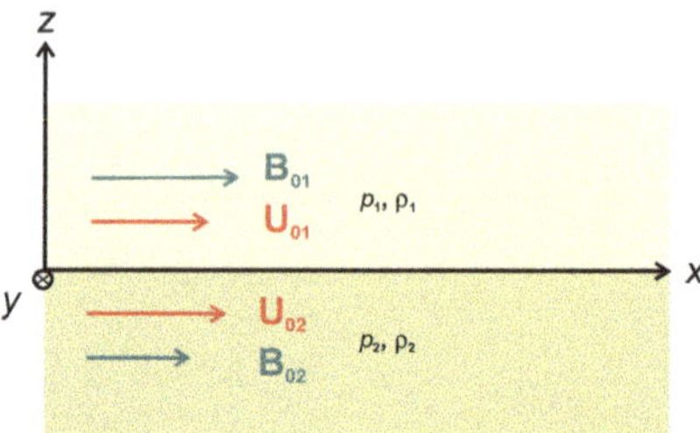

Fig. 3.8. Typical Kelvin–Helmholtz instability configuration.

(see Fig. 3.8). It occurs at a tangential discontinuity, which is the interface where an abrupt change in the flow velocity, density, temperature, and magnetic field is observed. When there is a perturbation of the boundary between the two fluids, it causes a constriction of one of the fluids. This leads to increased velocity and reduced pressure. For the other fluid, the boundary is expanding, and there is reduced flow and increased pressure. Because of this, a deformation of the boundary leads to a pressure gradient which creates a pressure force in the same direction as the deformation. Thus, a small perturbation on the boundary between the two fluids will grow. As fluid from one side moves to the other side of the boundary, it is then accelerated by the surrounding fluid, and it begins to roll up. This is what causes the vortices that characterize this instability.

Let us consider how the KHI develops on a tangential discontinuity between two homogeneous non-gravity half-spaces 1 ($z < 0$) and 2 ($z > 0$). At the tangential discontinuity all parameters change their values. The vector velocity U_0 and magnetic field B_0 change their values and directions lying in the plane parallel to the interface ($U_{0z} = B_{0z} = 0$). We assume that the perturbations of the equilibrium parameters are in the form

$$\tilde{f}(x, y, z, t) \propto f(z) \exp\left[i(k_x x + k_y y - \omega t)\right].$$

The linearization of ideal MHD equations for incompressible media

$$\nabla \cdot \tilde{\boldsymbol{v}} = 0,$$

$$\rho_0 \left(\frac{\partial}{\partial t} + \boldsymbol{U}_0 \cdot \nabla\right) \tilde{\boldsymbol{v}} = -\nabla\left(\tilde{p} + \frac{\boldsymbol{B}_0 \cdot \tilde{\boldsymbol{B}}}{\mu}\right) + \frac{1}{\mu}(\boldsymbol{B}_0 \cdot \nabla)\tilde{\boldsymbol{B}} = 0,$$

$$\frac{\partial \tilde{\boldsymbol{B}}}{\partial t} = \nabla \times \left(\tilde{\boldsymbol{v}} \times \boldsymbol{B}_0 + \boldsymbol{U}_0 \times \tilde{\boldsymbol{B}}\right)$$

yields the following ordinary differential equations for the Lagrangian vertical displacement ξ_z and the total pressure perturbation $p_{\text{tot}} = \tilde{p} + \boldsymbol{B}_0 \cdot \tilde{\boldsymbol{B}}/\mu$

(Mishin and Tomozov, 2016):

$$p_{\text{tot}} = \rho_0 \frac{\Omega^2}{k^2} \frac{\mathrm{d}\xi_z}{\mathrm{d}z}, \tag{3.38}$$

$$\frac{\mathrm{d}p_{\text{tot}}}{\mathrm{d}z} = \rho_0 \Omega^2 \xi_z. \tag{3.39}$$

Here

$$\Omega^2 = (\omega - \boldsymbol{k} \cdot \boldsymbol{U}_0)^2 - (\boldsymbol{k} \cdot \boldsymbol{v}_{\text{A}})^2, \quad \boldsymbol{k} = (k_x, k_y, 0), \quad \boldsymbol{v}_{\text{A}} = \boldsymbol{B}_0/\sqrt{\mu\rho_0},$$

and $\omega = \mathrm{Re}(\omega) + \mathrm{i}\,\mathrm{Im}(\omega)$ is the complex wave angular frequency. We note also that the fluid velocity perturbation $\tilde{\boldsymbol{v}}$ is given by the Lagrangian displacement $\boldsymbol{\xi}$ through the relation $\tilde{\boldsymbol{v}} = \partial\boldsymbol{\xi}/\partial t$.

By integrating Eqs. (3.38) and (3.39) through a thin layer across the interface, one obtains two boundary conditions for continuity of the vertical displacement and the continuity of the total pressure:

$$\xi_{z1} = \xi_{z2} \quad \text{at } z = 0,$$
$$\tilde{p}_1 + \boldsymbol{B}_{01} \cdot \tilde{\boldsymbol{B}} = \tilde{p}_2 + \boldsymbol{B}_{02} \cdot \tilde{\boldsymbol{B}} \quad \text{at } z = 0. \tag{3.40}$$

Labels 1 and 2 characterize the parameters on both sides of the interface. The set of Eqs. (3.38) and (3.39) has a damped solution in homogeneous domains:

$$f_1(z) = C_1 \exp(kz) \quad \text{for } z < 0,$$
$$f_2(z) = C_2 \exp(-kz) \quad \text{for } z > 0, \tag{3.41}$$

where C_1 and C_2 are constants. These solutions show that surface perturbations are generated and their amplitude decreases exponentially fast with distance away from the boundary. Substitution of solutions (3.41) in the boundary conditions (3.40) yields the wave dispersion equation on an incompressible tangential discontinuity:

$$\omega = \frac{\boldsymbol{k} \cdot (\rho_{01}\boldsymbol{U}_{01} + \rho_{02}\boldsymbol{U}_{02})}{\rho_{01} + \rho_{02}}$$
$$\pm \mathrm{i}\frac{\sqrt{\rho_{01}\rho_{02}(\boldsymbol{k} \cdot \Delta\boldsymbol{U}_0)^2 - (\rho_{01} + \rho_{02})(\boldsymbol{k} \cdot \boldsymbol{B}_{01})^2 + (\boldsymbol{k} \cdot \boldsymbol{B}_{02})^2/\mu}}{\rho_{01} + \rho_{02}}. \tag{3.42}$$

Here, $\Delta\boldsymbol{U}_0 = \boldsymbol{U}_{02} - \boldsymbol{U}_{01}$. It is seen that the wave becomes unstable if

$$(\boldsymbol{k} \cdot \Delta\boldsymbol{U}_0)^2 > \frac{(\rho_{01} + \rho_{02})(\boldsymbol{k} \cdot \boldsymbol{B}_{01})^2 + (\boldsymbol{k} \cdot \boldsymbol{B}_{02})^2/\mu}{\rho_{01}\rho_{02}}$$

and KHI instability growth rate is

$$\mathrm{Im}(\omega) = \frac{\sqrt{\rho_{01}\rho_{02}(\boldsymbol{k} \cdot \Delta\boldsymbol{U}_0)^2 - (\rho_{01} + \rho_{02})(\boldsymbol{k} \cdot \boldsymbol{B}_{01})^2 + (\boldsymbol{k} \cdot \boldsymbol{B}_{02})^2/\mu}}{\rho_{01} + \rho_{02}}. \tag{3.43}$$

In the homogeneous case, $\rho_{01} = \rho_{02}$ and $\boldsymbol{B}_{01} = \boldsymbol{B}_{02}$, we have

$$\mathrm{Im}(\omega) = \left[(\boldsymbol{k} \cdot \Delta \boldsymbol{U}_0)^2 - 4(\boldsymbol{k} \cdot \boldsymbol{v}_\mathrm{A})^2 \right]^{1/2},$$

and

$$\sqrt{\frac{(\rho_{01} + \rho_{02})(\boldsymbol{k} \cdot \boldsymbol{B}_{01})^2 + (\boldsymbol{k} \cdot \boldsymbol{B}_{02})^2/\mu}{\rho_{01}\rho_{02}}} = 2v_\mathrm{A}.$$

KHI takes place in cylindrical geometry too under the condition that the velocity shear at tube boundary exceeds some critical value. That will be demonstrated shortly in the next chapters. Here, we have considered the case of KHI in incompressible homogeneous flowing plasmas. The answer to the question how the gravity, the plasma compressibility, the plasma density inhomogeneity, the plasma resistivity, and the magnetic field inhomogeneity can influence the MHD wave propagation and the KHI the reader can find in Chandrasekhar (1961), Goedbloed and Poedts (2004), Goedbloed *et al.* (2010), Priest (2014) and Roberts (2019).

3.5.3 *Sausage and kink instabilities*

Consider a cylindrical plasma column (with radius a) that is confined (or pinched) by the toroidal magnetic field B_ϕ due to current I flowing along its surface or through its interior

$$B_\phi(r) = \frac{\mu I}{2\pi r}.$$

When plasma (at pressure p_0 and density ρ_0) contains no magnetic field (interior), the column is unstable to the interchange mode ($\boldsymbol{k} \perp \boldsymbol{B}$), since the confining field is concave to plasma. The place where it pinched (see Fig. 3.9), the toroidal field is increased and the radius is decreased. Therefore, magnetic pressure and tension increase which implies that the inward force is no longer balanced with the gas pressure and subsequently the perturbation grows. The place where the tube boundary bulges out, the toroidal field decreases and the radius increases. Therefore, the

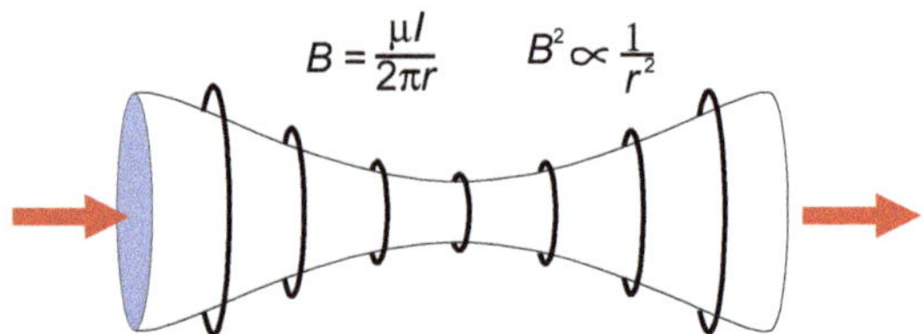

Fig. 3.9. Sausage instability configuration.

magnetic pressure and the tension decrease and the perturbation grows. This instability is so-called *sausage instability*—it corresponds to the $m = 0$ mode number, that is, to the sausage MHD mode. This instability is unstable in all wavelengths. The growth rate of sausage instability with the wavenumber $k \approx a^{-1}$ is

$$\mathrm{Im}(\omega) = \frac{\sqrt{2p_0/\rho_0}}{a}. \tag{3.44}$$

Note that the sausage wave is compressible one. The cylindrical plasma column can be stabilized against the sausage mode by the presence of a large enough axial magnetic field, B_{0z}. In that case, bearing in mind that the value of toroidal magnetic field at the interface is B_ϕ, the force balance on the interface gives

$$p_0 + \frac{B_{0z}^2}{2\mu} = \frac{B_\phi^2}{2\mu}.$$

The effect of Alfvén wave propagating along the axis with speed $v_{Az} = B_{0z}/\sqrt{\mu\rho_0}$ is to modify the dispersion relation to

$$\omega^2 = -\frac{2p_0}{\rho_0 a^2} + \frac{B_{0z}^2}{\mu\rho_0 a^2}. \tag{3.45}$$

Hence, the aforementioned force balance gives stability ($\omega^2 > 0$) when

$$B_{0z}^2 > \frac{1}{2}B_\phi^2. \tag{3.46}$$

Let us now consider the perturbation of kink kind to a cylindrical plasma column. Inside kinked plasma column, the magnetic pressure becomes strong, while outside of the kinked plasma column, the magnetic pressure becomes weak and subsequently the perturbation grows (becomes unstable) (see Fig. 3.10). This instability is so-called *kink instability*—it corresponds to the $m = 1$ mode number,

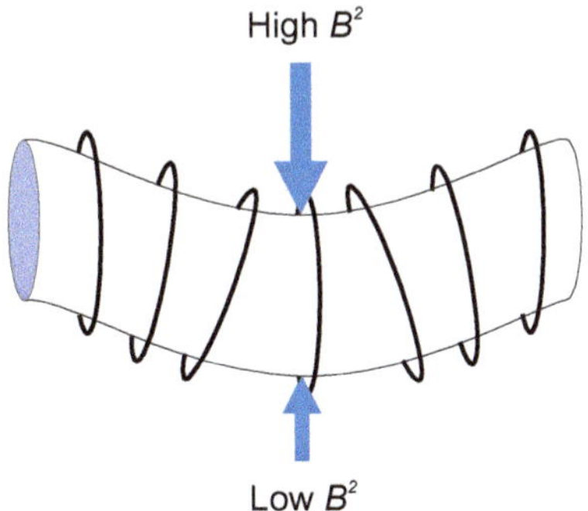

Fig. 3.10. Kink instability configuration.

that is, to the kink MHD mode. The fast kink mode is polarized quasi-perpendicular to the magnetic field and has $\cos\phi$ symmetry. It displaces the loop axis and is quasi-incompressible. An axial magnetic field in the cylindrical plasma column also affects and can stabilize this instability. The condition of stability for the kink mode is

$$\left(\frac{B_\phi}{B_z}\right)^2 < (ka)^2 = \left(\frac{2\pi a}{\lambda}\right)^2. \tag{3.47}$$

This inequality is the *Kruskal–Shafranov criterion for stability* (see Kruskal and Tuck, 1958; Shafranov, 1956). If the perturbed wavelength is long enough, the plasma column with helical magnetic field is unstable against the kink instability. In that case, the pitch of the perturbation follows the pitch of the helix, that is, the crests/troughs of the perturbations follow the field lines of the tube, $B_\phi/r + kB_z \geqslant 0$. Note also that the perturbation of the Lorentz force $\boldsymbol{j}_1 \times \boldsymbol{B}_0$ is zero.

Chapter 4

Normal Magnetohydrodynamic Modes in Solar Jets

4.1   Jet geometry, basic MHD equations, and wave dispersion relation

As we already have mentioned in the previous section, it is well established by observations that the solar atmosphere is magnetically structured: one can distinguish vertical or slightly inclined magnetic flux tubes having more often cylindrical shape. The plasma inside the flux tube can be at rest, but also moving with a velocity v_0 with respect to its environment. In the latter case, we talk for solar jets, such as spicules, plumes, dark mottles, macrospicules, extreme ultraviolet (EUV) jets, X-ray jets, coronal mass ejections (CMEs), etc. All jets, being themselves static or moving, support the propagation of the aforementioned wave modes. The question which immediately arises is how the spatially bounded structures will modify the principal MHD wave modes propagating in unbounded magnetized plasmas. To answer that question, we model a solar jet as a moving straight vertical cylinder (see Fig. 4.1) with radius a filled with ideal compressible plasma of density ρ_i and immersed in a constant magnetic field B_i directed along the z-axis. The most natural discontinuity, which occurs at the surface binding the cylinder, is the tangential one because it is the discontinuity that ensures an equilibrium total pressure balance. Moreover, it is worth noting that the jet is non-rotating and without twist—otherwise the centrifugal and the magnetic tension forces should be taken into account; later on, we will study how the twist and rotating will affect waves's propagation. The flow velocity, v_0, like the ambient magnetic field, is directed along the z-axis. The flow velocity, v_0, is an input parameter whose value determines the stability/instability status of the jet. Other input parameters are the density contrast $\eta = \rho_e/\rho_i$ (do not confuse with magnetic field diffusivity) and the ratio of external and internal equilibrium magnetic fields $b = B_e/B_i$.

51

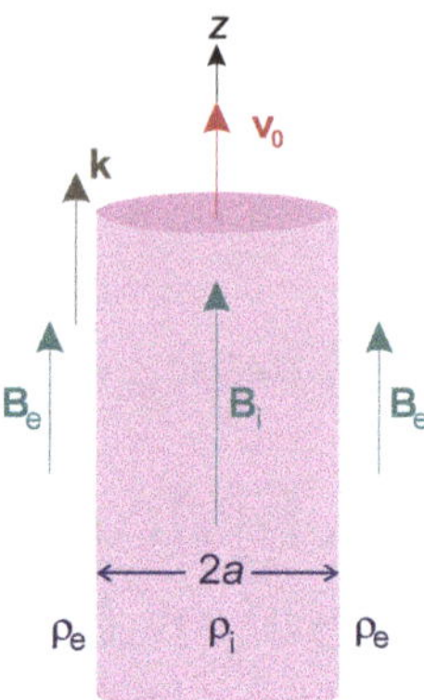

Fig. 4.1. Geometry of a solar flux tube containing flowing plasma with velocity v_0.

As we will study a linear wave propagation, the basic MHD variables can be presented in the form

$$\rho = \rho_0 + \rho_1, \quad p = p_0 + p_1, \quad v = v_0 + v_1 \quad \text{and} \quad \boldsymbol{B} = \boldsymbol{B}_0 + \boldsymbol{B}_1,$$

where ρ_0, p_0, and $\boldsymbol{B}_0$ are the equilibrium values in either medium, v_0 is the flow velocity, ρ_1, p_1, v_1, and $\boldsymbol{B}_1$ being the small perturbations of the basic MHD variables. For convenience, we chose the frame of reference to be attached to the ambient medium.

With the above assumptions, the basic set of MHD equations for the perturbations of the mass density, pressure, fluid velocity, and magnetic field become

$$\frac{\partial}{\partial t}\rho_1 + (\boldsymbol{v}_0 \cdot \nabla)\rho_1 + \rho_0 \nabla \boldsymbol{v}_1 = 0, \tag{4.1}$$

$$\rho_0 \frac{\partial}{\partial t}\boldsymbol{v}_1 + \rho_0 (\boldsymbol{v}_0 \cdot \nabla)\,\boldsymbol{v}_1 + \nabla \left(p_1 + \frac{1}{\mu}\boldsymbol{B}_0 \cdot \boldsymbol{B}_1\right) - \frac{1}{\mu}(\boldsymbol{B}_0 \cdot \nabla)\boldsymbol{B}_1 = 0, \tag{4.2}$$

$$\frac{\partial}{\partial t}\boldsymbol{B}_1 + (\boldsymbol{v}_0 \cdot \nabla)\boldsymbol{B}_1 - (\boldsymbol{B}_0 \cdot \nabla)\boldsymbol{v}_1 + \boldsymbol{B}_0 \nabla \cdot \boldsymbol{v}_1 = 0, \tag{4.3}$$

$$\frac{\partial}{\partial t}p_1 + (\boldsymbol{v}_0 \cdot \nabla)p_1 + \gamma\, p_0 \nabla \cdot \boldsymbol{v}_1 = 0, \tag{4.4}$$

$$\nabla \cdot \boldsymbol{B}_1 = 0. \tag{4.5}$$

We note that the gravity force term in momentum Eq. (4.2) has been omitted because one assumes that the mass density of the jet does not change appreciably in the limits of the jet's length.

From Eq. (4.1), we obtain that

$$\nabla \cdot \boldsymbol{v}_1 = -\frac{1}{\rho_0}\left[\frac{\partial}{\partial t}p_1 + (\boldsymbol{v}_0 \cdot \nabla)p_1\right].$$

(4.6)

Inserting this expression into Eq. (4.4), we get

$$\left[\frac{\partial}{\partial t} + (\boldsymbol{v}_0 \cdot \nabla)\right]p_1 - c_{\rm s}^2\left[\frac{\partial}{\partial t} + (\boldsymbol{v}_0 \cdot \nabla)\right]\rho_1 = 0,$$

which means that the pressure's and density's perturbations are related via the expression

$$p_1 = c_{\rm s}^2\rho_1, \quad \text{where } c_{\rm s} = (\gamma \, p_0/\rho_0)^{1/2}.$$

(4.7)

Assuming that each perturbation is presented, in cylindrical coordinates (r, ϕ, z), in the form $g(r)\exp i(-\omega t + m\phi + k_z z)$ with its amplitude $g(r)$ being just a function of r, and bearing in mind that in these coordinates the nabla operator has the form

$$\nabla \equiv \frac{\partial}{\partial r}\hat{r} + \frac{1}{r}\frac{\partial}{\partial \phi}\hat{\phi} + \frac{\partial}{\partial z}\hat{z},$$

Eq. (4.2) reads

$$-i\rho_0(\omega - \boldsymbol{k} \cdot \boldsymbol{v}_0)v_{1r} + \frac{d}{dr}\left(p_1 + \frac{1}{\mu}B_0 B_{1z}\right) - ik_z\frac{1}{\mu}B_0 B_{1r} = 0,$$

(4.8)

$$-\rho_0(\omega - \boldsymbol{k} \cdot \boldsymbol{v}_0)v_{1\phi} + \frac{m}{r}\left(p_1 + \frac{1}{\mu}B_0 B_{1z}\right) - k_z\frac{1}{\mu}B_0 B_{1\phi} = 0,$$

(4.9)

$$-\rho_0(\omega - \boldsymbol{k} \cdot \boldsymbol{v})v_{1z} + k_z\left(p_1 + \frac{1}{\mu}B_0 B_{1z}\right) - k_z\frac{1}{\mu}B_0 B_{1z} = 0.$$

(4.10)

Induction equation (4.3) gives

$$(\omega - \boldsymbol{k} \cdot \boldsymbol{v}_0)B_{1r} - k_z B_0 v_{1r} = 0,$$

(4.11)

$$(\omega - \boldsymbol{k} \cdot \boldsymbol{v}_0)B_{1\phi} - k_z B_0 v_{1\phi} = 0,$$

(4.12)

$$-i(\omega - \boldsymbol{k} \cdot \boldsymbol{v}_0)B_{1z} - ik_z B_0 v_{1z} + B_0\nabla \cdot \boldsymbol{v}_1 = 0.$$

(4.13)

Accordingly from Eq. (4.4) one obtains

$$-i(\omega - \boldsymbol{k} \cdot \boldsymbol{v}_0)p_1 + \gamma \, p_0\nabla \cdot \boldsymbol{v}_1 = 0.$$

(4.14)

Finally Eq. (4.5) yields

$$\frac{\mathrm{d}}{\mathrm{d}r} B_{1r} + \frac{1}{r} B_{1r} + \mathrm{i}\frac{m}{r} B_{1\phi} + \mathrm{i}k_z B_{1z} = 0. \tag{4.15}$$

From Eqs. (4.11)–(4.13), we express magnetic field perturbations via the fluid velocity perturbations to obtain that

$$B_{1r} = -\frac{k_z B_0}{\omega - \boldsymbol{k} \cdot \boldsymbol{v}_0} v_{1r}, \tag{4.16}$$

$$B_{1\phi} = -\frac{k_z B_0}{\omega - \boldsymbol{k} \cdot \boldsymbol{v}_0} v_{1\phi}, \tag{4.17}$$

$$B_{1z} = -\frac{k_z B_0}{\omega - \boldsymbol{k} \cdot \boldsymbol{v}_0} v_{1z} - \mathrm{i}\frac{B_0}{\omega - \boldsymbol{k} \cdot \boldsymbol{v}_0} \nabla \cdot \boldsymbol{v}_1. \tag{4.18}$$

On the other hand, from momentum component (4.8)–(4.10), one can express the velocity perturbations as functions of the magnetic field perturbations and the total pressure perturbation $p_{\mathrm{tot}} = p_1 + \frac{1}{\mu} B_0 B_{1z}$, to have

$$v_{1r} = -\frac{1}{\rho_0} \frac{1}{\omega - \boldsymbol{k} \cdot \boldsymbol{v}_0} \left[\mathrm{i}\frac{\mathrm{d}}{\mathrm{d}r} p_{\mathrm{tot}} + \frac{k_z B_0}{\mu} B_{1r} \right], \tag{4.19}$$

$$v_{1\phi} = \frac{1}{\rho_0} \frac{1}{\omega - \boldsymbol{k} \cdot \boldsymbol{v}_0} \left[\frac{m}{r} p_{\mathrm{tot}} - \frac{k_z B_0}{\mu} B_{1\phi} \right], \tag{4.20}$$

$$v_{1z} = \frac{1}{\rho_0} \frac{1}{\omega - \boldsymbol{k} \cdot \boldsymbol{v}_0} \left[k_z p_{\mathrm{tot}} - \frac{k_z B_0}{\mu} B_{1z} \right]. \tag{4.21}$$

From Eq. (4.14), we obtain that

$$\gamma\, p_0 \nabla \cdot \boldsymbol{v}_1 = \mathrm{i}\,(\omega - \boldsymbol{k} \cdot \boldsymbol{v}_0)\, p_1 + \mathrm{i}\,(\omega - \boldsymbol{k} \cdot \boldsymbol{v}_0)\, \frac{B_0 B_{1z}}{\mu} - \mathrm{i}\,(\omega - \boldsymbol{k} \cdot \boldsymbol{v}_0)\, \frac{B_0 B_{1z}}{\mu},$$

which can be rearranged in the form

$$\nabla \cdot \boldsymbol{v}_1 = \mathrm{i}\frac{\omega - \boldsymbol{k} \cdot \boldsymbol{v}_0}{\gamma\, p_0} \left(p_{\mathrm{tot}} - \frac{B_0 B_{1z}}{\mu} \right).$$

Multiplying and dividing the first multiplier by ρ_0 and similarly the magnetic pressure (inside the brackets) by B_0, we finally have

$$\nabla \cdot \boldsymbol{v}_1 = \mathrm{i}\frac{\omega - \boldsymbol{k} \cdot \boldsymbol{v}_0}{c_{\mathrm{s}}^2} \left[\frac{1}{\rho_0} p_{\mathrm{tot}} - \frac{v_{\mathrm{A}}^2}{B_0} B_{1z} \right]. \tag{4.22}$$

Going back to expressions (4.16)–(4.18) and inserting there the perturbations of the fluid velocity along with $\nabla \cdot \boldsymbol{v}_1$, we obtain the magnetic field perturbations expressed in terms of p_{tot}:

$$B_{1r} = \frac{i}{\rho_0} \frac{k_z B_0}{(\omega - \boldsymbol{k} \cdot \boldsymbol{v}_0)^2 - k_z^2 v_A^2} \frac{d}{dr} p_{\text{tot}}, \tag{4.23}$$

$$B_{1\phi} = -\frac{1}{\rho_0} \frac{k_z B_0}{(\omega - \boldsymbol{k} \cdot \boldsymbol{v}_0)^2 - k_z^2 v_A^2} \frac{m}{r} p_{\text{tot}}, \tag{4.24}$$

$$B_{1z} = \frac{B_0}{\rho_0} \frac{(\omega - \boldsymbol{k} \cdot \boldsymbol{v}_0)^2 - k_z^2 c_s^2}{(c_s^2 + v_A^2)\left[(\omega - \boldsymbol{k} \cdot \boldsymbol{v}_0)^2 - k_z^2 c_T^2\right]} p_{\text{tot}}, \tag{4.25}$$

where

$$c_T = \frac{c_s v_A}{\left(c_s^2 + v_A^2\right)^{1/2}} \tag{4.26}$$

is the defined in Section 3 *tube velocity* (Roberts and Webb, 1978). Now we insert these final expressions of magnetic field perturbations in terms of p_{tot} into Eq. (4.15) and obtain a second-order ordinary differential equation for the total pressure perturbation:

$$\left[\frac{d^2}{dr^2} + \frac{1}{r}\frac{d}{dr} - \left(\kappa^2 + \frac{m^2}{r^2}\right)\right] p_{\text{tot}} = 0. \tag{4.27}$$

Here, κ^2 is given by the expression

$$\kappa^2 = -\frac{\left[(\omega - \boldsymbol{k} \cdot \boldsymbol{v}_0)^2 - k_z^2 c_s^2\right]\left[(\omega - \boldsymbol{k} \cdot \boldsymbol{v}_0)^2 - k_z^2 v_A^2\right]}{(c_s^2 + v_A^2)\left[(\omega - \boldsymbol{k} \cdot \boldsymbol{v}_0)^2 - k_z^2 c_T^2\right]}, \tag{4.28}$$

where, recall, the tube velocity is given by Eq. (4.26). It is important to notice that both κ^2 (respectively κ) and the tube velocity, c_T, *have different values inside and outside the jet* due to the different sound and Alfvén speeds, which characterize correspondingly the jet and its surrounding medium.

As can be seen, Eq. (4.27) is the equation for the modified Bessel functions I_m and K_m of first and second kind, respectively, and, accordingly, its solutions in both media (the jet and its environment) are:

$$p_{\text{tot}}(r) = \begin{cases} \alpha_i I_m(\kappa_i r) & \text{for } r \leqslant a, \\ \alpha_e K_m(\kappa_e r) & \text{for } r \geqslant a, \end{cases}$$

where α_i and α_e are constants.

From Eq. (4.19), by inserting in it the expression of B_{1r} given by Eq. (4.23), one gets a formula relating v_{1r} with the first derivative (with respect to r) of p_{tot}

$$v_{1r} = -\frac{\mathrm{i}}{\rho_0}\frac{\omega - \boldsymbol{k}\cdot\boldsymbol{v}_0}{(\omega - \boldsymbol{k}\cdot\boldsymbol{v}_0)^2 - k_z^2 v_\mathrm{A}^2}\frac{\mathrm{d}}{\mathrm{d}r}p_{\text{tot}}. \tag{4.29}$$

It is clear that we have two different expressions of v_{1r}, which, bearing in mind the solutions to the ordinary second order differential Eq. (4.27), read

$$v_{1r}(r \leqslant a) = -\frac{\mathrm{i}}{\rho_\mathrm{i}}\frac{\omega - \boldsymbol{k}\cdot\boldsymbol{v}_0}{(\omega - \boldsymbol{k}\cdot\boldsymbol{v}_0)^2 - k_z^2 v_\mathrm{Ai}^2}\kappa_\mathrm{i}\alpha_\mathrm{i} I'_m(\kappa_\mathrm{i} r)$$

and

$$v_{1r}(r \geqslant a) = -\frac{\mathrm{i}}{\rho_\mathrm{e}}\frac{\omega}{\omega^2 - k_z^2 v_\mathrm{Ae}^2}\kappa_\mathrm{e}\alpha_\mathrm{e} K'_m(\kappa_\mathrm{e} r),$$

respectively. Here, the sign prime, $\prime$, implies differentiation of the Bessel functions with respect to their arguments. Now it is time to apply some boundary conditions, which link the solutions of total pressure perturbation, p_{tot}, and Lagrangian displacement ξ_r (obtainable from relation $v_{1r} = \partial\xi_r/\partial t$) at the interface $r = a$. The appropriate boundary conditions are as follows:

- p_{tot} has to be continuous across the interface,
- the transverse displacement, ξ_r, has also to be continuous (Chandrasekhar, 1961).

After applying the boundary conditions (we recall that for the ambient medium $v_0 = 0$) finally we arrive at the required dispersion relation of the normal MHD modes propagating along the jet (Terra-Homem *et al.*, 2003; Nakariakov, 2007; Zhelyazkov, 2012)

$$\frac{\rho_\mathrm{e}}{\rho_\mathrm{i}}(\omega^2 - k_z^2 v_\mathrm{Ae}^2)\kappa_\mathrm{i}\frac{I'_m(\kappa_\mathrm{i} a)}{I_m(\kappa_\mathrm{i} a)} - \left[(\omega - \boldsymbol{k}\cdot\boldsymbol{v}_0)^2 - k_z^2 v_\mathrm{Ai}^2\right]\kappa_\mathrm{e}\frac{K'_m(\kappa_\mathrm{e} a)}{K_m(\kappa_\mathrm{e} a)} = 0. \tag{4.30}$$

For the azimuthal mode number $m = 0$ the above equation describes the propagation of so-called *sausage* waves, while with $m = 1$ it governs the propagation of the *kink* waves (Edwin and Roberts, 1983). As we have already seen, the wave frequency, ω, is Doppler-shifted inside the jet. The two quantities κ_i and κ_e, whose squared magnitudes are given by Eq. (4.28) are termed *wave attenuation coefficients*. They characterize how quickly the wave amplitude having its maximal value at the interface, $r = a$, decreases as we go away in both directions. Depending on the specific sound and Alfvén speeds in a given medium, as well as on the density contrast, $\eta = \rho_e/\rho_i$, and the ratio of the embedded magnetic fields, $b = B_e/B_i$, the attenuation coefficients can be real or imaginary quantities. In the case when both κ_i and κ_e are real, we have a *pure surface wave*. The case κ_i imaginary and κ_e real corresponds to *pseudosurface waves* (or *body waves* according to Edwin and Roberts terminology (Edwin and Roberts, 1983)). In that case, the modified Bessel function inside the jet, I_0, becomes the spatially periodic Bessel function J_0. In the opposite situation, the wave energy is carried away from the flux tube—then the wave is called *leaky wave* (Cally, 1986).

For the kink waves, one defines the *kink speed* (Edwin and Roberts, 1983)

$$c_k = \left(\frac{\rho_i v_{Ai}^2 + \rho_e v_{Ae}^2}{\rho_i + \rho_e} \right)^{1/2} = \left(\frac{1 + b^2}{1 + \eta} \right)^{1/2} v_{Ai}, \tag{4.31}$$

which is independent of sound speeds and characterizes the propagation of transverse perturbations.

4.1.1 *Derivation of wave dispersion relation on using the operator coefficient techniques*

An alternative way for derivation of the wave dispersion relation is the so-called operator coefficient techniques developed by Goossens *et al.* (1992). These authors, starting from the basic MHD equations (in which the perturbed velocity v_1 is expressed via the Lagrangian displacement $\boldsymbol{\xi}$ in the form $v_1 = \partial \boldsymbol{\xi}/\partial t$), have obtained the following governing equations for the radial component ξ_r of the Lagrangian displacement $\boldsymbol{\xi}$ and for the total pressure perturbation p_{tot}:

$$D \frac{d}{dr}(r\xi_r) = C_1 r\xi_r - C_2 r p_{tot}, \tag{4.32}$$

$$D \frac{dp_{tot}}{dr} = C_3 \xi_r - C_1 p_{tot}. \tag{4.33}$$

The coefficient functions D, C_1, C_2, and C_3 depend on the equilibrium variables ρ_0, $\boldsymbol{B}_0$, $\boldsymbol{v}_0$ and on the Doppler-shifted frequency $\Omega = \omega - \boldsymbol{k} \cdot \boldsymbol{v}_0$. In Erdélyi and Fedun (2010) notation, these coefficient functions are as follows:

$$D = \rho_0(\Omega^2 - \omega_A^2)C_4, \tag{4.34}$$

$$C_1 = \frac{2B_{0\phi}}{\mu r}\left(\Omega^4 B_{0\phi} - \frac{m}{r} f_B C_4 \right), \tag{4.35}$$

$$C_2 = \Omega^4 - \left(k_z^2 + \frac{m^2}{r^2} \right) C_4, \tag{4.36}$$

$$C_3 = \rho_0 D \left[\Omega^2 - \omega_A^2 + \frac{2 B_{0\phi}}{\mu \rho_0} \frac{\mathrm{d}}{\mathrm{d}r} \left(\frac{B_{0\phi}}{r} \right) \right] + 4\Omega^4 \left(\frac{B_{0\phi}^2}{\mu r} \right)^2 - \rho_0 C_4 \frac{4 B_{0\phi}^2}{\mu r^2} \omega_A^2,$$

$$(4.37)$$

where

$$C_4 = (c_s^2 + c_A^2)(\Omega^2 - \omega_c^2), \tag{4.38}$$

and

$$f_B = \frac{m}{r} B_{0\phi} + k_z B_{0z}, \quad \omega_A^2 = \frac{f_B^2}{\mu \rho_0}, \quad \omega_c^2 = \frac{c_s^2}{c_s^2 + c_A^2} \omega_A^2.$$

Here ω_A is the Alfvén frequency and ω_c is the cusp frequency; the other notation is standard.

After eliminating ξ_r, the set of Eqs. (4.32) and (4.33) can be rewritten in the form of a well-known second-order ordinary differential equation (Hain and Lüst, 1958; Goedbloed, 1971; Sakurai *et al.*, 1991)

$$\frac{\mathrm{d}^2 p_{\text{tot}}}{\mathrm{d}r^2} + \left[\frac{C_3}{rD} \frac{\mathrm{d}}{\mathrm{d}r} \left(\frac{rD}{C_3} \right) \right] \frac{\mathrm{d} p_{\text{tot}}}{\mathrm{d}r} + \left[\frac{C_3}{rD} \frac{\mathrm{d}}{\mathrm{d}r} \left(\frac{rC_1}{C_3} \right) + \frac{1}{D^2} \left(C_2 C_3 - C_1^2 \right) \right] p_{\text{tot}}$$
$$= 0. \tag{4.39}$$

Having obtained the solutions to Eq. (4.39) in both media, and finding the corresponding expressions for ξ_r, we can merge these solutions through appropriate boundary conditions at the interface $r = a$ and derive the dispersion relation of the normal modes propagating along the magnetic flux tube.

For a untwisted magnetic flux tube, the coefficient $C_1 = 0$ and $C_3 = \rho_0 D(\Omega^2 - \omega_A^2)$; then Eq. (4.39) takes the form

$$\frac{\mathrm{d}^2 p_{\text{tot}}}{\mathrm{d}r^2} + \frac{1}{r} \frac{\mathrm{d} p_{\text{tot}}}{\mathrm{d}r} - \left(m_0^2 + \frac{m^2}{r^2} \right) p_{\text{tot}} = 0, \tag{4.40}$$

where

$$m_0^2 = -\frac{(\Omega^2 - k_z^2 c_s^2)(\Omega^2 - k_z^2 v_A^2)}{(c_s^2 + v_A^2)(\Omega^2 - \omega_c^2)}. \tag{4.41}$$

The cusp frequency, ω_c, is usually expressed via the tube speed, c_T, notably $\omega_c = k_z c_T$, where (Edwin and Roberts, 1983)

$$c_T = \frac{c_s v_A}{\sqrt{c_s^2 + v_A^2}} \quad \text{with } c_s = (\gamma p_0/\rho_0)^{1/2} \text{ and } v_A = B_0/\sqrt{\mu \rho_0}.$$

In the expression of sound speed, c_s, γ is the adiabatic index. The solutions for p_{tot} can be written in terms of modified Bessel functions: $I_m(m_{0i}r)$ inside the jet and $K_m(m_{0e}r)$ in its environment, that is,

$$p_{tot}(r) = \begin{cases} \alpha_i I_m(m_{0i}r) & \text{for } r \leqslant a, \\ \alpha_e K_m(m_{0e}r) & \text{for } r \geqslant a, \end{cases}$$

where α_i and α_i are constants. By expressing the radial component of the Lagrangian displacement, ξ_r, from Eq. (4.33) in the form

$$\xi_r = \frac{1}{\rho_0(\Omega^2 - \omega_A^2)} \frac{d p_{tot}}{dr}, \tag{4.42}$$

bearing in mind the above solutions for $p_{tot}(r)$, we obtain the following expressions for ξ_r in the two media:

$$\xi_r(r \leqslant a) = \frac{1}{\rho_i(\Omega^2 - k_z^2 v_{Ai}^2)} \alpha_i m_{0i} I_m'(m_{0i}r),$$

$$\xi_r(r > a) = \frac{1}{\rho_e(\omega^2 - k_z^2 v_{Ae}^2)} \alpha_e m_{0e} K_m'(m_{0e}r).$$

By applying the boundary conditions for continuity of p_{tot} and ξ_r across the interface, $r = a$, we arrive at the dispersion relation of normal MHD modes propagating in a flowing compressible jet surrounded by a static compressible plasma

$$\frac{\rho_e}{\rho_i}(\omega^2 - k_z^2 v_{Ae}^2) m_{0i} \frac{I_m'(m_{0i}a)}{I_m(m_{0i}a)} - \left[(\omega - \boldsymbol{k} \cdot \boldsymbol{v_0})^2 - k_z^2 v_{Ai}^2 \right] m_{0e} \frac{K_m'(m_{0e}a)}{K_m(m_{0e}a)}$$
$$= 0, \tag{4.43}$$

which not surprisingly coincides with the previously derived dispersion equation (4.30). Here, the wave attenuation coefficients, $m_{0i,e}$, have the notation used in Edwin and Roberts (1983) as well as in many articles on surface MHD waves. Recall that the wave frequency is Doppler-shifted inside the jet.

Since we are looking for unstable solutions to wave dispersion relations, assume the MHD modes are running along the flux tube having a real axial wavenumber k_z and a complex angular wave frequency $\omega \equiv \text{Re}\,\omega + i\,\text{Im}\,\omega$. Numerical results are usually presented as dependence of the complex wave phase velocity $v_{ph} = \omega/k_z$ on the k_z. For convenience, we normalize all velocities with respect to the Alfvén speed inside the tube, v_{Ai}, and the wavelength $\lambda = 2\pi/k_z$ with respect to the tube radius, a. Thus we have a complex dimensionless wave phase velocity v_{ph}/v_{Ai} and dimensionless axial wavenumber $k_z a$. The normalization of sound, c_s and tube, c_T, speeds or cusp frequencies, ω_c in the attenuation coefficients (4.28) or (4.41), contained in dispersion equation (4.30) or (4.43), requires the values of both the reduced plasma betas $\tilde{\beta}_{i,e} = c_{i,e}^2/v_{Ai,e}^2$ and the magnetic fields ratio $b = B_e/B_i$. Then, as the imaginary part of the complex frequency/phase velocity at a given wave number, k_z, and a critical jet speed, v_0^{cr}, has some non-zero positive value, one says that the wave becomes unstable—its amplitude begins to grow with time. In this case, the linear theory is no longer applicable and one ought to investigate the further wave propagation by means of a nonlinear theory. Our linear approach can determine just the instability threshold only. In the dimensionless form of the wave dispersion relation the jet speed is presented by the ratio v_0/v_{Ai} termed Alfvén Mach number M_A. An estimation of the critical Alfvén Mach number at which the KHI of the kink ($m = 1$) mode starts can be obtained by the inequality (Zaqarashvili *et al.*, 2014a):

$$M_A^2 > (1 + 1/\eta)(b^2 + 1). \tag{4.44}$$

4.2   An example for finding unstable solutions to the wave dispersion relation

Let us consider an example for finding unstable solutions to the wave dispersion relation (4.30) of the kink ($m = 1$) MHD mode. As a typical flowing magnetized plasma we take the solar wind. We assume that, according to daily observations, the constant magnetic field inside the jet is $B_i = 5 \times 10^{-5}\,\mathrm{G}$ and the electron number density is $n_i = 2.43 \times 10^6\,\mathrm{m}^{-3}$ at 1 AU, which implies that the Alfvén speed inside the wind is $v_{Ai} = 70\,\mathrm{km\,s}^{-1}$. Concerning the sound speeds, c_{si} and c_{se}, it is reasonable to take them both equal to $70\,\mathrm{km\,s}^{-1}$, and the Alfvén speed outside the wind to be $v_{Ae} = 100\,\mathrm{km\,s}^{-1}$. From these speeds, by using the total pressure balance equation (the sum of kinetic and magnetic pressures to be constant)

$$p_i + \frac{B_i^2}{2\mu} = p_e + \frac{B_e^2}{2\mu},$$

that can be rearranged in the form

$$\frac{\rho_e}{\rho_i} = \frac{v_{si}^2 + \frac{5}{6}v_{Ai}^2}{v_{se}^2 + \frac{5}{6}v_{Ae}^2}, \tag{4.45}$$

one can obtain the density contrast $\eta \equiv \rho_e/\rho_i = 0.679$. The magnetic field ratio, b, can be derived from the ratio v_{Ae}^2/v_{Ai}^2 and it is equal to

$$b \equiv \frac{B_e}{B_i} = \frac{v_{Ae}}{v_{Ai}}\sqrt{\frac{\rho_e}{\rho_i}} = 1.177. \tag{4.46}$$

With this value of b one finds that the external magnetic field is $B_e \cong 5.9 \times 10^{-5}\,\mathrm{G}$. The rest two input parameters for the numerical solving of the wave dispersion relation, namely the reduced plasma betas of both media, $\tilde{\beta}_{i,e}$, are easily obtainable and their values are 1.0 and 0.49, respectively.

Prior to starting the numerical task, we can predict: (i) the nature of propagating kink mode, (ii) the kink speed (4.31), and (iii) the threshold Alfvén Mach number for instability onset (4.44). The mode's nature is determined from the ordering of basic speeds (sound, Alfvén, and tube one) in both media, which in our case is as follows:

$$c_{Ti} < c_{Te} < c_{si} = c_{se} = v_{Ai} < v_{Ae}.$$

Note that the values of the tube speeds are $c_{Ti} = 49.5\,\mathrm{km\,s^{-1}}$ and $c_{Te} = 57.3\,\mathrm{km\,s^{-1}}$, respectively. According to Cally (1986) (see Table I there), at this ordering the kink ($m = 1$) mode, propagating in a static ($v_0 = 0$) magnetic flux tube, would be a pseudosurface/body wave of B_+^+-type. Moreover, the wave phase velocity, $v_{ph} = \omega/k_z$, must be higher than the internal Alfvén speed v_{Ai}, but lower than the external Alfvén speed v_{Ae}.

The kink speed, being the highest wave phase speed at $v_0 = 0$, has a magnitude of $83.4\,\mathrm{km\,s^{-1}}$, which is higher than v_{Ai}, but lower than v_{Ae}. Its normalized value is 1.1919.

According to the instability criterion (4.44), one should expect the KHI to start at $M_A > 2.4287$.

It is worth mentioning that finding the solutions to transcendental dispersion equation in complex variables is no a trivial task (Acton, 1990). One can use, in principal, three root-finding algorithms that can be generalized for complex numbers, notably the well-known Newton–Raphson and secant methods (see, for example, Woodfoord and Phillips, 1997), or the Muller's method (Muller, 1956).

After this short introduction, we can show the results of the numerical computations—they are displayed in Fig. 4.2. In searching for unstable solutions (non-zero imaginary parts) we have started with a little bit larger value of M_A than that predicted from the instability condition (4.44). Diminishing with small steps that M_A value, we finally have found that the instability should start at $M_A^{cr} = 2.4287$, which coincides exactly with the predicted value. We note that one can obtain a unstable solution even for $M_A = 2.4285$, however, the magnitudes of the $\mathrm{Im}(v_{ph}/v_{Ai})$ are extremely low: 0.000000012 at $k_z a = 0.005$ and 0.001473933 for $k_z a = 4.0$. That is why, a more acceptable value of the threshold M_A is 2.4287. This threshold value determines the critical solar wind speed for the instability onset—its value is $170.0\,\mathrm{km\,s^{-1}}$, which indeed is an accessible solar-wind speed. From these plots, we also can get the wave phase velocity and wave frequency growth rate for a fixed unstable wavelength. If we assume that the tube radius is $a = 1000\,\mathrm{km}$ (or equivalently $1\,\mathrm{Mm}$) and wish to know

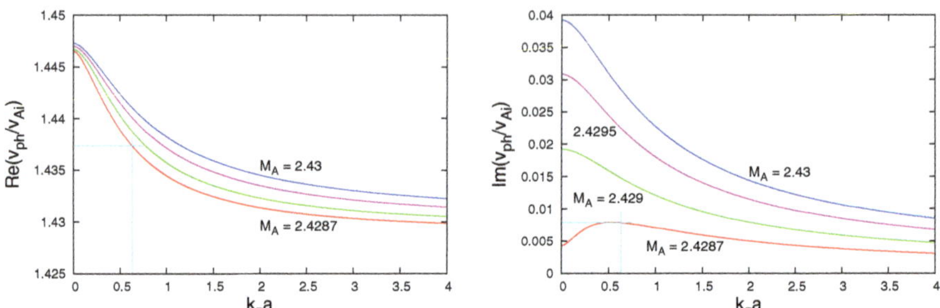

Fig. 4.2. (*Left panel*) Dispersion curves of forward propagating unstable kink waves at $\eta = 0.679$, $b = 1.177$ for $M_A = 2.4287$, 2.429, 2.4295, and 2.43. The red dispersion curve corresponds to the threshold Alfvén Mach number of 2.4287 at which the KHI emerges. (*Right panel*) Dimensionless growth rates of the unstable kink waves propagating along the flux tube at the same conditions.

what are the KHI parameters at $\lambda_{KH} = 10\,\text{Mm}$, we have to find out the values or normalized wave phase velocity and its normalized growth rate at $k_z a = 0.628$ (the dimensionless wavenumber that corresponds to $\lambda_{KH} = 10\,\text{Mm}$), from which to calculate those quantities in absolute units. The cyan lines in both plots fix the corresponding dimensionless values for the imaginary and real parts of the complex wave phase velocity, being equal to 0.007875829 and 1.437374269, respectively. These numbers yield

$$\gamma_{KH} = 0.3464 \times 10^{-3}\,\text{s}^{-1} \quad \text{and} \quad v_{ph} = 100.6\,\text{km s}^{-1}.$$

We see that the unstable kink mode with phase velocity of $100.6\,\text{km s}^{-1}$ is a super-Alfvénic wave. Having obtained the frequency growth rate, γ_{KH}, we can calculate the KHI developing/growth time

$$\tau_{KH} = \frac{2\pi}{\gamma_{KH}} = 18\,138.68\,\text{s} \cong 5.04\,\text{h}.$$

Obviously, this is a very slow Kelvin–Helmholtz instability, compared with the same instability in other jets, as we will see in the next chapters, but nevertheless its growth time is less the solar-wind lifetime from 0.5 to 2 days. Note also that a small increase in the Alfvén Mach number, M_A, can sensibly change the picture, that is, one can obtain shorter instability growth times.

The numerical calculations at $v_0 = 0$ ($M_A = 0$) reproduce the expected kink speed within four places behind the decimal point. That dispersion curve (not plotted here) possesses dimensionless wave phase velocity values from 1.19194 at $k_z a = 0.005$ to 1.11036 at $k_z a = 4.0$ in full agreement with Cally's (Cally, 1986) evaluations. Moreover, the computed attenuation coefficients confirm the prediction that the kink mode in a static tube is a pseudosurface/body wave. This is true for the stable waves with Alfvén Mach numbers less than M_A^{cr}. However, for all kink unstable waves with $M_A \geqslant M_A^{cr}$, the attenuation coefficients in both media are complex quantities which implies that they are neither pure surface, pseudosurface/body, or leaky waves—it is more logical to term them *generalized surface modes*. A more general approach in studying the evolution of wave dispersion curves depending on their Alfvén Mach numbers will be discussed shortly in the next chapter of the book.

Chapter 5

Kelvin–Helmholtz Instability in Solar Spicules

One of the most enduring mysteries in solar physics is why the Sun's outer atmosphere, or corona, is millions of kelvins hotter than its surface. Among the proposed theories for coronal heating is one that considers the role of spicules—first reported and hand drawn by the Jesuit Father Angelo Secchi (1877), solar spicules are often described as chromospheric plasma jets that protrude into the solar corona—in that process (see Athay and Holzer, 1982; Athay, 2000). For decades, it had been assumed that spicules might be sending heat into the corona. However, observational research in the 1980s found that spicule plasma did not reach coronal temperatures, and this line of study largely fell out of vogue. Kukhianidze *et al.* (2006) were the first to report the observation of kink waves in solar spicules—their wavelength was found to be $\sim$3500 km, and the period of waves was estimated to be in the range of 35–70 s. These authors argued that these waves may carry photospheric energy into the corona and therefore might be of importance to coronal heating. Zaqarashvili *et al.* (2007) analyzed consecutive height series of Hα spectra in solar limb spicules at the heights of 3800–8700 km above the photosphere, and detected Doppler-shift oscillations with periods of 20–25 and 75–110 s. According to authors, the oscillations could be caused by waves' propagation in thin magnetic flux tubes anchored in the photosphere. Moreover, observed waves can be used to improve our understanding of spicule seismology, and the magnetic field induction in spicules at the height of $\sim$6000 km above the photosphere is estimated to be 12–15 G. We note also that Ebadi *et al.* (2012) have reported the observations of standing kink waves in solar spicules. de Pontieu *et al.* (2007) identified a new class of spicules that moved much faster and were shorter lived than the traditional spicules, which have speeds of between 20 and 40 km s^{-1} and lifespans of from

3 to 7 minutes. These Type II spicules, observed in CaII 854.2 nm and Hα lines (see Sterling *et al.*, 2010), are much more dynamic: they form rapidly (in $\sim$10 s), are very thin ($\leqslant$200 km wide), have lifetimes of 10–150 s (at any one height), and shoot upwards at high speeds, often in excess of 100–150 km s^{-1}, before disappearing. The rapid disappearance of these jets had suggested that the plasma they carried might get very hot, but direct observational evidence of this process had not been reported. Both types of spicules are observed to carry Alfvén waves with significant amplitudes of order 20 km s^{-1}. De Pontieu *et al.* (2011) used new observations from the Atmospheric Imaging Assembly on NASA's launched *Solar Dynamics Observatory* and its Focal Plane Package for the Solar Optical Telescope (SOT) on the Japanese *Hinode* satellite. Their observations reveal "a ubiquitous coronal mass supply in which chromospheric plasma in fountainlike jets or spicules is accelerated upward into the corona, with much of the plasma heated to temperatures between $\sim$0.02 and 0.1 million kelvin (MK) and a small but sufficient fraction to temperatures above 1 MK. These observations provide constraints on the coronal heating mechanism(s) and highlight the importance of the interface region between photosphere and corona." Nevertheless, Moore *et al.* (2011) from *Hinode* observations of solar X-ray jets, Type II spicules, and granule-size emerging bipolar magnetic fields in quiet regions and coronal holes, advocate a scenario for powering coronal heating and the solar wind. In this scenario, Type II spicules and Alvfén waves are generated by the granule-size emerging bipoles in the manner of the generation of X-ray jets by larger magnetic bipoles. From observations and this scenario, the authors estimate that Type II spicules and their co-generated Alfvén waves carry into the corona an area-average flux of mechanical energy of $\sim$7 $\times$ 10^5 erg s^{-1} cm^{-2}. This is enough to power the corona and solar wind in quiet regions and coronal holes, hence indicates that the granule-size emerging bipoles are the main engines that generate and sustain the entire heliosphere. The upward propagation of high- and low-frequency Alfvén waves along spicules detected from SOT's observations on *Hinode* was also reported by He *et al.* (2009) and Tavabi *et al.* (2011). He *et al.* (2009) found in four cases that the spicules are modulated by high-frequency ($\geqslant$0.02 Hz) transverse fluctuations. These fluctuations are suggested to be Alfvén waves that propagate upwards along the spicules with phase speeds ranging from 50 to 150 km s^{-1}. Three of the modulated spicules display clear wave-like shapes with short wavelengths shorter than 8 Mm. We note that at the same time, Kudoh and Shibata (1999) presented a torsional Alfvén-wave model of spicules (actually the classical Type I spicules) and discussed the possibility of wave coronal heating—they estimated that the energy flux transported into corona is of about 3 $\times$ 10^5 erg s^{-1} cm^{-2}, i.e., roughly half of the flux carried by the Alfvén waves running on Type II spicules (see Moore *et al.*, 2011). Tavabi *et al.* (2011)

performed a statistical analysis of the SOT/*Hinode* observations of solar spicules and their wave-like behavior, and argued that there is a possible upward propagation of Alfvén waves inside a doublet spicule with a typical wave's period of 110 s. Using high resolution observations of spicules obtained with the Swedish 1 m Solar Telescope, Pereira *et al.* (2016) have found that that spicules display a rich and detailed spatial structure, and show a distribution of transverse velocities that, when aligned with the line of sight, can make them appear at different Hα wing positions. They become more abundant at positions closer to the line core, reflecting a distribution of Doppler shifts and widths. In Hα width maps they stand out as bright features both on disk and off limb, reflecting their large Doppler motions and possibly higher temperatures than in the typical Hα formation region. The authors also found that the sudden appearance of spicules can be explained by Doppler shifts from their transverse motions.

Recently, there have been performed numerical simulations of solar spicules. Iijima and Yokoyama (2017) have offered a 3D MHD simulation of the formation of solar chromospheric jets with twisted magnetic field lines. Kuźma *et al.* (2017a) using the PLUTO code, numerically solved adiabatic and non-adiabatic MHD equations in 2D cylindrical geometry. They followed the evolution of spicules triggered by pulses that are launched in a vertical velocity component from the upper chromosphere. The numerical results reveal that the velocity pulse is steepened into a shock that propagates upward into the corona. The chromospheric cold and dense plasma follows the shock and rises into the corona with the mean speed of 20–25 km s^{-1}. In a subsequent article Kuźma *et al.* (2017b) using the newly developed JOANNA code, numerically solved two-fluid (for ions + electrons and neutrals) equations in 2D Cartesian geometry. They followed the evolution of a spicule triggered by the time-dependent signal in ion and neutral components of gas pressure launched in the upper chromosphere. The authors used the potential magnetic field, which evolves self-consistently, but mainly plays a passive role in the dynamics. The numerical results recover the occurrence of spicules with aforementioned velocities. In addition, it was established that the formed spicule exhibits the upflow/downfall of plasma during its total lifetime of around 3–4 minutes, and it follows the typical characteristics of a classical spicule, which is modeled by magnetohydrodynamics. Martínez-Sykora *et al.* (2017) compared magnetohydrodynamic simulations of spicules with observations from the Interface Region Imaging Spectrograph and the Swedish 1-m Solar Telescope. Spicules are shown to occur when magnetic tension is amplified and transported upward through interactions between ions and neutrals or ambipolar diffusion. The tension is impulsively released to drive flows, heat plasma (through ambipolar diffusion), and generate Alfvénic waves.

Since spicules support Alfvén (or more generally magnetohydrodynamic) wave propagation, it is of great importance to determine their dispersion characteristics and more specifically their stability/instability status. If while propagating along the jets, MHD waves become unstable and the expected instability is of the Kelvin–Helmholtz type, that instability can trigger the onset of wave turbulence leading to an effective plasma jet heating and the acceleration of the charged particles. We note that the Alfvénic turbulence is considered to be the most promising source of heating in the chromosphere and extended corona (van Ballegooijen *et al.*, 2011). That possibility, based on high-resolution observations of the chromosphere and transition region (TR) with the *Interface Region Imaging Spectrograph* (*IRIS*) and of the corona with the Atmospheric Imaging Assembly (AIA) on board the *Solar Dynamics Observatory* (*SDO*), is discussed in detail by Srivastava *et al.* (2017) and De Pontieu *et al.* (2017).

Studies for the occurrence of KHI in solar spicules were carried out by Zhelyazkov (2012), Ajabshirizadeh *et al.* (2015), Ebadi (2016), and Antolin *et al.* (2018). Here, we outline the essence of finding and numerically exploring the dispersion curves of kink and sausage modes for the MHD waves traveling primarily along the Type II spicules for various values of the jet speed. In studying wave propagation characteristics, we assume that the axial wave number k_z ($\hat{z}$ is the direction of the embedded constant magnetic fields in the two media) is real, while the angular wave frequency, ω, is complex. The imaginary part of that complex frequency is the wave growth rate when a given mode becomes unstable. All of our analysis is based on a linearized set of equations for the adopted form of magnetohydrodynamics. We show that the stability/instability status of the traveling waves depends entirely on the magnitudes of the flow velocities and the values of two important control parameters, namely the so-called density contrast (the ratio of the mass density inside to that outside the flux tube) and the ratio of the background magnetic field of the environment to that of the spicules themselves.

5.1   Geometry and the wave dispersion relations

The simplest model of spicules is a straight vertically moving with velocity U cylinder with radius a filled with ideal compressible plasma of density $\rho_i \sim 3 \times 10^{-13}\,\mathrm{g\,cm^{-3}}$ (Sterling, 2000) and immersed in a constant magnetic field B_i directed along the z-axis. Our flux tube can be considered as that in Fig. 3 in Matsumoto and Shibata (2010) starting at the height of $2\,\mathrm{Mm}$ from the tube footpoint. The mass density of the environment, ρ_e, is much, say 50–100 times, less than that of the spicule, while the magnetic field induction B_e might be of the order of or less than $B_i \sim 10$–$15\,\mathrm{G}$. We note that while the parameters of classical Type I

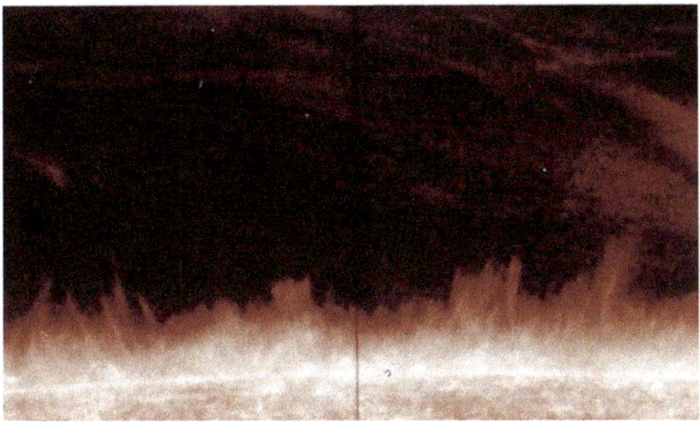

Fig. 5.1. Solar spicules observed by the Interface Region Imaging Spectrograph. *Credit to: NASA.*

spicules are well-documented (Beckers, 1968, 1972), those of Type II spicules are generally disputed—Senteno *et al.* (2010), for example, on using a novel inversion code for Stokes profiles caused by the joint action of atomic level polarization and the Hanle and Zeeman effects to interpret the observations, claim that magnetic fields as strong as $\sim$50 G were detected in a very localized area of the slit, which might represent a lower field strength for the organized network spicules. Moreover, it was reported using AIA/*SDO* and *IRIS* that fast 40–100 km s^{-1} Type II spicules that are only briefly visible (1–2 minutes) in chromospheric observables such as Ca II H 3968 Å (see de Pontieu *et al.*, 2007; Pereira *et al.*, 2012) and in which a fraction of the plasma appears to be heated to at least TR temperatures (see, for example, Pereira *et al.*, 2014; Rouppe van der Voort *et al.*, 2015; Skogsrud *et al.*, 2015) before returning to the surface (Fig. 5.1).

Assuming that the density contrast—the ratio of equilibrium plasma density outside to that inside of spicule—is $\eta = 0.01$, our choice of the sound and Alfvén speeds in the jet is $c_{\rm si} = 10$ km s^{-1} and $v_{\rm Ai} = 70$ km s^{-1}, respectively, while those speeds in the environment are correspondingly $c_{\rm se} \cong 605$ km s^{-1} and $v_{\rm Ae} = 250$ km s^{-1}. All these values are in agreement with the condition for the balance of total pressures at the flux tube interface—that condition can be expressed in the form

$$p_{\rm i} + \frac{B_{\rm i}^2}{2\mu} = p_{\rm e} + \frac{B_{\rm e}^2}{2\mu},$$

which yields (Edwin and Roberts, 1983)

$$\frac{\rho_{\rm e}}{\rho_{\rm i}} = \frac{c_{\rm si}^2 + \frac{\gamma}{2} v_{\rm Ai}^2}{c_{\rm se}^2 + \frac{\gamma}{2} v_{\rm Ae}^2}.$$

Here $c_s = (\gamma p_0/\rho_0)^{1/2}$ is the sound speed, where $\gamma = 5/3$ is the polytropic index, $v_A = B_0/(\mu\rho_0)^{1/2}$ is the Alfvén speed, and μ is the plasma permittivity. For simplicity in notation, p_0, ρ_0, and B_0 denote the equilibrium pressure, mass density, and magnetic field in either medium.

We assume that the magnetohydrodynamic, say Alfvén, waves are generated by random, turbulent-like, motions in the photosphere and that they propagate upwards in the flux tubes modeling the spicules. The basic equations governing the small-amplitude wave propagation are the linearized MHD equations for the perturbations of the fluid velocity v, density ρ, pressure p, and magnetic field B. These equations have been already displayed in Sec. 4.1 (see Eqs. (4.1)–(4.5)), which finally yield the wave dispersion relation given by Eq. (4.30) there. We can use here that equation just by replacing the jet velocity v_0 by U:

$$\frac{\rho_e}{\rho_i}(\omega^2 - k_z^2 v_{Ae}^2)\kappa_i \frac{I_m'(\kappa_i a)}{I_m(\kappa_i a)} - \left[(\omega - k \cdot U)^2 - k_z^2 v_{Ai}^2 \right]\kappa_e \frac{K_m'(\kappa_e a)}{K_m(\kappa_e a)} = 0, \quad (5.1)$$

where the squared wave attenuation coefficients $\kappa_{i,e}$ are given by the expression

$$\kappa^2 = -\frac{\left[(\omega - k \cdot U)^2 - k_z^2 c_s^2\right]\left[(\omega - k \cdot U)^2 - k_z^2 v_A^2\right]}{(c_s^2 + v_A^2)\left[(\omega - k \cdot U)^2 - k_z^2 c_T^2\right]}. \quad (5.2)$$

Recall that the tube speed, c_T, is equal to $c_s v_A/\sqrt{c_s^2 + v_A^2}$. For the azimuthal mode number $m = 0$, the above equation describes the propagation of so-called *sausage* waves, while with $m = 1$ it governs the propagation of the *kink* waves (Edwin and Roberts, 1983). It is also seen that the wave frequency, ω, is Doppler-shifted inside the jet. As we have already discussed in Sec. 4.1, depending on the values of wave attenuation coefficients $\kappa_{i,e}$, the propagating MHD mode can be pure surface, pseudosurface/body, or leaky wave. In our case for the specified sound and Alfvén speeds, both modes turn out to be generally pseudosurface (body) waves that can, however, at some flow speeds become pure surface waves. The two tube speeds, respectively, are $c_{Ti} = 9.9\,\mathrm{km\,s^{-1}}$ and $c_{Te} = 231\,\mathrm{km\,s^{-1}}$. A specific speed for the kink ($m = 1$) MHD mode is the kink speed, which can be presented in the form (see Eq. (4.31)):

$$c_k = \left(\frac{1 + b^2}{1 + \eta}\right)^{1/2} v_{Ai},$$

where b is the ratio of the embedded magnetic fields, B_e/B_i, and η is the density contrast. The kink speed is independent of sound speeds and characterizes the propagation of transverse perturbations. Its value in our case is $74\,\mathrm{km\,s^{-1}}$.

It is obvious that dispersion Eq. (5.1) of either mode can be solved only numerically. Before starting that job, we normalize all velocities to the Alfvén speed

v_{Ai} inside the jet thus defining the non-dimensional phase velocity $V_{\mathrm{ph}} = v_{\mathrm{ph}}/v_{\mathrm{Ai}}$ and the *Alfvén Mach number* $M_{\mathrm{A}} = U/v_{\mathrm{Ai}}$. The wavelength is normalized to the tube radius a which means that the non-dimensional wave number is $K = k_z a$. The calculation of wave attenuation coefficients requires the introduction of three numbers, notably the two ratios $\bar{\beta} = c_s^2/v_{\mathrm{A}}^2$ correspondingly in the jet and its environment, and the ratio of the background magnetic field outside to that inside the flow, b, in addition to the density contrast, η. We recall that the two $\bar{\beta}$s are 1.2 times smaller than the corresponding plasma betas in both media—the latter are given by the expressions $\beta_{\mathrm{i,e}} = 2\bar{\beta}_{\mathrm{i,e}}/\gamma$. Thus, with listed above the specific speeds in both media and $\eta = 0.01$, the input parameters in the numerical procedure are

$$\eta = 0.01, \quad \bar{\beta}_{\mathrm{i}} \cong 0.02, \quad \bar{\beta}_{\mathrm{e}} \cong 5.86, \quad b \cong 0.36, \quad \text{and} \quad M_{\mathrm{A}}.$$

The value of the Alfvén Mach number, M_{A}, naturally depends on the value of the streaming velocity U. Our choice of this value is $100\,\mathrm{km\,s^{-1}}$, which yields $M_{\mathrm{A}} = 1.43$. With these input values, we will calculate the dispersion curves of first kink waves and then sausage ones.

5.1.1 *Dispersion diagrams of kink waves*

We start by calculating the dispersion curves of kink waves assuming that the angular wave frequency ω is real. As a reference, we first assume that the plasma in the flux tube is static, i.e., $M_{\mathrm{A}} = 0$. The dispersion curves, which present the dependence of the normalized wave phase velocity on the normalized wave number, are in this case shown in Fig. 5.2. One can recognize three types of waves: a sub-Alfvénic slow magnetoacoustic wave (in red color) labeled c_{Ti} (which is actually the normalized value of c_{Ti} to v_{Ai}), an almost Alfvén wave labeled c_{k} (the green

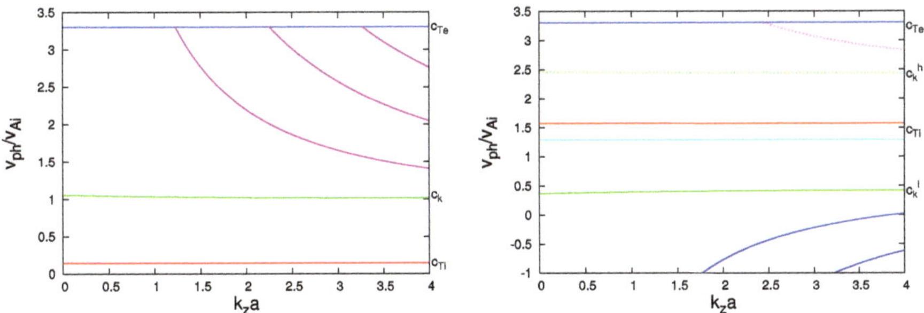

Fig. 5.2. (*Left panel*) Dispersion curves of kink waves propagating along the flux tube at $\eta = 0.01$, $b = 0.36$ and $M_{\mathrm{A}} = 0$. (*Right panel*) The same but for $M_{\mathrm{A}} = 1.43$.

curve), and a family of super-Alfvénic waves (the magenta dispersion curves). We note that one can obtain by numerically solving Eq. (5.1) the mirror images (with respect to the zeroth line) of the c_k-labeled dispersion curve, as well as of the fast super-Alfvénic waves—both being backward propagating modes that are not plotted in the left panel of Fig. 5.2. The right panel of Fig. 5.2 shows how all these dispersion curves change when the plasma inside the tube flows. One sees that the flow first shifts upwards the almost Alfvén wave now labeled c_k^h, as well as high-harmonic super-Alfvénic waves—where both of the altered types of waves are depicted with points style lines in the left panel of Fig. 5.2. Second, the slow magnetoacoustic wave (with label c_{Ti} in the left panel of Fig. 5.2) is replaced by two, now, super-Alfvénic waves, whose dispersion curves (in cyan and orange colors) are collectively labeled c_{Ti}. These two waves have practically constant normalized phase velocities equal to 1.289 and 1.571, respectively, which are the $(M_A \mp c_{Ti}^0)$-values, where c_{Ti}^0 is the normalized magnitude of the slow magnetoacoustic wave at $M_A = 0$. Unsurprisingly, one gets a c_k^l-labeled curve which is the mirror image of the c_k^h-labeled curve. That is why this curve is plotted in green and, as can be seen, it is now a forward propagating wave that has, however, a smaller normalized phase velocity than that of its sister c_k^h-labeled points-style-line dispersion curve. Moreover, there appears to be a family of generally backward propagating waves (below the c_k^l-labeled curve) plotted in blue color that can similarly be considered as a mirror image of the high-harmonic super-Alfvénic waves. The most interesting waves especially for the Type II spicules seems to be the waves labeled c_k. It would be interesting to see whether these modes can become unstable at some, say critical, value of the Alfvén Mach number M_A. To study this, we have to assume that the wave frequency is complex, i.e., $\omega \to \omega + i\bar{\gamma}$, where $\bar{\gamma}$ is the expected instability growth rate. Thus, the dispersion equation becomes complex (complex wave phase velocity and real wave number) and finding the solutions to a transcendental complex equation is generally a difficult numerical task (Acton, 1990).

Before starting to derive a numerical solution of the complex version of Eq. (5.1), we can simplify that equation. Bearing in mind that the plasma beta inside the jet is very small ($\beta_i \cong 0.024$) and that of the surrounding medium quite high (of order 7), we can treat the jet as a cool plasma and the environment as a hot incompressible fluid. We point out that according to the numerical simulation of spicules by Matsumoto and Shibata (2010) the plasma beta at heights greater than 2 Mm is of the order of 0.03–0.04 (consider Fig. 4 of their paper). For cool plasma $c_s \to 0$, hence the normalized wave attenuation coefficient $\kappa_i a = \left[1 - (V_{ph} - M_A)^2\right]^{1/2} K$, while in the incompressible environment $c_s \to \infty$ and the corresponding attenuation coefficient is simply equal to k_z, i.e.,

$\kappa_e a = K$. Under these circumstances, the simplified dispersion equation of kink waves takes the form

$$(V_{ph}^2\eta - b^2)\big[1 - (V_{ph} - M_A)^2\big]^{1/2}\frac{I_1'(\kappa_i a)}{I_1(\kappa_i a)} - \big[(V_{ph} - M_A)^2 - 1\big]\frac{K_1'(K)}{K_1(K)} = 0,$$

$$(5.3)$$

where we recall that $\kappa_i a = \big[1 - (V_{ph} - M_A)^2\big]^{1/2} K$, and the normalized wave phase velocity V_{ph} is a complex number. We note that this simplified version of the dispersion equation of the kink waves closely reproduces the dispersion curves labeled c_k in Figs. 5.2 and 5.3.

To investigate the stability/instability status of the kink waves, we numerically solve Eq. (5.3) using the Muller method (Muller, 1956) to find the complex roots at fixed input parameters $\eta = 0.01$ and $b = 0.36$, and varying Alfvén Mach numbers M_A from zero to some reasonable numbers. Before starting any numerical procedure for solving the aforementioned dispersion equation, we note that for each input value of M_A one can get two c_k-dispersion curves one of which (for relatively small magnitudes of M_A) has a normalized phase velocity roughly equal to $M_A - 1$ and a second dispersion curve associated with dimensionless phase velocity equal to $M_A + 1$. These curves are similar to the dispersion curves labeled c_k^l and c_k^h in the right panel of Fig. 5.2. The numerical solutions to Eq. (5.3) are shown in Fig. 5.3. For $M_A = 0$, except for the dispersion curve with normalized phase velocity approximately equal to 1, one can find a dispersion curve with normalized phase velocity close to -1—that curve is not plotted in Fig. 5.3. Similarly, for $M_A = 2$ one obtains a curve at $V_{ph} = 1$ and another at $V_{ph} = 3$, and so on. With

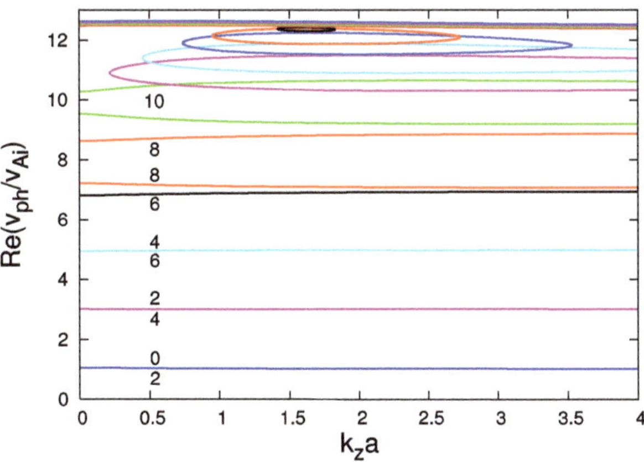

Fig. 5.3. Dispersion curves of kink waves propagating along the flux tube at $\eta = 0.01$, $b = 0.36$, and various values of M_A. *Reprinted with permission.*

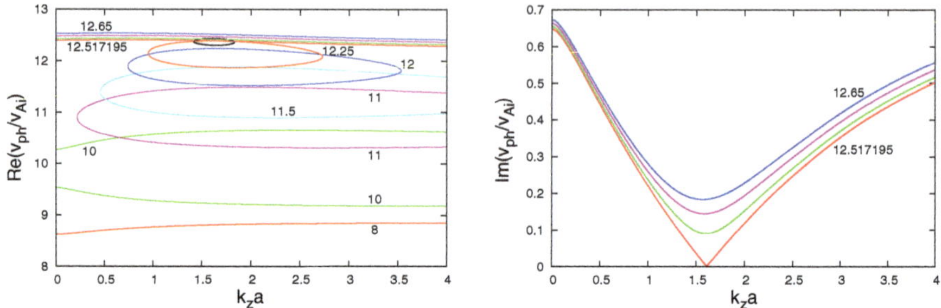

Fig. 5.4. (*Left panel*) Dispersion curves of kink waves propagating along the flux tube at $\eta = 0.01$, $b = 0.36$, and for relatively large values of M_A. The red dispersion curve corresponds to the threshold Alfvén Mach number of 12.517195 at which the KHI emerges. (*Right panel*) Growth rates of unstable kink waves propagating along the flux tube at $\eta = 0.01$, $b = 0.36$, and values of M_A equal to 12.517195, 12.55, 12.6, and 12.65, respectively. *Reprinted with permission.*

increasing the magnitude of the Alfvén Mach number, the kink waves change their structure—for small numbers being pseudosurface (body) waves and for $M_\mathrm{A} > 4$ becoming pure surface modes. Another effect associated with the increase in M_A, is that, for instance at $M_\mathrm{A} \geqslant 8$, the shapes of pairs of dispersion curves begin to visibly change as can be seen in Fig. 5.3 and in the left panel of Fig. 5.4. The most interesting observation is that for $M_\mathrm{A} \geqslant 11$ both curves start to merge and at $M_\mathrm{A} = 12$ they form a closed dispersion curve. The ever increasing of M_A yields yet smaller closed dispersion curves—the smallest one depicted in Fig. 5.4 corresponds to $M_\mathrm{A} = 12.5$. All these dispersion curves present stable propagation of the kink waves. However, for $M_\mathrm{A} \geqslant 12.517$, we obtain a new family of wave dispersion curves that correspond to an unstable wave propagation. We plot in the left panel of Fig. 5.4 four curves of that kind which have been calculated for $M_\mathrm{A} = 12.517195$, 12.55, 12.6, and 12.65, respectively. The growth rates of the unstable waves are shown in the right panel of Fig. 5.4. The instability that arises is of the Kelvin–Helmholtz type. We recall that the KHI occurs when velocity shear is presented within a continuous fluid or when there is a sufficient velocity difference across the interface between two fluids (Chandrasekhar, 1961). In our case, we have the second option and the relative jet velocity U plays the role of the necessary velocity difference across the interface between the spicule and its environment.

The main outstanding question is whether one can really observe such an instability in spicules. The answer to that question is obviously negative—to register the onset of KHI of kink waves traveling on a Type II spicule, one would need to observe jet velocities of the order of or higher than $876 \, \mathrm{km \, s^{-1}}$! If we assume that the density contrast, η, possesses a value of 0.02 and the ratio of the background magnetic fields, b, is equal to 0.35 (which may be deduced from a slightly different set of characteristic sound and Alfvén speeds in both media), the critical Alfvén Mach number at which the instability starts is a little bit lower (equal to 8.9) but the corresponding jet speed is still too high ($=712\,\text{km s}^{-1}$) to be registered in a spicule. The value of 712 was computed under the assumption that the Alfvén speed inside the jet is $80\,\text{km s}^{-1}$. Very similar dispersion curves and growth rates of unstable kink waves were obtained for cylindrical jets when both media were treated as incompressible fluids. In that case, dispersion Eq. (5.3) becomes a quadratic equation that provides solutions for the real and imaginary parts of the normalized wave phase velocity in closed forms (see Zhelyazkov, 2010). It is astonishing that for our jet with $b = 0.36$ and $\eta = 0.01$ the quadratic dispersion equation yields a critical Alfvén Mach number for the onset of a KHI equal to 12.5, which is slightly lower than its magnitude obtained from Eq. (5.3). With this new critical Alfvén Mach number, the required jet speed for the instability onset is $875\,\text{km s}^{-1}$. It is worth mentioning that for the same $\eta = 0.01$, but for $b = 1$ (equal background magnetic fields), the quadratic equation yields a much higher critical Alfvén Mach number ($=15.65$), which means that the critical jet speed grows up to $1096\,\text{km s}^{-1}$. This consideration shows that both the density contrast, η, and the ratio of the constant magnetic fields, b, are equally important in determining the critical Alfvén Mach number. Moreover, since Eq. (5.3) and its simplified form of a quadratic equation yield almost similar results (both for dispersion curves and growth rates when kink waves become unstable) firmly corroborates the correctness of the numerical solutions to the complex dispersion Eq. (5.3).

Ajabshirizadeh *et al.* (2015) exploring the KH instability of the kink ($m = 1$) mode in Type II spicules performed an interesting plasma and magnetic field parameter study for the instability onset. In the beginning, as an illustration, they investigated the stability/instability status of kink waves traveling along a spicule with a density contrast $\eta = 0.1$ (too low for these kind of chromospheric jets) and a ratio of axial magnetic fields, $b = 0.25$. They modeled the spicule as a cool plasma surrounded by incompressible magnetized plasma. Dispersion curves and growth rates of unstable kink mode for various Alfvén Mach numbers (that is, jet velocities) plotted in their Fig. 1 in many aspects are similar to our Fig. 5.4. It is necessary, however, to emphasize that the marginal growth rate curve should be refined, notably it is naturally to consider as a marginal growth rate curve the one that touches (at levels in the range of 0.003–0.005) the $k_z a$-axis—notably that curve defines the actual threshold Alfvén Mach number. We have also spotted that

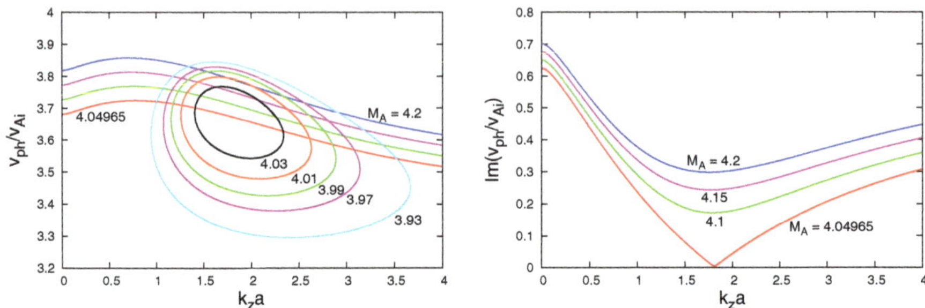

Fig. 5.5. (*Left panel*) Dispersion curves of unstable kink ($m = 1$) MHD mode propagating on a moving untwisted flux tube of cool plasma surrounded by incompressible magnetized plasma at $\eta = 0.1$ and $b = 0.25$ for four values of the Alfvén Mach number $M_A = 4.04965, 4.1, 4.15,$ and 4.2. Closed dispersion curves correspond to stable kink wave propagation. (*Right panel*) The normalized growth rates for the same values of M_A. Red curves in both plots correspond to the onset of KH instability.

the growth rate curves in Fig. 1 in Ajabshirizadeh *et al.* (2015) corresponding to Alfvén Mach numbers equal to 3.93 and 3.97 are not adequate—our computations with their input parameters show that the dispersion curves associated with these two numbers depict stable kink mode propagation (see the left panel in Fig. 5.5). The marginal growth rate curve was obtained at $M_A^{\mathrm{cr}} = 4.04965$ that yields with a reference Alfvén speed of $40 \, \mathrm{km \, s^{-1}}$ a critical jet speed of $162 \, \mathrm{km \, s^{-1}}$, being higher than the maximal Type II spicules' speed of $150 \, \mathrm{km \, s^{-1}}$. On the other hand, their curves plotted in Fig. 4 in Ajabshirizadeh *et al.* (2015) seem to be correct. One can see that according to the curve corresponding to $\eta = 0.1$ for $b = 0.25$ the critical Alfvén Mach number, M_A^{cr}, must be a little higher than 4, in fine agreement with our computations. We have checked also that with $\eta = 0.01$ the critical Alfvén Mach numbers corresponding to b equal to 0, 0.36, and 1 are 11.991254, 12.517196, and 15.64737, respectively, and all they are in good agreement with the highest curve in Fig. 4 in Ajabshirizadeh *et al.* (2015). Another interesting observation is the circumstance that the minima of all computed by us "v"-shaped (see the red curve in the right panel of Fig. 5.5) growth rate curves were grouped around $k_z a = 1.6$, more specifically 1.609 at $b = 0$, 1.611 for $b = 0.36$, and 1.623 with $b = 1$.

Andries and Goossens (2001) were the first to derive a criterion for predicting the rising of KH instability in untwisted magnetic flux tubes (in the limit of cold plasma, $\beta_i = 0$), notably it (instability) occurs if (in our notation)

$$M_A > 1 + b/\sqrt{\eta}. \tag{5.4}$$

In their paper, Ajabshirizadeh *et al.* (2015), on using asymptotic presentations of modified Bessel functions, claim that the KH instability will emerge if

$$M_{\mathrm{A}} > \left[\frac{\eta + 1}{\eta} \left(b^2 + 1 \right) \right]^{1/2} . \qquad (5.5)$$

We note that this criterion generalizes the criterion of Holzwarth *et al.* (2007) applicable for untwisted tube with non-magnetic environment, that is, $B_{\mathrm{e}} = 0$; see Eq. (5) in Holzwarth *et al.* (2007). It is intriguing to see how accurately these two criteria predict the threshold/critical Alfvén Mach number for KHI emergence in our and Ajabshirizadeh *et al.* (2015)'s jet. In Table 5.1, we present the input data (η and b) for computing the critical Alfvén Mach numbers using the two criteria and compare the predicted critical Alfvén Mach numbers with those found from the numerical solving of dispersion Eq. (5.3). The new criterion (5.5) of Ajabshirizadeh *et al.* (2015) is in any way superior to that of Andries and Goossens (2001)—it yields values of the Alfvén Mach number rather close to those found from the solutions to Eq. (5.3). Nevertheless, we should not forget that the obtaining of a correct threshold/critical Alfvén Mach number, $M_{\mathrm{A}}^{\mathrm{cr}}$, can be done by carefully studying the evolution of the pair of kink-speed dispersion curves with gradually increasing M_{A} to find the marginal growth rate curve—this procedure sometimes requires many takes during the computation. The final criterion, of course, is the product of $M_{\mathrm{A}}^{\mathrm{cr}}$ and the reference Alfvén speed that defines the critical flow velocity for the KH instability onset.

Recently, Ebadi (2016) performed a purely numeric 2D simulations of compressible, adiabatic, inviscid, ideal MHD equations to show that KH-type instability may take place in spicule conditions. He considers a spicule of radius $a = 500\,\mathrm{km}$, length $L = 8000\,\mathrm{km}$, steady flow $v_0 = 25\,\mathrm{km\,s^{-1}}$, magnetic field $B_0 = 10\,\mathrm{G}$, density contrast $\eta = 0.01$, kinetic pressure $p_0 = 3.7 \times 10^{-2}\,\mathrm{N\,m^{-2}}$, adiabatic index $\gamma = 1.4$, Alfvén speed $v_{\mathrm{A}0} = 50\,\mathrm{km\,s^{-1}}$, transit Alfvén time $\tau = a/v_{\mathrm{A}0} = 10\,\mathrm{s}$, and dimensionless wavenumber $ka = \pi/2$. The perturbations of the fluid velocity v and magnetic field B were assumed to be independent of y and being two-dimensional (lying in the xz-plane). The number of mesh grid points was set as 200×800. In addition, the time step was chosen as $10^{-5}\,\mathrm{s}$, and the system lengths in x and z dimensions (simulation box sizes) were set to be $(-1, 1)$

Table 5.1. Predicted and computed critical Alfvén Mach numbers.

Input data	M_{A} from Eq. (5.4)	M_{A} from Eq. (5.5)	M_{A} computed
$\eta = 0.01, b = 0.36$	4.6	10.681	12.517196
$\eta = 0.10, b = 0.25$	1.791	3.419	4.04965

and $(0, 8)$. The results of the computations of the density variation for different times (from 0 to 227.5 s) are depicted in Figs. 1 and 2 in Ebadi (2016). Initial density profile is plotted in the top left panel of Fig. 1 and it has been shown that spicule environment density is lower than the spicule density by two orders of magnitude. The top right panel of Fig. 1 in Ebadi (2016) shows the onset of instability and transition to turbulent flow in fluids of different densities, moving at various speeds. It should be noted that there is no observational evidence about KH instability in solar spicules to authors' knowledge until now. It is assumed that if such instability can take place in spicules, it needs to be observed with high resolutions. Another point that should be emphasized here is that the simulations, performed in this study show some different schemes for the KH instability which may be related to the selected equilibrium flow (it is assumed to be zero in the spicule environment). For more quantitative comparisons between schemes, time evolution of the total magnetic energy and total energy has been plotted in Figs. 3 and 4 in Ebadi (2016). The development of KH instability leads to momentum and energy transport, dissipation, and mixing of fluids. When magnetic fields are involved, the possibility of field amplification can take place. The linear stage of KH instability, like other macro-instabilities in the framework of MHD, is characterized by exponential growth of time: $\exp(i\omega t)$, where the negative imaginary part of complex angular frequency ω yields this growth. The analytical growth rate for the same geometry the reader can find in Cavus and Kazkapan (2013).

Very recently, Antolin *et al.* (2018) having analyzed *Hinode* and *IRIS* observations of several spicules, they found different behaviors in terms of their Doppler velocity evolution and collective motion of their sub-structure. Some have a Doppler shift sign change that is rather fixed along the spicule axis, and lack coherence in the oscillatory motion of strand-like structure, matching rotation models, or long-wavelength torsional Alfvén waves. Others exhibit a Doppler shift sign change at maximum displacement and coherent motion of their strands, suggesting a collective MHD wave. By comparing with an idealized 3D MHD simulation performed with the CIP-MOCCT scheme and combined with radiative transfer modeling, the authors analyze the role of transverse MHD waves and associated instabilities in spicule-like features. They find that transverse wave induced Kelvin–Helmholtz (TWIKH) rolls lead to coherence of strand-like structure in imaging and spectral maps, as seen in some observations. The rapid transverse dynamics and the density and temperature gradients at the spicule boundary lead to ring-shaped Mg II k at 2796.35 Å and Ca II H at 3969.59 Å source functions in the transverse cross-section, potentially allowing *IRIS* to capture the KH instability dynamics. Twists and currents propagate along the spicule at Alfvénic speeds, and the temperature variations within TWIKH rolls, produce the sudden appearance/disappearance of strands seen in Doppler velocity and in Ca II H intensity. However, only a mild intensity increase in higher-temperature lines is obtained, suggesting there is an additional heating mechanism at work in spicules.

The numerical model of Antolin *et al.* (2018) considered a typical spicule with a dense and cold core with $n_i = 50n_e$ (that is, $\eta = 0.02$) and $T_i = \frac{1}{100}T_e$, where n denotes the total number density. The authors take initially $T_i = 10^4$ K, $n_i = 6 \times 10^{10}$ cm^{-3}, values commonly found in spicules. The density is given by $\rho = \mu m_p n$, where μ is the mean molecular weight and m_p is the proton mass. To keep pressure balance throughout the atmosphere, the magnetic field varies slightly from $B_i = 14.5$ G to $B_e = 14.43$ G, and the plasma β is 0.02 everywhere. The MHD equations exclude gravity, loop expansion, and curvature since Antolin *et al.* (2018) focus on the transverse dynamics observed in spicules, for which the authors assumed they were second-order factors. Furthermore, the effects of radiative cooling and thermal conduction were not included. The numerical grid contains $512 \times 256 \times 100$ points in the x, y, and z directions, respectively. Thanks to the symmetric properties of the kink mode only half the plane in y and half the loop were modeled (from $z = 0$ to $z = 50$ Mm), and the authors set symmetric boundary conditions in all boundary planes except for x, where periodic boundary conditions were imposed. To minimize the influence from lateral boundary conditions, the spatial grids in x and y were uniform with a resolution of 7.8 km where the loop dynamics are produced, and increase further away leading to a maximum distance of ≈ 8 Mm from the loop center. The simulation was close to ideal with no explicit resistivity or viscosity. The work of Antolin *et al.* (2018) opens a new page in studying KHI in solar spicules because for the first time there is a signature for twist in the jets. Moreover, this higher-resolution model indicates that the small-scale processes triggered by dynamic instabilities associated with transverse MHD waves may also play a significant role in the observed spicule features.

5.1.2 *Dispersion diagrams of sausage waves*

The dispersion curves of sausage waves ($m = 0$) both in a static and in a flowing plasma shown in Fig. 5.6 are very similar to those of kink waves (compare with Fig. 5.2). The latter curves were calculated from dispersion Eq. (5.1) with azimuthal mode number $m = 0$ for the same input parameters as in the case of kink waves. The main difference is that the c_k-labeled green dispersion curve is replaced by a curve corresponding to the Alfvén wave inside the jet. We note that the dispersion curve in the right panel of Fig. 5.6 corresponding to a normalized

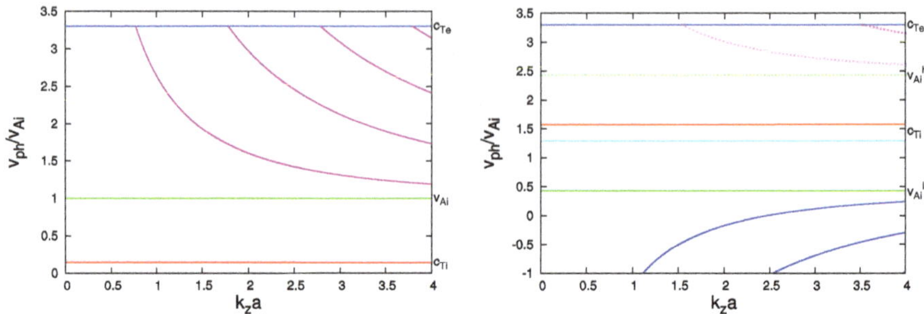

Fig. 5.6. (*Left panel*) Dispersion curves of sausage waves propagating along the flux tube at $\eta = 0.01$, $b = 0.36$ and $M_A = 0$. (*Right panel*) The same but for $M_A = 1.43$. *Reprinted with permission.*

phase velocity of the order of 0.5 (more exactly of 0.43) is labeled v_{Ai}^l because it can be considered as the one dispersion curve of the (1.43 ± 1)-curves that can be derived from the dispersion equation. As in the case of kink waves, the dispersion curve corresponding to the higher speed has the label v_{Ai}^h. Here, we also get two almost dispersionless curves collectively labeled c_{Ti} (in the same colors, cyan and orange, as in the right panel of Fig. 5.2) with normalized wave phase velocities equal to 1.289 and 1.571. One of those curve starting at 1.2886 decreases to 1.2875, while the other slightly increases starting at 1.5714 and finishing at 1.5725. The same holds for the two similar curves in the right panel of Fig. 5.2.

When examining the stability properties of sausage waves as a function of the Alfvén Mach number M_A, we use Eq. (5.3) while changing the order of the modified Bessel functions from 1 to 0. As in the case of kink waves, we are interested primarily in the behavior of the waves whose phase velocities are multiples of the Alfvén speed. The results of numerical calculations to the complex dispersion equation are shown in Fig. 5.7. It turns out that for all reasonable Alfvén Mach numbers the waves are stable. This is unsurprising because the same conclusion was drawn by solving precisely the complex dispersion equation governing the propagation of sausage waves in incompressible flowing cylindrical plasmas (see Zhelyazkov, 2010). In Fig. 5.7, almost all dispersion curves have two labels: one for the $(M_A - 1)$-labeled curve at given M_A (the label is below the curve), and second for the $(M_A' + 1)$-labeled curve associated with the corresponding $(M_A' = M_A - 2)$-value (the label is above the curve). This labeling is quite complex because for all M_A we find dispersion curves that overlap: for instance, the higher-speed dispersion curve (i.e., that associated with the $(M_A + 1)$-value) for $M_A = 0$ coincides with the lower-speed dispersion curve (i.e., that associated with the $(M_A - 1)$-value) for $M_A = 2$. In contrast to the kink waves that for

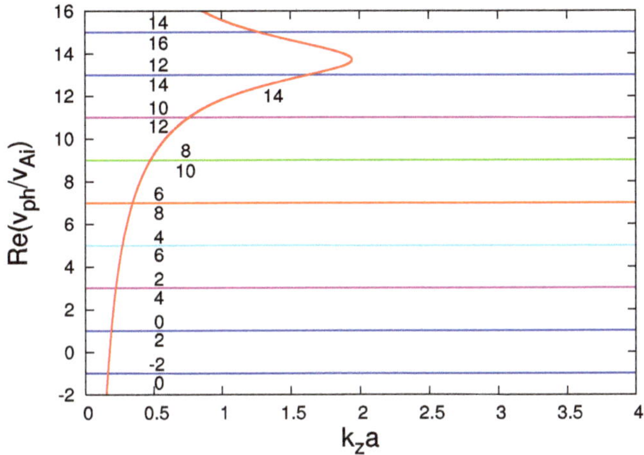

Fig. 5.7. Dispersion curves of sausage waves propagating along the flux tube at $\eta = 0.01$, $b = 0.36$, and for various values of M_A. *Reprinted with permission.*

$M_A > 4$ are pure surface modes, the sausage waves can be both pseudosurface (body) and pure surface modes, or one of the pair can be a surface mode while the other is a pseudosurface one. For example, all dispersion curves for $M_A = 0$ and 8 correspond to the pseudosurface (body) waves, while the curves' pair associated with $M_A = 4$ describes the dispersion properties of pure surface waves. For the other Alfvén Mach–numbers, one of the waves is a pseudosurface and the other is a pure surface. However, there is a "rule": if, for instance, the higher-speed wave with $M_A = 10$ is a pseudosurface mode, the lower-speed wave for $M_A = 12$ is a pure surface wave. We finish the discussion of sausage waves with the following conclusion: with increasing Alfvén Mach number M_A, the initially independent high-harmonic waves and their mirroring counterparts begin to merge—this is clearly seen in Fig. 5.7 for $M_A = 14$—the resulting dispersion curve is in red color. A similar dispersion curve can be obtained for $M_A = 12$, too; the merging point of the corresponding two high-harmonic dispersion curves lies, however, at around $k_z a = 4.1$, i.e., beyond the limits of the plot. It is also evident that in the long wavelength limit the bottom part of the red-colored dispersion curve describes a backward propagating sausage pseudosurface (body) wave. Another peculiarity of the same dispersion curve is the circumstance that for the range of dimensionless wave numbers between 1 and 2, one can have two different wave phase velocities. Which one is detected, the theory cannot predict.

In conclusion to this chapter, we can say the all studies of the kink ($m = 1$) MHD mode propagation characteristics in solar spicules show that the KHI can occur at relatively high velocity flows which are inaccessible even for fast Type II spicules. However, a twist in the magnetic field of the flux tube or its environment may have the effect of lowering the instability threshold as that has been demonstrated by Antolin *et al.* (2018) and eventually lead to the triggering of observable Kelvin–Helmholtz instability in solar spicules.

Chapter 6

Kelvin–Helmholtz Instability in Solar Photospheric Twisted Flux Tubes

6.1 Introduction

With the wealth of recent high-resolution observations from *SDO*, *Hinode*, *Stereo*, *TRACE*, *SOHO*, *RHESSI*, and *IRIS* as well as from ground-based observations, the concepts of the magnetic flux tube and coronal loops have become increasingly important for understanding explosive phenomena such as solar flares and eruptive prominences and also for the problem of the solar corona heating mechanism. The observations reveal a wide range of shapes and sizes, ranging from the small-size magnetic flux tubes found in the granular network, sunspots, coronal loops, and arcades with many loops to huge loops such as prominences. The origin and basic properties of magnetic flux tubes are thoroughly discussed in Ryutova (2018) and especially for photospheric ones in Gent *et al.* (2013), Dominguez *et al.* (2014), and Inoue *et al.* (2018). The registered mass flow speeds in magnetic tubes range from a few tens of kilometers per second in the photosphere/chromosphere (see, e.g., Shibata *et al.*, 2007; Katsukawa *et al.*, 2007; de Pontieu *et al.*, 2007) up to hundreds or thousands of kilometers per second in coronal holes and X-ray jets.

An additional feature of the flux tube is its twist. There is observational evidence for a twist in the solar atmosphere. Aschwanden (1999) confirmed the existence of twisted loops in the corona by detecting a post-flare loop system in extreme ultraviolet wavelengths (171 Å) with the *Transition Region and Coronal Explorer* (*TRACE*). Rotational movement along a loop as observed in the active region NOAA 8668 of the photosphere by Chae *et al.* (2001) indicates that there is a twist in the magnetic field lines. Srivastava *et al.* (2010), using multi-wavelength observations of *SOHO*/MDI, SOT-*Hinode*/blue-continuum (4504 Å), *G* band (4305 Å),

83

CaII H (3968 Å), and *TRACE* 171 Å presented the observational signature of a strongly twisted magnetic loop in AR 10960 during the period 04:43 UT–04:52 UT on June 4, 2007.

As we already know, the hydromagnetic flows are generally unstable against the Kelvin–Helmholtz instability when the flow speed exceeds a certain threshold/critical value (Chandrasekhar, 1961). This turns out to be true for photospheric jets, too (Rae, 1983; Karpen *et al.*, 1993). On the other hand, the twisted magnetic field itself can cause the so-called *kink instability* even when there is no flow. Magnetic tubes are subject to the kink instability when the twist exceeds a critical value (see, for example, Lundquist, 1951; Hood and Priest, 1979). Oscillations and waves and their stability in twisted magnetic flux tubes without flow have been studied in the framework of the normal mode analysis in many works treating primarily MHD waves in coronal magnetic flux tubes/coronal loops.

An open question is how a flow along a twisted magnetic flux tube will change the dispersion properties of the propagating modes and their stability. It turns out that the flow may decrease the threshold for the kink instability, as was tested experimentally in a laboratory twisted plasma column (Furno *et al.*, 2007). This observation was theoretically confirmed by Zaqarashvili *et al.* (2010). The authors studied the influence of axial mass flows on the stability of an isolated twisted magnetic tube of incompressible plasma embedded in a perfectly conducting unmagnetized plasma. Two main results were found. First, the axial mass flow reduces the threshold of the kink instability in twisted magnetic tubes. Second, the twist of the magnetic field leads to the Kelvin–Helmholtz instability of sub-Alfvénic flows for the harmonics with a sufficiently large azimuthal mode number m. The effect is more significant for photospheric magnetic flux tubes than for coronal ones. Díaz *et al.* (2011) also studied the equilibrium and stability of twisted magnetic flux tubes with mass flows, but for flows along the field lines. The authors focused on the stability and oscillatory modes of magnetic tubes with a uniform twist in a zero-beta plasma surrounded by a uniform cold plasma embedded in a purely longitudinal magnetic filed. Regarding the equilibrium, the authors claimed that the only value of the flow that satisfies the equations for their magnetic field configuration is a super-Alfvénic one. The main conclusion is that the twisted tube is subject to the kink instability unless the magnetic field pitch is very high, since the Lundquist criterion is significantly lowered. This is caused by the requirement of having an Alfvén Mach number greater than 1, so the magnetic pressure balances the magnetic field tension and fluid inertia. The authors suggest that this type of instability might be observed in some solar atmospheric structures, such as surges.

Our aim is to investigate the effect of a twisted magnetic field on the stability of kink ($m = 1$) waves propagating on a cylindrical incompressible radially homogeneous photospheric tube with axial mass flow, assuming that the waves are subject, under certain conditions, to the Kelvin–Helmholtz instability (KHI). To evaluate that effect, we compare the critical flow speeds at which the instability occurs with those in the untwisted magnetic tubes. It is necessary to point out that our simplified model of radially homogeneous plasma density inside the tube rules out the resonant absorption of MHD modes that can modify both the wave frequency spectrum and conditions under which the KHI occurs. The radial density inhomogeneity is usually introduced as a non-uniform transition layer that continuously connects the internal density to the external one. Owing to this transverse inhomogeneous transitional layer, wave modes with $m \neq 1$ are spatially damped by resonant absorption. The influence of a twisted magnetic field on the stability status of kink waves propagating along radially inhomogeneous solar cylindrical photospheric/chromospheric jets is beyond the scope of our, and as was mentioned above, we limit ourselves to considering radially homogeneous flowing plasmas only.

6.2 Geometry, the basic MHD equations, and the wave dispersion relation

We consider a magnetic flux tube with radius a and density ρ_i embedded in a uniform field environment with density ρ_e. Both media (inside and outside the tube) are supposed to be incompressible. The magnetic field inside the tube is helicoidal, $\boldsymbol{B}_i = (0, B_{i\phi}(r), B_{iz}(r))$, while outside the magnetic field is uniform and directed along the z-axis, $\boldsymbol{B}_e = (0, 0, B_e)$. Both ρ_i and ρ_e are assumed to be homogeneous. We consider the mass flow $\boldsymbol{v}_0 = (0, 0, v_0)$ directed along the z-axis; thus the equilibrium mass flow is not field-aligned. No mass flow is present outside the tube, therefore the surrounding photospheric medium is considered to be uniformly magnetized ($B_e \hat{z}$ is constant), uniform (ρ_e constant), and lacking mass flow at equilibrium.

In cylindrical equilibrium, the magnetic field and plasma pressure satisfy the equilibrium condition in the radial direction

$$\frac{\mathrm{d}}{\mathrm{d}r}\left(p_i + \frac{B_i^2}{2\mu}\right) = -\frac{B_{i\phi}^2}{\mu r}. \tag{6.1}$$

Here, $B_i(r) = (B_{i\phi}^2 + B_{iz}^2)^{1/2} = |\boldsymbol{B}_i|$ denotes the strength of the equilibrium magnetic field, and μ is the magnetic permeability. We note that in Eq. (6.1) the total (plasma plus magnetic) pressure gradient is balanced by the tension force (the right-hand side of Eq. (6.1)) in the twisted field. We consider the special case of an equilibrium with uniform twist, i.e., the one for which $B_{i\phi}(r)/r B_{iz}(r)$ is a constant. Accordingly, the background magnetic field is assumed to be

$$\boldsymbol{B}(r) = \begin{cases} (0, Ar, B_{\text{iz}}) & \text{for } r \leqslant a, \\ (0, 0, B_{\text{e}}) & \text{for } r > a, \end{cases} \tag{6.2}$$

where A, B_{iz}, and B_{e} are constant. Then, the equilibrium condition (6.1) gives the equilibrium plasma pressure $p_{\text{i}}(r)$ as

$$p_{\text{i}}(r) = p_0 - \frac{A^2 r^2}{\mu},$$

where p_0 is the plasma pressure at the center of the tube. We note that the jump in the value of $B_\phi(r)$ across $r = a$ implies a surface current there.

The plasma motion is governed by the set of linearized MHD equations for an ideal incompressible plasma:

$$\rho \frac{\partial}{\partial t} \boldsymbol{v}_1 + \rho (\boldsymbol{v}_0 \cdot \nabla) \boldsymbol{v}_1 + \nabla \left(p_1 + \frac{\boldsymbol{B}_0 \cdot \boldsymbol{B}_1}{\mu} \right)$$

$$- \frac{1}{\mu} (\boldsymbol{B}_0 \cdot \nabla) \boldsymbol{B}_1 - \frac{1}{\mu} (\boldsymbol{B}_1 \cdot \nabla) \boldsymbol{B}_0 = 0, \tag{6.3}$$

$$\frac{\partial}{\partial t} \boldsymbol{B}_1 - \nabla \times (\boldsymbol{v}_1 \times \boldsymbol{B}_0) - \nabla \times (\boldsymbol{v}_0 \times \boldsymbol{B}_1) = 0, \tag{6.4}$$

$$\nabla \cdot \boldsymbol{v}_1 = 0, \tag{6.5}$$

$$\nabla \cdot \boldsymbol{B}_1 = 0. \tag{6.6}$$

Here, the index "0" denotes equilibrium values of the fluid velocity and the medium magnetic field, and the index "1" their perturbations. Below, the sum $p_1 + \boldsymbol{B}_0 \cdot \boldsymbol{B}_1/\mu$ in Eq. (6.3) will be replaced by p_{tot}, which represents the total pressure perturbation.

Assuming that all perturbations are proportional to $g(r) \exp\left[i(-\omega t + m\phi + k_z z)\right]$ with $g(r)$ being just a function of r, and that in cylindrical coordinates the nabla operator has the form

$$\nabla \equiv \frac{\partial}{\partial r} \hat{r} + \frac{1}{r} \frac{\partial}{\partial \phi} \hat{\phi} + \frac{\partial}{\partial z} \hat{z},$$

from the above set of equations one can obtain a second-order differential equation for the total pressure perturbation p_{tot}

$$\left[\frac{d^2}{dr^2} + \frac{1}{r} \frac{d}{dr} - \left(\kappa^2 + \frac{m^2}{r^2} \right) \right] p_{\text{tot}} = 0, \tag{6.7}$$

as well as an expression for the radial component v_{1r} of the fluid velocity perturbation $\boldsymbol{v}_1$ in terms of p_{tot} and its first derivative

$$v_{1r} = -\mathrm{i}\frac{1}{\rho}\frac{1}{Y}\frac{\omega - \boldsymbol{k}\cdot\boldsymbol{v}_0}{(\omega - \boldsymbol{k}\cdot\boldsymbol{v}_0)^2 - \omega_{\text{A}}^2}\left(\frac{\mathrm{d}}{\mathrm{d}r}p_{\text{tot}} - Z\frac{m}{r}p_{\text{tot}}\right). \qquad (6.8)$$

In Eq. (6.7), κ is the so-called *wave attenuation coefficient*, which characterizes the space structure of the wave and whose squared magnitude is given by the expression

$$\kappa^2 = k_z^2\left(1 - \frac{4A^2\omega_{\text{A}}^2}{\mu\rho\left[(\omega - \boldsymbol{k}\cdot\boldsymbol{v}_0)^2 - \omega_{\text{A}}^2\right]^2}\right), \qquad (6.9)$$

where

$$\omega_{\text{A}} = \frac{1}{\sqrt{\mu\rho}}\,(mA + k_z B_z) \qquad (6.10)$$

is the so-called *local Alfvén frequency* (Bennett *et al.*, 1999). The numerical coefficients Z and Y in the expression of v_{1r} (see Eq. (6.8)) are, respectively,

$$Z = \frac{2A\omega_{\text{A}}}{\sqrt{\mu\rho}\left[(\omega - \boldsymbol{k}\cdot\boldsymbol{v}_0)^2 - \omega_{\text{A}}^2\right]} \quad \text{and} \quad Y = 1 - Z^2.$$

As seen from the expressions for the wave attenuation coefficient and the radial fluid velocity component perturbation, the two quantities have different values inside and outside the twisted flux tube owing to the different spatial structure of the magnetic field in both media, and to the different mass densities (ρ_{i} and ρ_{e}, respectively). It is important to note that the attenuation coefficient κ_{e} outside the tube is simply equal to k_z, whilst that inside the tube can become a purely imaginary quantity, thus allowing the propagation of pseudosurface (body) waves along the tube. Notably the twisted magnetic field inside the tube enables the existence of these waves, which otherwise are pure surface modes (Bennett *et al.*, 1999).

Equation (6.7) has two different solutions inside and outside the jet

$$p_{1\text{tot}}(r) = \begin{cases} \alpha_{\text{i}} I_m(\kappa_{\text{i}} r) & \text{for } r \leqslant a, \\ \alpha_{\text{e}} K_m(k_z r) & \text{for } r > a. \end{cases}$$

Here I_m and K_m are the modified Bessel functions of the first and second kind, and α_{i} and α_{e} are arbitrary constants.

The boundary conditions have to ensure that the normal component of the interface perturbation

$$\xi_r = -\frac{v_{1r}}{\mathrm{i}(\omega - \boldsymbol{k}\cdot\boldsymbol{v}_0)}$$

remains continuous across the unperturbed tube boundary $r = a$, and also that the total Lagrangian pressure is conserved across the perturbed boundary. This leads to the conditions (Bennett *et al.*, 1999)

$$\xi_r{}^{\text{inner}} = \xi_r{}^{\text{outer}}, \tag{6.11}$$

$$\left[p_{\text{tot}} - \frac{B_\phi^2}{\mu a} \xi_r \right]^{\text{inner}} = \left[p_{\text{tot}} - \frac{B_\phi^2}{\mu a} \xi_r \right]^{\text{outer}}. \tag{6.12}$$

Applying boundary conditions (6.11) and (6.12) to our solutions of p_{tot} and v_{1r} (and respectively ξ_r), we obtain after some algebra the dispersion relation of the normal modes propagating along a twisted magnetic tube with axial mass flow $\boldsymbol{v}_0$

$$\frac{\left[(\omega - \boldsymbol{k} \cdot \boldsymbol{v}_0)^2 - \omega_{\text{Ai}}^2 \right] F_m(\kappa_i a) - 2m\omega_{\text{Ai}} \frac{A}{\sqrt{\mu\rho_i}}}{\left[(\omega - \boldsymbol{k} \cdot \boldsymbol{v}_0)^2 - \omega_{\text{Ai}}^2 \right]^2 - 4\omega_{\text{Ai}}^2 \frac{A^2}{\mu\rho_i}}$$
$$= \frac{P_m(k_z a)}{\frac{\rho_e}{\rho_i}(\omega^2 - \omega_{\text{Ae}}^2) + \frac{A^2}{\mu\rho_i} P_m(k_z a)}, \tag{6.13}$$

where

$$F_m(\kappa_i a) = \frac{\kappa_i a I_m'(\kappa_i a)}{I_m(\kappa_i a)} \quad \text{and} \quad P_m(k_z a) = \frac{k_z a K_m'(k_z a)}{K_m(k_z a)}.$$

This dispersion relation is a generalization of Eq. (23) in the work by Bennett *et al.* (1999) that is applicable to a twisted magnetic flux tube without flow and of Eq. (13) in the work by Zaqarashvili *et al.* (2010) that is applicable to a twisted magnetic flux tube with a flow embedded in a non-magnetic environment. As we have already seen, the wave frequency ω is Doppler-shifted inside the jet. In the above equation, a prime (′) denotes the derivative of a Bessel function to its dimensionless argument, and the local Alfvén frequencies inside and outside the tube are correspondingly

$$\omega_{\text{Ai}} = \frac{1}{\sqrt{\mu\rho_i}}(mA + k_z B_{\text{iz}}) \quad \text{and} \quad \omega_{\text{Ae}} = \frac{k_z B_e}{\sqrt{\mu\rho_e}}.$$

The wave attenuation coefficient inside the tube, according to Eq. (6.9), is given by

$$\kappa_i = k_z \left(1 - \frac{4A^2 \omega_{\text{Ai}}^2}{\mu\rho_i \left[(\omega - \boldsymbol{k} \cdot \boldsymbol{v}_0)^2 - \omega_{\text{Ai}}^2 \right]^2} \right)^{1/2}.$$

Assuming no twist, when $B_{i\phi} = 0$, the dispersion relation (6.13) reduces to

$$\frac{\rho_e}{\rho_i}\left(\omega^2 - k_z^2 v_{Ae}^2\right)\frac{I'_m(k_z a)}{I_m(k_z a)} - \left[(\omega - \boldsymbol{k}\cdot\boldsymbol{v}_0)^2 - k_z^2 v_{Ai}^2\right]\frac{K'_m(k_z a)}{K_m(k_z a)} = 0. \qquad (6.14)$$

The Alfvén speeds in this equation are expressed in terms of the background magnetic fields in both media and the corresponding mass densities via the well-known definitions

$$v_{Ai} = \frac{B_i}{\sqrt{\mu\rho_i}} \quad\text{and}\quad v_{Ae} = \frac{B_e}{\sqrt{\mu\rho_e}},$$

respectively. Equation (6.14) is akin to a dispersion relation in the same form as was previously obtained by Edwin and Roberts (1983) for untwisted magnetic tubes without flow. We also recall that for the kink mode ($m = 1$) one defines the so-called *kink speed* (Edwin and Roberts, 1983)

$$c_k = \left(\frac{\rho_i v_{Ai}^2 + \rho_e v_{Ae}^2}{\rho_i + \rho_e}\right)^{1/2} = \left(\frac{v_{Ai}^2 + (\rho_e/\rho_i)v_{Ae}^2}{1 + \rho_e/\rho_i}\right)^{1/2}, \qquad (6.15)$$

which characterizes the propagation of transverse perturbations.

6.3 Numerical solutions and wave dispersion diagrams

We focus our study on the propagation of the kink mode, i.e., for $m = 1$. It is obvious that Eqs. (6.13) and (6.14) can be solved only numerically. The first step is to define the input parameters that characterize the twisted magnetic tube and the axial flow, and to normalize all variables in the dispersion equations. The density contrast between the tube and its environment is characterized by the parameter $\eta = \rho_e/\rho_i$. In investigating the kink instability in twisted magnetic tubes with axial and field-aligned flows with their equilibrium magnetic fields in the form $\boldsymbol{B}(r) = (0, Ar, B_0)$, Zaqarashvili *et al.* (2010) and Díaz *et al.* (2011) chose as twist characteristics the so-called *dimensionless pitch* $k_z p$, in which the specific length p is defined (in their notation) as

$$p = \frac{B_0}{A} \quad\text{or more generally as}\quad p = \frac{B_z}{A}.$$

That dimensionless pitch can be presented in the form $k_z p = 2\pi p/\lambda$, where $2\pi p$ is the magnetic field pitch in its common sense. As we showed, this parameter is associated with a fixed wavelength λ. To study the Kelvin–Helmholtz instability in these tubes we prefer to specify the twist by the ratio of the two magnetic field components, B_ϕ and B_z, evaluated at the inner boundary of the tube, $r = a$, i.e., via B_ϕ/B_z, where $B_\phi = Aa$. This choice is symbolically equivalent to the replacement of the wavelength λ by the tube radius a. Following Aschwanden (2005), the magnetic twist parameter B_ϕ/B_z can be estimated as $B_\phi/B_z = \tan\theta$, where θ is the shear angle between the untwisted and twisted field line. The geometric shear angle, θ, can observationally be measured in twisted photospheric magnetic tubes and in twisted coronal loops. We note that our choice of the twist parameter allows us to immediately compare the dispersion curves of kink waves and the growth rates when the waves become unstable in twisted magnetic tubes with those in untwisted ones.

As usual, we normalize the velocities to the Alfvén speed $v_{\mathrm{Ai}} = B_{\mathrm{i}}/(\mu\rho_{\mathrm{i}})^{1/2}$, where $B_{\mathrm{i}} = (B_{\mathrm{i}\phi}^2 + B_{\mathrm{i}z}^2)^{1/2}$ is evaluated at the inner boundary of the tube, $r = a$. Thus, we introduce the dimensionless wave phase velocity $V_{\mathrm{ph}} = v_{\mathrm{ph}}/v_{\mathrm{Ai}}$ and the *Alfvén Mach number* $M_{\mathrm{A}} = v_0/v_{\mathrm{Ai}}$, the latter characterizing the axial mass flow. The wavelength is normalized to the tube radius a, which means that the dimensionless wave number is $K = k_z a$. It is worth to discuss the normalization of the local Alfvén frequency ω_{Ai}. After multiplying both sides of Eq. (6.10) by the tube radius a and extracting a B_{i} from the brackets, we derive that

$$a\omega_{\mathrm{Ai}} = \frac{B_{\mathrm{i}}}{\sqrt{\mu\rho_{\mathrm{i}}}}\left(m\frac{B_{\mathrm{i}\phi}}{B_{\mathrm{i}}} + k_z a \frac{B_{\mathrm{i}z}}{B_{\mathrm{i}}}\right), \quad \text{where} \quad \frac{B_{\mathrm{i}z}}{B_{\mathrm{i}}} = \left(1 - \frac{B_{\mathrm{i}\phi}^2}{B_{\mathrm{i}}^2}\right)^{1/2}.$$

To simplify the numerical code, and largely for clarity, we can redefine our twist parameter as $B_{\mathrm{i}\phi}/B_{\mathrm{i}} \equiv \varepsilon$, and consequently the normalized local Alfvén frequency is expressed in terms of the Alfvén speed v_{Ai}, the dimensionless wave number K, and ε:

$$a\omega_{\mathrm{Ai}} = v_{\mathrm{Ai}}(m\varepsilon + Kb_z),$$

where the dimensionless axial magnetic field component $b_z = (1 - \varepsilon^2)^{1/2}$. The new twist parameter, ε, is related to the real twist B_ϕ/B_z via the simple relation

$$\varepsilon = \frac{B_\phi/B_z}{\sqrt{1 + (B_\phi/B_z)^2}}. \tag{6.16}$$

Thus, if we have an observationally measured shear angle θ of some solar twisted magnetic tube, then $B_\phi/B_z = \tan\theta$ can be inserted into expression (6.16) to obtain the adopted twist ε of that specific magnetic tube.

Before starting the numerical calculation, we have to specify the values of the input parameters. First we will examine, as in the work of Zaqarashvili *et al.* (2010), the case of an isolated twisted magnetic tube with axial mass flow, i.e., with $B_{\mathrm{e}} = 0$, and hence for parameter $b \equiv B_{\mathrm{e}}/B_{\mathrm{i}} = 0$. Then, we will study the influence of the external magnetic field on the instability onset by choosing b equal to 0.1 and 0.5, respectively. Concerning the density contrast, η, we consider a typical for photospheric tube $\eta = 2$. To satisfy the Kruskal–Shafranov stability criterion (see Kruskal and Tuck, 1958; Shafranov, 1956) for a kink instability, we assume that the azimuthal component of the magnetic field is smaller than its axial component, i.e., we will choose our twist parameter ε to be always less than 1. In particular, we will study the dispersion diagrams of kink waves and their growth rates (when the waves are unstable) for three fixed values of ε: 0 (untwisted magnetic tube), 0.1, and 0.4. The Alfvén Mach number M_A will take values from zero (no flow) to any reasonable number.

We start by calculating the dispersion curves of kink waves assuming that the twist parameter $\varepsilon = 0.1$, and that the angular wave frequency ω is real. As a reference, we first assume that the plasma in the flux tube is static, i.e., $M_A = 0$. The dispersion curves, which present the dependence of the normalized wave phase velocity, $v_{\rm ph}/v_{\rm Ai}$, on the normalized wave number, $k_z a$, are in this case shown in Fig. 6.1. One can recognize two types of waves: a sub-Alfvénic wave labeled c_k (the green curve), and a family of super-Alfvénic waves accommodated in a narrowband depicted by the two purple dispersion curves. The green dispersion curve is related to the kink speed defined by Eq. (6.15). For a weakly twisted magnetic tube ($\varepsilon = 0.1$) with a density contrast $\eta = 2$ the kink speed is equal to $0.5774v_{\rm Ai}$— our numerical code yields in the long wavelength limit, $c_k = 0.5738v_{\rm Ai}$, which agrees well with the value calculated from Eq. (6.15). If we assume that the Alfvén speed inside the tube is typically $10\ {\rm km\,s}^{-1}$, then the kink speed in the twisted magnetic tube is equal to $5.7\ {\rm km\,s}^{-1}$, or approximately to 6 kilometers per second. We

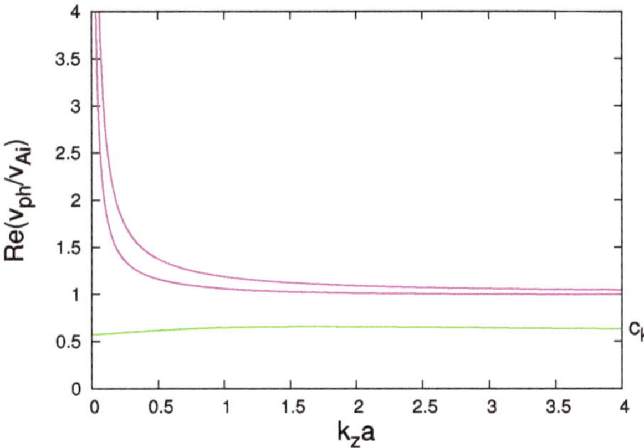

Fig. 6.1. Dispersion curves of kink waves propagating along a twisted magnetic flux tube at $\eta = 2$, $B_e/B_i = 0$, $B_{i\phi}/B_i = 0.1$, and $M_A = 0$. The green curve is the dispersion curve of the wave associated with the kink speed c_k. The two purple curves depict a band that contains nine dispersion curves of similar form, not plotted here. *Reprinted with permission.*

recall that the expression of c_k was defined for untwisted magnetic tubes and there is no guarantee that it will be valid for twisted tubes as well.

Bennett *et al.* (1999), in studying the kink wave propagation in a twisted magnetic tube with parameters similar to ours, had discovered a band that contains an infinite number of pseudosurface (body) wave harmonics (Fig. 5(a) in their paper). Among those, generally super-Alfvénic, waves we were able to easily calculate only 11 dispersion curves from Eq. (6.13) with search steps (in decreasing order) from 0.00025 to 0.000000025. They are not plotted in Fig. 6.1 because there is no room for all of them in that narrowband, but they will be shown shortly in a similar plot for a tube with $\varepsilon = 0.4$. With diminishing search step, taking it equal to 0.0000000025 or smaller, one can extract a few (3) additional curves, but finding them is an extremely hard numerical task—for such small search steps the data are very noisy. However, these high-harmonics pseudosurface kink waves are not too important to us since we are primarily interested in the evolution of the kink mode associated with the kink speed c_k when the plasma starts to flow. It is worth noticing that the c_k-labeled dispersion curve for small dimensionless wave numbers, till $k_z a = 0.1228$, describes a pseudosurface (body) wave, while behind that value the wave is a pure surface mode. These types of waves are sometimes called *hybrid waves* (Bennett *et al.*, 1999).

It is essential to emphasize that the dispersion equation (6.13) also yields solutions for negative values of the wave phase velocity (i.e., for backward-propagating waves), which are a mirror image of the solutions in the positive semi-space. Including the flow, the whole pattern shifts en bloc upwards, changing of course both the c_k-labeled dispersion curve and the wave harmonics dispersion curves. For each M_A we have *two* c_k-dispersion curves that initially, for small Alfvén Mach numbers, are independent, but at certain magnitudes of M_A begin to interact; for example, they form semi-closed or closed dispersion curves by merging. These dispersion curves are an indication that we are in a region in which the kink wave may become unstable.

To study the stability/instability status of the c_k-kink wave, we have to assume that the wave frequency is complex, i.e., $\omega \rightarrow \omega + i\gamma$, where γ is the expected instability growth rate. Thus, the dispersion equation becomes complex (complex wave phase velocity and real wave number) and deriving the solutions to a transcendental complex equation is generally a hard task (see Acton, 1990). We numerically solve Eq. (6.13) using the Muller method (Muller, 1956) to find the complex roots at fixed input parameters $\eta = 2$, $b = 0$, and $\varepsilon = 0.1$, and varying Alfvén Mach numbers M_A from zero to generally reasonable values. The evolution of the two kink waves with increasing flow velocity (or equivalently M_A) is shown in Fig. 6.2. We note that while for rest plasma or small Alfvén Mach

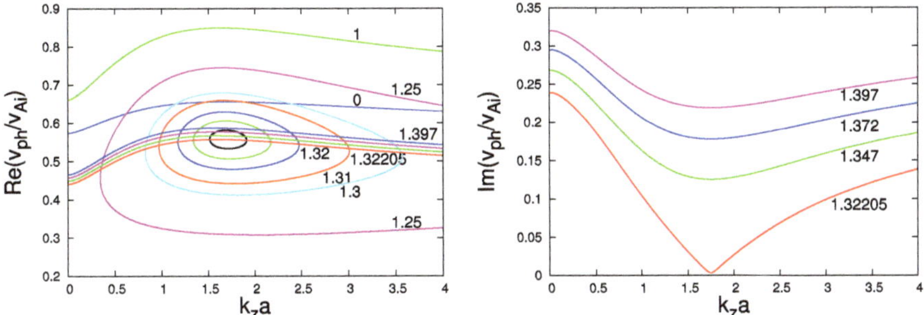

Fig. 6.2. (*Left panel*) Dispersion curves of forward propagating stable and unstable kink waves in a twisted magnetic flux tube with the same input parameters as in Fig. 6.1 for various values of the Alfvén-Mach number M_A. The non-labeled green closed curve corresponds to $M_A = 1.325$, and the black one—to $M_A = 1.328$. (*Right panel*) Growth rates of the unstable kink waves for $M_A = 1.32205, 1.347, 1.372,$ and 1.397. *Reprinted with permission.*

numbers the lower kink wave (i.e., the one that is a mirror image of the green curve in Fig. 6.1) is a backward one, for $M_A = 1$ it becomes a forward wave. At the next increase of the Alfvén Mach number the two dispersion curves change their form and structure—initially being hybrid modes, for $M_A > 1.23$ they are pure surface waves. The most interesting observation is that for $M_A \geqslant 1.2173$ both curves begin to merge and for $M_A \geqslant 1.2933$ they form a closed dispersion curve. The ever increasing M_A yields yet smaller closed dispersion curves—the smallest one depicted in the left panel of Fig. 6.2 corresponds to $M_A = 1.328$. All these dispersion curves present a stable propagation of the kink waves. However, for $M_A \geqslant 1.32205$ we obtain a new family of wave dispersion curves that correspond to an unstable wave propagation. We plot in Fig. 6.2 four curves of that kind which were calculated for $M_A = 1.32205, 1.347, 1.372,$ and 1.397, respectively. The growth rates of the unstable waves are shown in the right panel of the same figure. The instability that arises is of the Kelvin–Helmholtz type. Note the shape of red curve labeled 1.32205: the growth rate has relatively high values for small normalized wave numbers, then passes through a minimum (its relative value there is ~ 0.003) after which it starts to grow. That specific curve is to some extent a marginal curve—for values of M_A lower than 1.32205 the kink wave is stable whilst at $M_A \geqslant 1.32205$ the wave is definitely unstable. This means that the instability onset at the critical M_A will prevent the occurrence of the neutral dispersion curve with $M_A = 1.328$.

It is interesting to see how the twist of the magnetic field has changed the dispersion curves and growth rates of unstable kink waves calculated for an untwisted magnetic tube. In that case, the dispersion curves and growth rates can be obtained

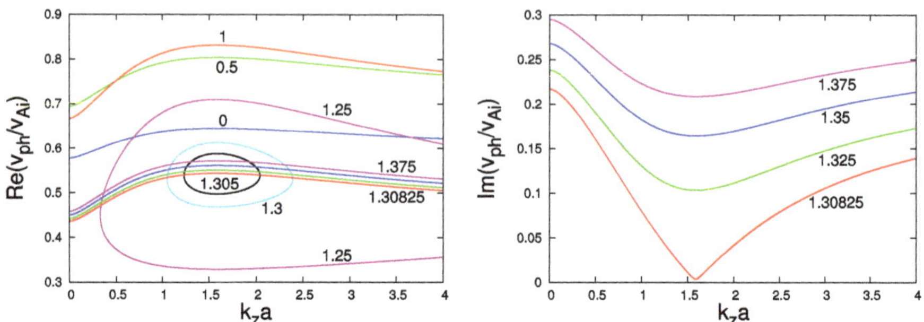

Fig. 6.3. (*Left panel*) Dispersion curves of forward propagating stable and unstable kink waves in an untwisted magnetic flux tube at $\eta = 2$ and $B_e / B_i = 0$ for various values of the Alfvén Mach number M_A. (*Right panel*) Growth rates of the unstable kink waves for $M_A = 1.30825$, 1.325, 1.35, and 1.375. *Reprinted with permission.*

through the exact numerical solutions of Eq. (6.14)—in that case the dispersion relation is a quadratic equation. The results are shown in Fig. 6.3. Comparing Figs. 6.2 and 6.3, we immediately see that the dispersion diagrams and growth rates of the unstable kink waves are very similar. There are differences, of course; firstly the magnetic field twist visibly extends the closed dispersion curves in the horizontal direction, and secondly, it slightly increases the threshold Alfvén Mach number for the onset of the Kelvin–Helmholtz instability—for the untwisted magnetic tube it is equal to 1.30825, whilst for the twisted one it is slightly higher—the corresponding value is 1.32205. The similarities between the dispersion diagrams and the growth rates are important to us—they indirectly confirm the correctness of our numerical solutions derived by solving the complex dispersion equation (6.13).

Figure 6.4 shows the dispersion curves of kink waves when the twist parameter $\varepsilon = 0.4$ and there is no flow. Now one can see all 11 easily derived dispersion curves of the wave harmonics. They have similar shapes, and as seen in Figs. 6.4 and 6.5, one can distinguish two sub-groups of four and five curves, respectively. We note that the eleventh (the highest) dispersion curve is not visible in the left panel of Fig. 6.5 because it simply begins at about $k_z a = 0.23$, which is far beyond the right side of the plot. Another important observation is that the kink speed is markedly lower than expected from Eq. (6.15) value—here its normalized magnitude is equal to 0.5167, which corresponds to a kink speed of 5 kilometers per second. The kink wave itself is a hybrid mode—till $k_z a = 0.5623$ it is a pseudosurface (body) wave, becoming a pure surface wave for the next values of the normalized wave number. When plasma in the tube is flowing, the kink waves are pure surface mode for $M_A > 1.197$.

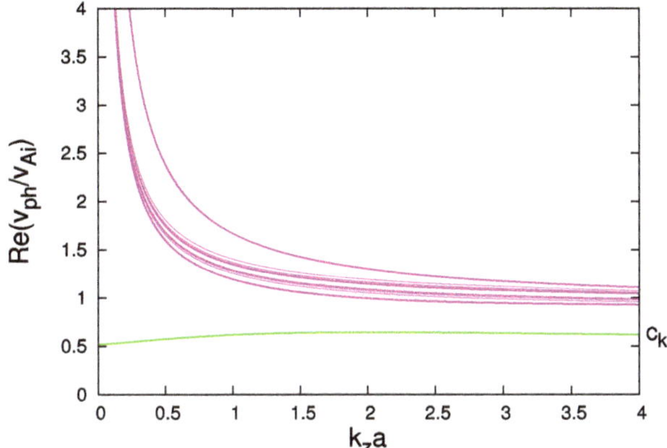

Fig. 6.4. Dispersion curves of kink waves propagating along a twisted magnetic flux tube at $\eta = 2$, $B_e/B_i = 0$, $B_{i\phi}/B_i = 0.4$, and $M_A = 0$. The green curve is the dispersion curve of the wave associated with the kink speed c_k. The purple curves depict all 11 dispersion curves; 9 of them are divided into two sub-groups. The "evolution" of those curves when increasing $k_z a$ can be seen in the next figure, which shows them near the left top and right bottom corners of the plot. *Reprinted with permission.*

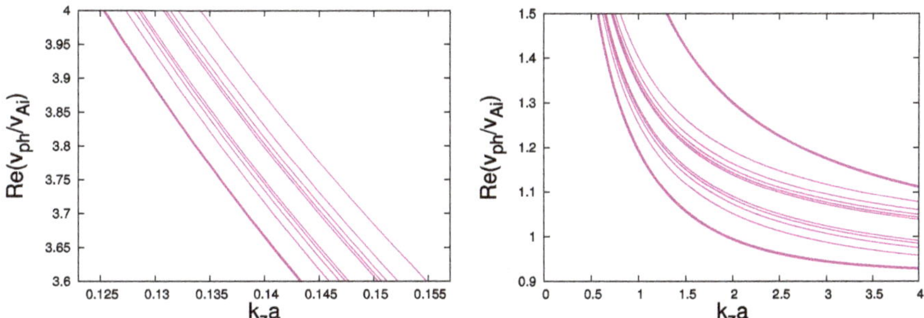

Fig. 6.5. Zoomed-in parts of the 11 dispersion curves near the left top (*Left panel*) and right bottom (*Right panel*) corners of the plot in Fig. 6.4. *Reprinted with permission.*

The dispersion curves and growth rates of kink waves for this higher value of the magnetic field twist are given in Fig. 6.6. At the left panel of this figure, we see dramatic changes in the shape and position of the closed dispersion curves. The most striking changes, however, are the circumstances that the onset of unstable kink waves, like in Fig. 6.2, starts at a value of the Alfvén Mach number lower than that of the narrowest closed dispersion curve—1.250075 vs. 1.29! In untwisted flux tubes, the kink mode appears at Alfvén Mach numbers higher than those of the smallest closed dispersion curves, as seen in Fig. 6.3 (cf. also Fig. 4 in Zhelyazkov,

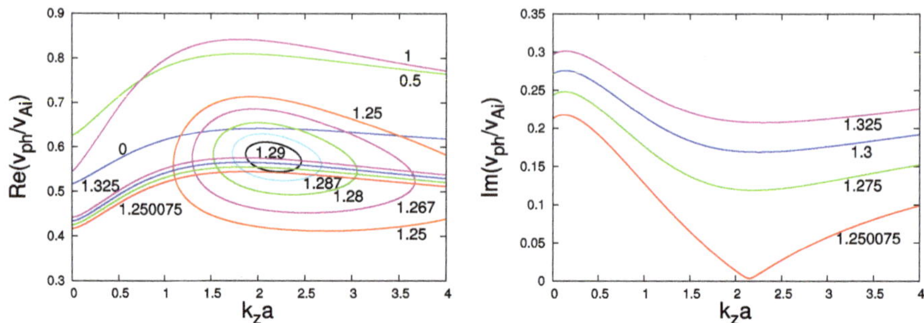

Fig. 6.6. (*Left panel*) Dispersion curves of forward-propagating stable and unstable kink waves in a twisted magnetic flux tube with the same input parameters as in Fig. 6.4 for various values of the Alfvén Mach number M_A. (*Right panel*) Growth rates of the unstable kink waves for $M_\mathrm{A} = 1.250075, 1.275, 1.3$, and 1.325. *Reprinted with permission.*

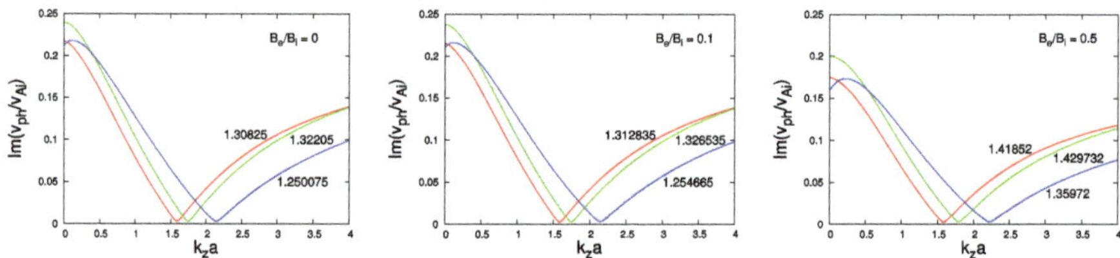

Fig. 6.7. Specific growth rates of kink waves propagating along an untwisted/twisted magnetic flux tube at $\eta = 2$, $B_\mathrm{e}/B_\mathrm{i} = 0$ (*Left panel*), $B_\mathrm{e}/B_\mathrm{i} = 0.1$ (*Middle panel*), and $B_\mathrm{e}/B_\mathrm{i} = 0.5$ (*Right panel*) for various values of the twist parameter $B_{\mathrm{i}\phi}/B_\mathrm{i}$, equal to 0 (red curve), 0.1 (green curve), and 0.4 (blue curve), respectively. The curve labels denote the threshold Alfvén Mach number. *Reprinted with permission.*

2012). In reality, of course, the onset of a Kelvin–Helmholtz instability in a twisted tube will "reject" the co-existence of stable kink waves like those labeled 1.267, 1.28, 1.287, and 1.29 in Fig. 6.6.

Similar calculations were performed for the two external magnetic fields specified by $b = 0.1$ and 0.5. As expected, the environment's magnetic field stabilizes the kink wave propagation (Bennett *et al.*, 1999)—the instability onset occurs at slightly higher threshold Alfvén Mach numbers, but the shapes of the dispersion and growth rate curves are similar to those of an isolated twisted flux tube as depicted in Figs. 6.2, 6.3, and 6.6. The influence of the magnetic field twist, ε, at a fixed external magnetic field, b, and that of the environment's magnetic field at a fixed tube twist on the specific growth rate curves are illustrated in Figs. 6.7 and 6.8. As seen in Fig. 6.7, at a fixed magnetic field configuration, the tube twist

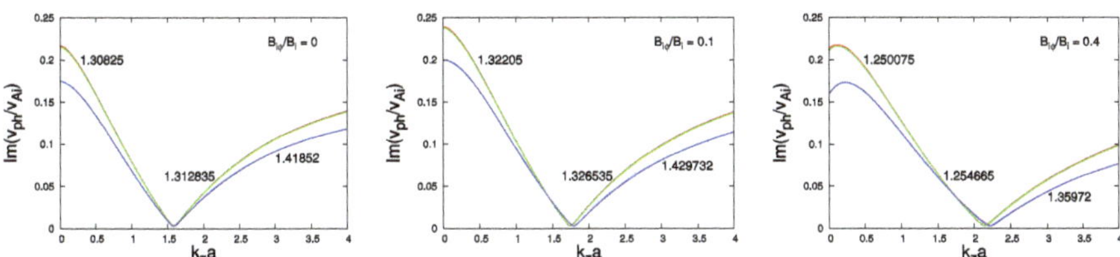

Fig. 6.8. Specific growth rates of kink waves propagating along an untwisted/twisted magnetic flux tube at $\eta = 2$, $B_{i\phi}/B_i = 0$ (*Left panel*), $B_{i\phi}/B_i = 0.1$ (*Middle panel*), and $B_{i\phi}/B_i = 0.4$ (*Right panel*) for various values of the environment's magnetic field B_e/B_i, equal to 0 (red curve), 0.1 (green curve), and 0.5 (blue curve), respectively. The curve labels denote the threshold Alfvén Mach number. *Reprinted with permission.*

shifts to the right the minimum of the specific growth rate curve. This is most pronounced at $b = 0.5$. A twist of 0.4 notably decreases the wave growth rate for each b and yields the lowest threshold Alfvén Mach number for a given b. Thus, for a photospheric twisted magnetic tube with $\varepsilon = 0.4$ flow speeds in the range 12.5–13.6 km s^{-1} can ensure the occurrence of a Kelvin–Helmholtz instability of the kink waves propagating along that tube. These flow speeds can be observed/detected in the solar photosphere. It becomes clear from Fig. 6.8 that an external magnetic field visibly increases the threshold Alfvén Mach number and diminishes the wave growth rate—the latter effect is strongest for a magnetic tube with twist parameter $\varepsilon = 0.4$.

To summarize, finding the conditions under which Kelvin–Helmholtz instability emerges in twisted magnetic flux tube requires a thorough exploration of the evolution of wave dispersion curves with gradually increasing the Alfvén Mach number from zero to larger values. A rough estimation for the threshold M_A at which the instability occurs can be obtained by the instability criterion (Zaqarashvili *et al.*, 2014a)

$$|m|M_A^2 > \left(1 + \frac{1}{\eta}\right)\left(|m|\frac{B_{ez}^2}{B_{iz}^2} + 1\right).$$

For the kink mode, i.e., $m = 1$, this inequality gives $M_A > 1.2247$ for $\eta = 2$ and $b \equiv B_{ez}/B_{iz} = 0$ (isolated flux tube). Our numerical findings of the critical Alfvén Mach numbers agree with this criterion. Kelvin–Helmholtz instability in twisted flux tubes was also numerically modeled (Murawski *et al.*, 2016) and numerically obtained properties of the Kelvin–Helmholtz instability confirm the analytical predictions for the occurrence of the instability.

The next steps in studying the Kelvin–Helmholtz instability of kink waves in twisted magnetic tubes are obviously (i) to investigate the role of the transverse density inhomogeneity inside the tube and associated with it resonant wave absorption on the kink waves propagation and their stability/instability status, and (ii) to consider a field-aligned mass flow instead of an axial one.

Chapter 7

Kelvin–Helmholtz Instability in Solar Surges and Dark Mottles

7.1 Kelvin–Helmholtz instability in solar surges

Surges are phenomena in which dark dense mass are ejected in the solar atmosphere from chromospheric into coronal heights. Usually, they appear as straight or slightly curved ejective structures, and they often recur (Roy, 1973b,a; Švestka, 1976; Foukal, 2004). At first, they were studied in Hα by Newton (1942) and by McMath and Mohler (1948). They have a typical size of 38 000–220 000 km, a transverse velocity of 30–200 km s^{-1}, and a lifetime of 10–20 min. The rotational or helical motions were, on occasions, also observed in surge activity (Gu *et al.*, 1994; Canfield *et al.*, 1996). Surges were also observed as emission in He II 304 Å solar images (Bohlin *et al.*, 1975; Georgakilas *et al.*, 1999), obtained by the slitless XUV spectrograph on a *Skylab* mission and with the Extreme Ultraviolet Imaging Telescope on board the *Solar and Heliospheric Observatory* (*SOHO*; Domingo *et al.*, 1995), respectively. With developing observational instruments and spacecrafts, chromospheric ejections were detected by using data from the 50 cm Swedish Vacuum Solar Telescope and the *Transition Region and Coronal Explorer* (*TRACE*) (Brooks *et al.*, 2007), as well as from the Big Bear Solar Observatory (Chen *et al.*, 2008) and with the Solar Ultraviolet Measurements of Emitted Radiation (SUMER; Wilhelm *et al.*, 1995) spectroscopic observations on board the *SOHO* (Chen *et al.*, 2008).

Based on observational studies, it is now accepted that the driving mechanism of mass ejection in surges is magnetic reconnection at chromospheric heights. Kurokawa and Kawai (1993) found that cool surges occur at the very beginning phases of magnetic-flux emergence, and suggested that surges are produced by magnetic reconnection between the emerging flux and the pre-existing magnetic flux. Canfield *et al.* (1996) reported that circumstances favorable to magnetic reconnection are produced by moving satellite spots in a surge-productive region. Over the years, using various observational spacecrafts there were detected chromospheric surges. Tziotziou *et al.* (2005), on the basis of an analysis of high temporal and spatial resolution CaⅡ H chromospheric limb observations obtained with the Dutch Open Telescope (DOT) along with the *TRACE* satellite in the 195 Å and 1600 Å passbands, observed a solar surge in active region AR10486 located near the solar limb. Uddin *et al.* (2012) presented a multi-wavelength study of recurrent surges originated due to the photospheric reconnections. Recurrent surges with rotational motion at AR 10930 on the west limb were also observed by *Hinode* Solar Optical Telescope (SOT) continuously from 11:21 ut on December 18 to 09:58 ut on December 19, 2006, using the Ca Ⅱ H broadband filter (Bong *et al.*, 2014). Chandra *et al.* (2015) reported a multi-wavelength study of solar jets on 2010 December 11 using the *Solar Dynamics Observatory* (*SDO*; Pesnell *et al.*, 2012) data. They found an increase in the amplitude of oscillations close to their footpoints of the observed jets, which provides the evidence for the wave-induced reconnection as a mechanism for jets triggering. A surge associated with an M3.7 flare was observed on 2011 September 25 at N12 E33 in active region 11302 by *Hinode*/EIS, *RHESSI*, and *SDO*/AIA (Doschek *et al.*, 2015). Canfield *et al.* (1996) also showed that a high-temperature X-ray jet and a cool untwisted surge can coexist located side by side at the site (see Fig. 10c in their paper). Kamio *et al.* (2010) using the EUV Imaging Spectrometer (EIS) on *Hinode* and SUMER on *SOHO* reported an X-ray jet associated with a cool macrospicule/surge eruption in He Ⅱ 304 Å above the limb. Shibata *et al.* (1992a) and Yokoyama and Shibata (1995) succeeded to reproduce surge mass ejections from chromospheric heights by magnetic reconnection between an emerging flux and a pre-existing magnetic field. This numerical result was later confirmed by Nindos *et al.* (1998) who studied the radio properties of 18 X-ray coronal jets as observed by the *Yohkoh* SXT (Tsuneta *et al.*, 1991) using Nobeyama Radioheliograph 17 GHz data. From the SXT images, Nindos *et al.* (1998) computed the coronal plasma parameters at the location of the surge. At the time of maximum surge activity, they found electron temperature $T_e = 2.8 \times 10^6$ K and emission measure $EM = 5.0 \times 10^{45}$ cm^{-3}, as well as derived constraints on the ejecta electron number density, notably $n_e < 6.1 \times 10^{10}$ cm^{-3}. A close correlation among cool surges, magnetic flux cancellations, and UV brightenings found at the edge of an emerging flux region was reported by Yoshimura *et al.* (2003). The authors showed that the surge activities in Hα and the brightenings in *TRACE* 1600 Å images correlate well in both time and space with the cancellation of magnetic fluxes around an emerging flux region. These facts are consistent with the magnetic reconnection model. The released energy through magnetic reconnection, which is estimated to be 10^{28} erg, is sufficiently large to produce surge activities.

Cool solar surges and associated soft X-ray loops were also studied by Schmieder *et al.* (1994) who performed simultaneous observations of NOAA AR 6850 on 1991 October 7, made with the MSDP spectrograph operating on the solar tower in Meudon and with the *Yohkoh* SXT. By measuring the volume emission measures of the two flaring loops (northern and southern ones) and the surge region (mid-part of the surge), the authors concluded that at 10:24–10:30 UT the temperature was $(3\text{--}4) \times 10^{6}$ K. Successive chromospheric surges ejected from a filament in active region NOAA AR 10407 on 2003 July 14, as observed by the THEMIS telescope in the center of the Hα line during the time interval 9:20–9:40 UT, were reported by Zuccarello *et al.* (2011). An eruption of a miniature Hα filament that took the form of a surge was also detected by Yang *et al.* (2012) on 2001 March 27 in NOAA AR 9401 by high-resolution Hα observations from the Big Bear Solar Observatory. Kayshap *et al.* (2013) have observed a solar surge in NOAA AR 11271 using the *SDO* data on 2011 August 25, possibly triggered by chromospheric activity. They also measured the temperature and density distribution of the observed surge during its maximum rise and found an average temperature and a number density of 2×10^{6} K and 4.17×10^{9} cm^{-3}, respectively.

As seen, the electron number density and the temperature of solar surges can vary in rather wide limits, from $\sim 10^{12}$ cm^{-3} and 10^{4} K for cool surges to $\sim 10^{9}$ cm^{-3} and 10^{6} K for high-temperature EUV surges. Since each surge is a jet in a well-defined magnetic flux tube, it is naturally to expect that the magnetohydrodynamic waves propagating along the magnetized plasma flow can become unstable against the KH instability. Note, however, that except the existence of a longitudinal velocity shear, leading to KH instability, we can also have transverse velocity shear initiating this instability. Ofman *et al.* (1994), in numerically studying the coronal heating by the resonant absorption of Alfvén waves in an inhomogeneous slab, have shown that a shear Alfvén wave can trigger a KH instability and its onset depends on the phase shift between azimuthal velocity and magnetic field perturbations. Anyway, it is interesting to establish whether MHD waves traveling along the surge jet can become unstable within its velocity range of 20–200 km s^{-1}.

7.1.1 *Surge models, basic parameters, and governing equations*

Here we explore one of the four surges observed by Brooks *et al.* (2007) in the solar active region NOAA AR 8227 (N26°, E09°) on 1998 May 30 from 7:50 to

16:50 UT. The electron number density derived from the emission measure analysis of the *TRACE* Fe IX 171 Å images is equal to $n_i = 3.83 \times 10^{10}\,\mathrm{cm}^{-3}$, and the jet velocity is of the order of 45–50 km s^{-1}. Estimated chromospheric magnetic field is $B_{\mathrm{foot}} = 25$ G and from the conservation of magnetic flux between the reconnection region and the photosphere one finds that the coronal magnetic field is 7–10 G. Brooks *et al.* (2007) claim that their observations confirm the emerging flux regions model of Kurokawa and Kawai (1993) and Shibata *et al.* (1994); moreover the clear evidence of spatiotemporal correlations between chromospheric (Ca II K, Hα) and coronal brightenings (*TRACE* Fe IX 171 Å) indicates that the chromosphere is heated up to coronal temperatures at the surge footpoint by the energy release during magnetic field reconnection, as shown in the numerical simulation of Yokoyama and Shibata (1996).

It is clear that the main body of the aforementioned surge is positioned at the TR/lower corona region. Thus, we can take the electron number density of the surge environment to be equal to $n_e = 2 \times 10^9\,\mathrm{cm}^{-3}$, and its temperature reasonably can be $T_e = 2 \times 10^6$ K. Concerning the surge temperature, we suppose it to be equal to $T_i = 2 \times 10^5$ K. Then with a background magnetic field $B_e = 7$ G and a density contrast $\eta = \rho_e/\rho_i = 0.052$ (we assume that plasma densities in both media are homogeneous), we have the following characteristic sound and Alfvén speed inside the surge and in the surrounding magnetized plasma: $c_{si} = 52.5$ km s^{-1}, $v_{Ai} \cong 67.1$ km s^{-1} (more exactly 67.093 km s^{-1}, which value determines the magnetic field inside the jet to be equal to $B_i = 6.02$ G), and $c_{se} \cong 166$ km s^{-1}, $v_{Ae} = 341.2$ km s^{-1}, respectively. Thus, the plasma betas of the two media are correspondingly $\beta_i = 0.794$ and $\beta_e = 0.284$. Incompressible plasma is a good approximation to study the KH instability, though the assumed values in the environment are not favorable for the incompressibility. Nevertheless, in the following we consider incompressible plasma media inside and outside the surge.

We model the surge as a vertically moving with a velocity v_0 cylindrical flux tube with radius $a = \Delta\ell/2$ (see Fig. 7.1), where $\Delta\ell = 7$ Mm is the surge width. Our frame of reference is attached to the TR/coronal plasma which implies that v_0 is the relative jet velocity with respect to its environment. We must mention that because the density contrast, η, is relatively high, in such a case, like in spicules, the occurrence of a KH instability, for instance of kink ($m = 1$) waves, becomes possible at generally high Alfvén Mach numbers (the Alfvén Mach number is defined as the ratio of jet velocity to Alfvén speed inside the jet, $M_A = v_0/v_{Ai}$) and respectively high critical flow velocities being far beyond the speeds accessible for surges/spicules in the solar atmosphere (Zhelyazkov, 2012; Zhelyazkov and Zaqarashvili, 2012). We note that this requirement applies for KH instability

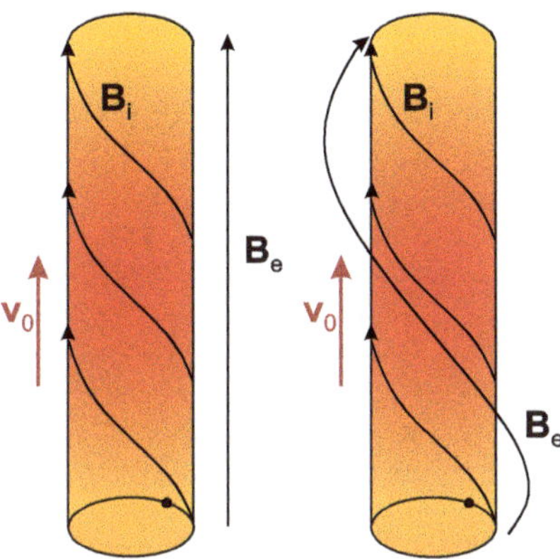

Fig. 7.1. Equilibrium magnetic field geometries of a cool solar surge.

occurring in the longitudinal direction (subject to longitudinal shear flow) and it is less strict and far more plausible for that occurring in the transfer direction (subject to transverse shear flow). Therefore, in our case of longitudinal shear flow the only possible way for emerging a KH instability in surges is the excitation of higher MHD harmonics that can become unstable at sub-Alfvénic flow velocities in twisted tubes (Zaqarashvili *et al.*, 2010). Here, we consider two possible magnetic field geometries: (i) a moving twisted magnetic flux tube embedded in untwisted magnetic field $\boldsymbol{B}_e$ (the left tube in Fig. 7.1), and (ii) a moving twisted magnetic flux tube surrounded by plasma with twisted magnetic field lines (the right tube in Fig. 7.1). In our cylindrical coordinate system (r, ϕ, z), the magnetic field has the following form: $\boldsymbol{B} = (0, B_\phi(r), B_z(r))$, and the flow profile inside the tube is $\boldsymbol{v}_0 = (0, 0, v_0)$. In general, v_0 can be a function of r, but we consider the simplest homogeneous case. The unperturbed magnetic field $\boldsymbol{B}$ and the pressure p satisfy the pressure balance equation

$$\frac{\mathrm{d}}{\mathrm{d}r}\left(p + \frac{B_\phi^2 + B_z^2}{2\mu}\right) = -\frac{B_\phi^2}{\mu r}, \tag{7.1}$$

where μ is the magnetic permeability.

As the unperturbed parameters depend on the r coordinate only, the perturbations can be Fourier analyzed with $\exp\left[-\mathrm{i}(\omega t - m\phi - k_z z)\right]$. The equation governing the incompressible plasma dynamics is (Goossens *et al.*, 1992) as follows:

$$\frac{\mathrm{d}^2 p_{\text{tot}}}{\mathrm{d}r^2} + \left[\frac{C_3}{rD}\frac{\mathrm{d}}{\mathrm{d}r}\left(\frac{rD}{C_3}\right)\right]\frac{\mathrm{d}p_{\text{tot}}}{\mathrm{d}r} + \left[\frac{C_3}{rD}\frac{\mathrm{d}}{\mathrm{d}r}\left(\frac{rC_1}{C_3}\right) + \frac{1}{D^2}(C_2 C_3 - C_1^2)\right]p_{\text{tot}}$$
$$= 0. \tag{7.2}$$

where

$$D = \rho(\Omega^2 - \omega_{\rm A}^2), \quad C_1 = -\frac{2m B_\phi}{\mu r^2}\left(\frac{m}{r}B_\phi + k_z B_z\right),$$

$$C_2 = -\left(\frac{m^2}{r^2} + k_z^2\right),$$

$$C_3 = D^2 + D\frac{2B_\phi}{\mu}\frac{\rm d}{{\rm d}r}\left(\frac{B_\phi}{r}\right) - \frac{4B_\phi^2}{\mu r^2}\rho\omega_{\rm A}^2;$$

$$\omega_{\rm A} = \frac{\boldsymbol{k} \cdot \boldsymbol{B}}{\sqrt{\mu\rho}} = \frac{1}{\sqrt{\mu\rho}}\left(\frac{m}{r}B_\phi + k_z B_z\right)$$

$$\tag{7.3}$$

is the local Alfvén frequency,

$$\Omega = \omega - \boldsymbol{k} \cdot \boldsymbol{v}_0 \tag{7.4}$$

is the Doppler-shifted frequency, and $p_{\rm tot}$ is total (thermal + magnetic) pressure perturbation. Radial displacement ξ_r is expressed through the total pressure perturbation as

$$\xi_r = \frac{D}{C_3}\frac{{\rm d}p_{\rm tot}}{{\rm d}r} + \frac{C_1}{C_3}p_{\rm tot}. \tag{7.5}$$

The solution to this equation depends on the magnetic field and density profile. To obtain the dispersion relation of MHD modes, we find the solutions to Eqs. (7.2) and (7.5) inside and outside the tube and merge the solutions through boundary conditions.

7.1.2 *Wave dispersion relations*

Prior to obtaining the wave dispersion relations for the two magnetic configurations, we first specify the twist of the magnetic flux tube to be a uniform one, i.e.,

$$\boldsymbol{B}_{\rm i} = (0, Ar, B_{\rm iz}), \tag{7.6}$$

where A and $B_{\rm iz}$ are constant. We specify the twist of the magnetic field, ε, by the ratio $B_{\rm i\phi}/B_{\rm iz}$, where $B_{\rm i\phi}$ is evaluated at the inner surface of the tube, that is, $\varepsilon = Aa/B_{\rm iz}$. In the following, we only consider $\varepsilon \ll 1$, therefore the configuration is stable for kink and sausage instabilities. In the simpler case (left flux tube in Fig. 7.1), the magnetic field outside the tube is homogeneous, $\boldsymbol{B}_{\rm e} = (0, 0, B_{\rm e})$, and the solution to Eq. (7.2) inside the tube bounded at the tube axis is (see Zaqarashvili *et al.*, 2014a)

$$p_{\rm tot}(r \leqslant a) = \alpha_{\rm i} I_m(\kappa_{\rm i}r), \tag{7.7}$$

where I_m is the modified Bessel function of order m and α_i is a constant. Transverse displacement can be written using Eq. (7.5) as where I_m is the modified Bessel function of order m and α_i is a constant. Transverse displacement can be written using Eq. (7.5) as

$$\xi_{ir} = \frac{\alpha_i}{r} \left\{ \frac{(\Omega^2 - \omega_{Ai}^2)\kappa_i r\, I'_m(\kappa_i r)}{\rho_i(\Omega^2 - \omega_{Ai}^2)^2 - 4A^2\omega_{Ai}^2/\mu} - \frac{2m A\omega_{Ai} I_m(\kappa_i r)/\sqrt{\mu\rho_i}}{\rho_i(\Omega^2 - \omega_{Ai}^2)^2 - 4A^2\omega_{Ai}^2/\mu} \right\},$$

(7.8)

where the attenuation coefficient κ_i and Alfvén frequency ω_{Ai} are given by

$$\kappa_i = k_z \left[1 - 4A^2\omega_{Ai}^2/\mu\rho_i(\Omega^2 - \omega_{Ai}^2)^2 \right]^{1/2},$$

(7.9)

$$\omega_{Ai} = \frac{m A + k_z B_{iz}}{\sqrt{\mu\rho_i}},$$

(7.10)

and prime sign means a differentiation by the Bessel function argument.

The solution to Eq. (7.2) outside the flux tube bounded at infinity is

$$p_{\text{tot}}(r > a) = \alpha_e K_m(k_z r),$$

(7.11)

where K_m is the modified Bessel function of order m and α_e is a constant. Transverse displacement can be written as

$$\xi_{er} = \frac{\alpha_e}{r} \frac{k_z r\, K'_m(k_z r)}{\rho_e(\omega^2 - \omega_{Ae}^2)},$$

(7.12)

where, as before, the prime sign means a differentiation by the Bessel function argument, and the local Alfvén frequency is

$$\omega_{Ae} = \frac{k_z B_e}{\sqrt{\mu\rho_e}} = k_z v_{Ae}.$$

(7.13)

Here, $v_{Ae} = B_e/\sqrt{\mu\rho_e}$ is the Alfvén speed in the tube environment.

The boundary conditions which merge the solutions inside and outside the twisted magnetic flux tube are the continuity of the radial component of the Lagrangian displacement

$$\xi_{ir}|_{r=a} = \xi_{er}|_{r=a}$$

(7.14)

and the total pressure perturbation (Bennett *et al.*, 1999)

$$\left. p_{\text{tot}\,i} - \frac{B_{i\phi}^2}{\mu a}\xi_{ir} \right|_{r=a} = p_{\text{tot}\,e}|_{r=a},$$

(7.15)

where total pressure perturbations $p_{\text{tot}\,i}$ and $p_{\text{tot}\,e}$ are given by Eqs. (7.7) and (7.11), respectively. Applying these boundary condition, after some algebra we finally derive the dispersion relation of the normal MHD modes propagating along a twisted magnetic flux tube with axial mass flow $\boldsymbol{v}_0$ surrounded by plasma embedded in a homogeneous magnetic field

$$\frac{(\Omega^2 - \omega_{\text{Ai}}^2)F_m(\kappa_i a) - 2mA\omega_{\text{Ai}}/\sqrt{\mu\rho_i}}{(\Omega^2 - \omega_{\text{Ai}}^2)^2 - 4A^2\omega_{\text{Ai}}^2/\mu\rho_i} = \frac{P_m(k_z a)}{\frac{\rho_e}{\rho_i}(\omega^2 - \omega_{\text{Ae}}^2) + A^2 P_m(k_z a)/\mu\rho_i},$$

$$(7.16)$$

where, recall, $\Omega = \omega - \boldsymbol{k}\cdot\boldsymbol{v}_0$ is the Doppler-shifted wave frequency in the moving flux tube, and

$$F_m(\kappa_i a) = \frac{\kappa_i a I_m'(\kappa_i a)}{I_m(\kappa_i a)} \quad\text{and}\quad P_m(k_z a) = \frac{k_z a K_m'(k_z a)}{K_m(k_z a)}.$$

The reader can see a derivation of Eq. (7.16) starting from the basic equations of ideal magnetohydrodynamics in Chapter 6.

In the case when the outside magnetic field is also twisted (the right flux tube in Fig. 7.1), we consider that magnetic field, $\boldsymbol{B}_e$, has the form (Zaqarashvili *et al.*, 2014a)

$$\boldsymbol{B}_e = \left(0,\, B_{e\phi}\frac{a}{r},\, B_{ez}\left(\frac{a}{r}\right)^2\right),$$

$$(7.17)$$

and the density is presented as $\rho = \rho_e(a/r)^4$, so that the Alfvén frequency

$$\omega_{\text{Ae}} = \frac{mB_{e\phi} + k_z a B_{ez}}{\sqrt{\mu\rho_e}a}$$

$$(7.18)$$

is constant, which allows us to find an analytical solution to the governing Eq. (7.2). The total pressure perturbation outside the tube is governed by the Bessel-type equation

$$\frac{d^2 p_{\text{tot}}}{dr^2} + \frac{5}{r}\frac{d p_{\text{tot}}}{dr} - \left(\frac{n^2}{r^2} + \kappa_e^2\right)p_{\text{tot}} = 0,$$

$$(7.19)$$

where

$$n^2 = m^2 - \frac{4m^2 B_{e\phi}^2}{\mu\rho_e a^2(\omega^2 - \omega_{\text{Ae}}^2)} + \frac{8m B_{e\phi}\omega_{\text{Ae}}}{\sqrt{\mu\rho_e}a(\omega^2 - \omega_{\text{Ae}}^2)},$$

$$(7.20)$$

and

$$\kappa_e^2 = k_z^2\left(1 - \frac{4B_{e\phi}^2\omega^2}{\mu\rho_e(\omega^2 - \omega_{\text{Ae}}^2)^2 a^2}\right).$$

$$(7.21)$$

A solution to Eq. (7.19) bounded at infinity is

$$p_{\text{tot}}(r > a) = \alpha_{\text{e}} \frac{a^2}{r^2} K_\nu(\kappa_{\text{e}} r), \tag{7.22}$$

where $\nu = \sqrt{4 + n^2}$ and α_{e} is a constant.

Transversal displacement can be written as

$$\xi_{er} = \alpha_{\text{e}} \frac{r(\omega^2 - \omega_{\text{Ae}}^2)\kappa_{\text{e}} r K_\nu'(\kappa_{\text{e}} r)}{a^2 \rho_{\text{e}}(\omega^2 - \omega_{\text{Ae}}^2)^2 - 4B_{\text{e}\phi}^2 \omega^2/\mu}$$

$$-\alpha_{\text{e}} \frac{r}{a} \left\{ \frac{2a(\omega^2 - \omega_{\text{Ae}}^2)}{a^2 \rho_{\text{e}}(\omega^2 - \omega_{\text{Ae}}^2)^2 - 4B_{\text{e}\phi}^2 \omega^2/\mu} \right.$$

$$\left. + \frac{2m B_{\text{e}\phi}\omega_{\text{Ae}}/\sqrt{\mu\rho_{\text{e}}}}{a^2 \rho_{\text{e}}(\omega^2 - \omega_{\text{Ae}}^2)^2 - 4B_{\text{e}\phi}^2 \omega^2/\mu} \right\} K_\nu(\kappa_{\text{e}} r). \tag{7.23}$$

By applying boundary conditions (7.14) and (7.15), where the transversal displacements are given by Eqs. (7.12) and (7.23), and total pressure perturbations $p_{\text{tot}\,i}$ and $p_{\text{tot}\,e}$, accordingly, by Eqs. (7.7) and (7.22), we obtain the dispersion relation of the normal MHD modes propagating along a twisted magnetic flux tube with axial mass flow $\boldsymbol{v}_0$ surrounded by plasma embedded in twisted magnetic field (Bogdanova *et al.*, 2018)

$$\frac{(\Omega^2 - \omega_{\text{Ai}}^2)F_m(\kappa_i a) - 2m A\omega_{\text{Ai}}/\sqrt{\mu\rho_i}}{(\Omega^2 - \omega_{\text{Ai}}^2)^2 - 4A^2\omega_{\text{Ai}}^2/\mu\rho_i} = \frac{a^2(\omega^2 - \omega_{\text{Ae}}^2)Q_\nu(\kappa_{\text{e}} a) - G}{L - Ha^2(\omega^2 - \omega_{\text{Ae}}^2)Q_\nu(\kappa_{\text{e}} a) - G}, \tag{7.24}$$

where

$$Q_\nu(\kappa_{\text{e}} a) = \frac{\kappa_{\text{e}} a K_\nu'(\kappa_{\text{e}} a)}{K_\nu(\kappa_{\text{e}} a)}, \quad L = a^2 \rho_{\text{e}}(\omega^2 - \omega_{\text{Ae}}^2)^2 - \frac{4B_{\text{e}\phi}^2 \omega^2}{\mu},$$

$$H = \frac{B_{\text{e}\phi}^2}{\mu a^2} - \frac{A^2}{\mu}, \quad G = 2a^2(\omega^2 - \omega_{\text{Ae}}^2) + \frac{2ma B_{\text{e}\phi}\omega_{\text{Ae}}}{\sqrt{\mu\rho_{\text{e}}}}.$$

Note that the left-hand sides of dispersion equations (7.16) and (7.24) are identical (this is not surprising), but the right-hand sides are completely different due to the very different magnetic field environments.

7.1.3 *Numerical calculations and results*

The main goal of our study is to determine under which conditions the MHD waves propagating along the moving flux tube can become unstable. To conduct this investigation, it is necessary to assume that the wave frequency ω is a complex quantity, that is, $\omega \to \omega + i\gamma$, where γ is the instability growth rate (do not confuse it with the adiabatic index), while the longitudinal wave number k_z is a real variable in the wave dispersion relation. The occurrence of the expected KH instability is determined primarily by the jet velocity and in searching for a critical or threshold value of it, we will gradually change velocity magnitude from zero to that critical value (and beyond). Thus, we have to solve dispersion relations in complex variables, obtaining the real and imaginary parts of the wave frequency, or as is commonly accepted, of the wave phase velocity $v_{\rm ph} = \omega/k_z$, as functions of k_z at various values of the velocity shear between the surge and its environment, v_0.

Before starting the numerical task, we have to normalize all variables and to specify the input parameters. The wave phase velocity, $v_{\rm ph}$, and the other speeds are normalized to the Alfvén speed inside the jet, $v_{\rm Ai}$, which is calculated using the axial magnetic fields $B_{\rm iz}$. The wavelength, $\lambda = 2\pi/k_z$, is normalized to the tube radius, a, that is equivalent to introducing a dimensionless wavenumber $k_z a$. For normalizing the Alfvén frequency in the environment, $\omega_{\rm Ae}$, except the density contrast η and the tube radius a, we have to additionally specify the ratio of the axial magnetic field components in both media, $b = B_{\rm ez}/B_{\rm iz}$. For our surge and its environment that ratio is equal to 1.1622, we shall simply take $b = 1$. In the dimensionless analysis, the flow speed, v_0, will be presented by the Alfvén Mach number $M_{\rm A} = v_0/v_{\rm Ai}$.

We first begin with the numerical solving Eq. (7.16) for the fluting-like $m = -2$ mode and one obtains four instability windows on the $k_z a$-axis whose position and width depend upon the magnetic field twist parameter ε (see Fig. 7.2). The critical jet speeds for emergence of a KH instability at $\varepsilon = 0.025, 0.1, 0.2$ and 0.4 correspondingly are $59.6\,{\rm km\,s^{-1}}$, $58.0\,{\rm km\,s^{-1}}$, $56.3\,{\rm km\,s^{-1}}$, and $54.7\,{\rm km\,s^{-1}}$. These velocities are higher than the speeds registered by Brooks *et al.* (2007), but they (the velocities) are generally accessible for cool surges. The three narrow windows corresponding to $\varepsilon = 0.025$, $\varepsilon = 0.1$, and $\varepsilon = 0.2$ are practically inapplicable to our surge–environment configuration: the wavelength, $\lambda = \pi\,\Delta\ell/k_z a$, of unstable $m = -2$ harmonics becomes comparable to the surge's height; for instance, for $k_z a = 0.3$ (the middle of the second instability window) the wavelength of the unstable $m = -2$ harmonic at $\varepsilon = 0.1$ is $\lambda_{\rm KH} = 73\,{\rm Mm}$. Actually only in the fourth instability window (at $\varepsilon = 0.4$), one can have an observable unstable $m = -2$ MHD mode: for instance, at $k_z a = \pi/2$ the wavelength is $\lambda_{\rm KH} = 14\,{\rm Mm}$, and the corresponding dimensionless mode phase velocity growth rate is equal to 0.20388 which implies a mode growth rate $\gamma_{\rm KH} = 0.0061\,{\rm s^{-1}}$. The instability growth time $\tau_{\rm KH} \equiv 2\pi/\gamma_{\rm KH} \cong 1023\,{\rm s} = 17\,{\rm min}$ is in the limit of surge's lifetime.

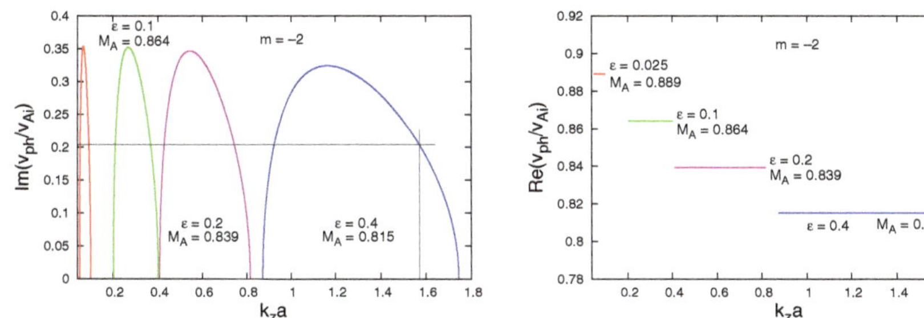

Fig. 7.2. (*Left panel*) Growth rates of the unstable $m = -2$ MHD mode propagating on incompressible jets in four different twisted internal magnetic fields (with $\varepsilon = 0.025$, 0.1, 0.2, and 0.4) at $\eta = 0.052$, $b = 1$, and corresponding critical M_A numbers. For $k_z a = \pi/2$, the wavelength of the unstable $m = -2$ harmonic for $\varepsilon = 0.4$ is $\lambda_{KH} = 14\,\text{Mm}$, and the mode growth rate is $\gamma_{KH} = 0.0061\,\text{s}^{-1}$. (*Right panel*) Marginal dispersion curves of the unstable $m = -2$ MHD mode for the critical M_A numbers as functions of the magnetic field twist parameter ε. At $k_z a = \pi/2$, the critical jet velocity for $\varepsilon = 0.4$ is $v_0^{cr} = 54.7\,\text{km s}^{-1}$.

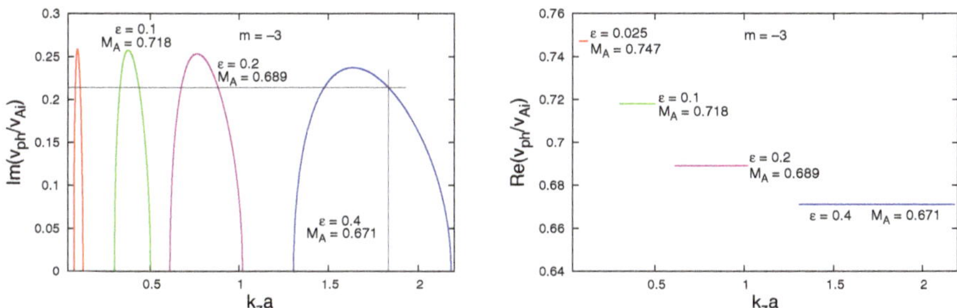

Fig. 7.3. (*Left panel*) Similar as Fig. 7.2 but for the case $m = -3$. For $k_z a = 1.832$, the wavelength of the unstable $m = -3$ harmonic at $\varepsilon = 0.4$ is $\lambda_{KH} = 12\,\text{Mm}$, and the mode growth rate is $\gamma_{KH} = 0.0075\,\text{s}^{-1}$. (*Right panel*) Marginal dispersion curves of the unstable $m = -3$ MHD mode for the critical M_A numbers as functions of the magnetic field twist parameter ε. At $k_z a = 1.832$ and $\varepsilon = 0.4$, the critical jet velocity is $v_0^{cr} = 45.0\,\text{km s}^{-1}$.

Figures 7.3 and 7.4 show the results for the $m = -3$ and $m = -4$ MHD modes. The first observation, comparing Figs. 7.2–7.4, is that with increasing the modulus of the MHD harmonic, the range of wave propagation (for a set of four magnetic filed twist parameters equal to 0.025, 0.1, 0.2, and 0.4) becomes wider. On the other hand, the instability windows do not recover all dimensionless wavenumber's values—there appears gaps between the windows. That is well-pronounced for the $m = -4$ MHD mode. Those gaps in Figs. 7.3 and 7.4 can be filled in by performing the calculation for some other values of the magnetic field twist parameter, ε, say for $\varepsilon = 0.3$. The second observation is that with increasing the modulus of the

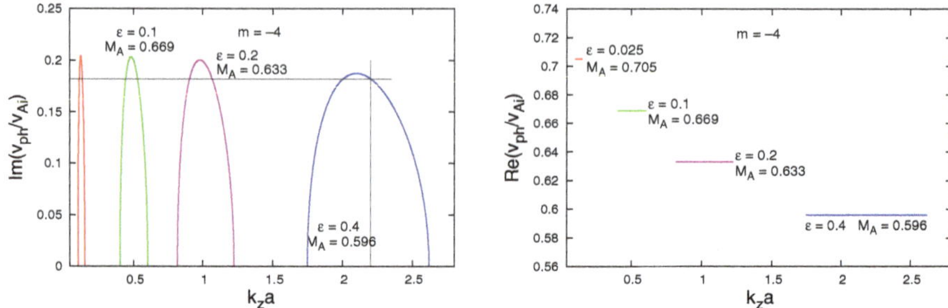

Fig. 7.4. (*Left panel*) Similar as Fig. 7.2 but for the case $m = -4$. For $k_z a = 2.199$, the wavelength of the unstable $m = -4$ harmonic for $\varepsilon = 0.4$ is $\lambda_{\text{KH}} \cong 10\,\text{Mm}$, and the mode growth rate is $\gamma_{\text{KH}} = 0.0077\,\text{s}^{-1}$. (*Right panel*) Marginal dispersion curves of the unstable $m = -4$ MHD mode for the critical M_A numbers as functions of the magnetic field twist parameter ε. At $k_z a = 2.199$ and $\varepsilon = 0.4$, the critical jet velocity is $v_0^{\text{cr}} = 40.0\,\text{km s}^{-1}$.

MHD mode the maximal values of the normalized mode phase velocity growth rate decrease. Common in all cases is the circumstance that one can really have unstable MHD harmonics (with reasonable wavelengths) only in the fourth window for $\varepsilon = 0.4$. Particularly, for the $m = -4$ MHD mode, one could also expect an unstable wavelength of $\lambda_{\text{KH}} = 14\,\text{Mm}$ at $\varepsilon = 0.3$.

Let us now give some numbers for critical velocities, unstable wavelengths and corresponding mode growth rates of the $m = -3$ and $m = -4$ MHD modes. For the $m = -3$ harmonic (see Fig. 7.3), the critical jet velocities for emergence of a KH instability at $\varepsilon = 0.025$, 0.1, 0.2, and 0.4 correspondingly are $50.1\,\text{km s}^{-1}$, $48.2\,\text{km s}^{-1}$, $46.2\,\text{km s}^{-1}$, and $45.0\,\text{km s}^{-1}$ as the lowest value is obtained in the fourth instability window with $\varepsilon = 0.4$. If we pick out a $k_z a = 1.832$, that is, an unstable $\lambda_{\text{KH}} = 12\,\text{Mm}$, the corresponding dimensionless mode phase velocity growth rate at $\varepsilon = 0.4$ is equal to 0.2139 which implies a mode growth rate $\gamma_{\text{KH}} = 0.0075\,\text{s}^{-1}$ and growth time $\tau_{\text{KH}} = 836.7$; $\text{s} = 13.9\,\text{min}$. If we shift to the left at $k_z a = 1.64$ (with maximal normalized mode phase velocity growth rate), the wavelength is $\lambda_{\text{KH}} = 13.4\,\text{Mm}$, and the corresponding mode growth rate is of the same order, namely $\gamma_{\text{KH}} = 0.0074\,\text{s}^{-1}$ or $\tau_{\text{KH}} = 14.1\,\text{min}$. With exciting the $m = -4$ harmonic, one can have relatively shorter unstable wavelengths as well as lower critical flow velocities—the latter are, in descending order, equal to $47.3\,\text{km s}^{-1}$, $44.9\,\text{km s}^{-1}$, $42.5\,\text{km s}^{-1}$, and $40.0\,\text{km s}^{-1}$, respectively— as seen in the range of deduced by Brooks *et al.* (2007) threshold speeds. For $k_z a = 2.199$ (see Fig. 7.4), $\lambda_{\text{KH}} = 10\,\text{Mm}$ and the corresponding $\text{Im}(v_{\text{ph}}/v_{\text{Ai}})$ value is 0.181754 that yields $\gamma_{\text{KH}} = 0.0077\,\text{s}^{-1}$, which implies instability developing time of $13.7\,\text{min}$. If we choose $k_z a \cong 2.094$—the value at which the dimensionless mode phase velocity growth rate is maximal—the corresponding unstable wavelength is $\lambda_{\mathrm{KH}} = 10.5\,\mathrm{Mm}$, and for $\mathrm{Im}(v_{\mathrm{ph}}/v_{\mathrm{Ai}})_{\mathrm{max}} = 0.187083$ the mode growth rate is $\gamma_{\mathrm{KH}} = 0.0075\,\mathrm{s}^{-1}$ or $\tau_{\mathrm{KH}} \cong 13.9\,\mathrm{min}$, being of the same order as that at $\lambda_{\mathrm{KH}} = 10\,\mathrm{Mm}$. Note that for each unstable MHD harmonic the normalized mode phase velocity on given dispersion curve (see the right panels in Figs. 7.2 to 7.4) is equal to its label $\mathsf{M_A}$. Therefore, the unstable perturbations are frozen in the flow and consequently they are vortices rather than waves. This observation is consistent with the KH instability in the hydrodynamics that deals with unstable vortices.

A surge implies both an upward and a downward motion of plasma in an often ballistic fashion. These two stages therefore introduce two regimes, one with $v_0 > 0$ and one with $v_0 < 0$ for the flow. If the waves are being continuously generated somewhere in the chromosphere and thus keep propagating upward during the entire lifetime of the surge, the Doppler-shifted frequency of the wave, $\omega - \boldsymbol{k} \cdot \boldsymbol{v}_0$, will change drastically at the moment of flow velocity reversal. Computations show, however, that the only change is associated with a reverse of the wave phase velocity, $\mathrm{Re}(\omega)/k_z$—the instability growth rates, $\mathrm{Im}(\omega)/k_z$, are the same as in an upward plasma motion. The wave dispersion curves in a downward plasma flow are the mirror images of the dispersion curves in upward flowing plasma.

One obtains a much more interesting picture when studying the more complicated case of a twisted magnetic flux tube surrounded by plasma embedded in a twisted background magnetic field. Our choice for the twist characteristics of the two magnetic fields (internal and external ones) are $\varepsilon_{\mathrm{i}} = 0.3$ and $\varepsilon_{\mathrm{e}} = 0.01$, respectively. Numerical solving Eq. (7.24) for the three mode numbers $m = -2$, $m = -3$, and $m = -4$ gives for each mode number two instability windows: one of them with relatively high maximal growth rate, and a second window, next to the former, with one order lower maximal growth rate. These two families of instability windows, for clarity, are presented in Figs. 7.5–7.6, respectively. As seen from Fig. 7.5, one can observe a KH instability mostly for the $m = -4$ MHD mode and partly for the $m = -3$ harmonic. The wave growth rates of unstable modes are of the same order like those illustrated in Figs. 7.2–7.4, that is, few inverse milliseconds. The critical flow velocities for the $m = -2$, $m = -3$, and $m = -4$ modes are correspondingly equal to $46\,\mathrm{km\,s^{-1}}$, $41\,\mathrm{km\,s^{-1}}$, and $37\,\mathrm{km\,s^{-1}}$. The second family of instability windows, shown in Fig. 7.6, has two distinct peculiarities: first, the instability windows are shifted to the right-hand side of the $k_z a$-axis, that is, the propagation range of unstable MHD modes is extended, and second, the marginal dispersion curves are not constant—the normalized wave phase velocities gradually decrease with increasing wavenumber. If we fix the normalized wavenumber to be $k_z a = 2$, the wavelength of the unstable $m = -4$ mode is $\lambda_{\mathrm{KH}} = 11\,\mathrm{Mm}$, and at $\mathrm{Im}(v_{\mathrm{ph}}/v_{\mathrm{Ai}}) = 0.02457$ the mode growth rate is $\gamma_{\mathrm{KH}} = 0.00094\,\mathrm{s^{-1}}$,

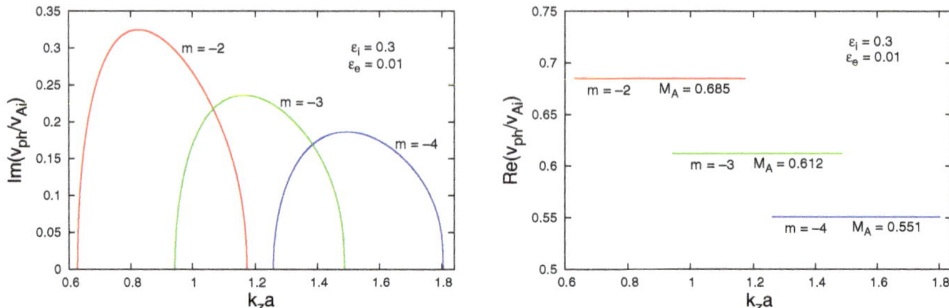

Fig. 7.5. (*Left panel*) Growth rates of the unstable $m = -2$, $m = -3$, and $m = -4$ MHD modes propagating on incompressible twisted jets with $\varepsilon_i = 0.3$ in a twisted external magnetic field with $\varepsilon_e = 0.01$ at $\eta = 0.052$, $b = 1$, and critical M_A numbers equal to 0.685, 0.612, and 0.551, respectively. (*Right panel*) Marginal dispersion curves of the unstable $m = -2$, $m = -3$, and $m = -4$ MHD modes for the critical M_A numbers. The critical surge velocities of these modes are correspondingly equal to $46\,\mathrm{km\,s^{-1}}$, $41\,\mathrm{km\,s^{-1}}$, and $37\,\mathrm{km\,s^{-1}}$.

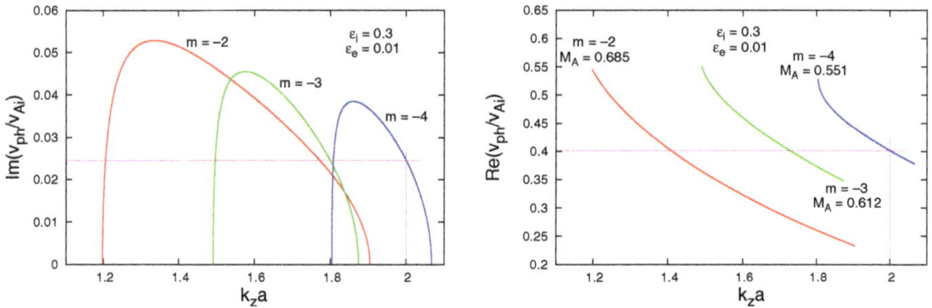

Fig. 7.6. (*Left panel*) Similar as Fig. 7.5 but for the second family of instability windows. For $k_z a = 2$, the wavelength of the unstable $m = -4$ harmonic is $\lambda_{\mathrm{KH}} = 11\,\mathrm{Mm}$, and the mode growth rate is $\gamma_{\mathrm{KH}} = 0.00094\,\mathrm{s^{-1}}$. (*Right panel*) The critical surge velocity of the $m = -4$ harmonic at $k_z a = 2$ is equal to $26.9\,\mathrm{km\,s^{-1}}$. In contrast to the instability windows plotted in the right panel of Fig. 7.5, in this family of instability windows the normalized phase velocities are not constant.

much lower than the growth rates in the first family of instability windows. Note also, that the critical jet velocity for emerging a KH instability now is remarkably lower—its normalized value is 0.401042 that implies $v_0^{\mathrm{cr}} = 26.9\,\mathrm{km\,s^{-1}}$, that is, approximately the half of the surge speed evaluated by Brooks *et al.* (2007). Thus, one can conclude that high-harmonic MHD modes can become unstable against the KH instability for accessible sub-Alfvénic velocities—this is more pronounced in the case when both magnetic fields are twisted.

Among the MHD modes which propagate along the moving tube only the high harmonics with negative azimuthal mode numbers can become unstable at accessible critical flow velocities. It is known that the perturbations with wave vector nearly perpendicular to the magnetic field, that is, $\boldsymbol{k} \cdot \boldsymbol{B} \approx 0$, are the most unstable ones (Pataraya and Zaqarashvili, 1995; Zaqarashvili *et al.*, 2014b). These modes are pure vortices in the incompressible limit, therefore they have the strongest growth rate due to KH instability. This condition yields opposite signs for azimuthal mode (m) and axial (k_z) wave numbers. Here we consider positive k_z, therefore the harmonics with the negative m have maximal growth rates. If one takes negative k_z, then the harmonics with positive m will have maximal growth rates. Negative k_z means the harmonics propagating opposite to the flow, hence a superposition of unstable harmonics with positive and negative m may lead to a less velocity of KH vortices compared to the flow speed.

To sum up, one can observe practically a developing KH instability only if the parameter ε characterizing the magnetic field twist is large enough. For instance, at the first magnetic field configuration (the left flux tube in Fig. 7.1) the $m = -3$ harmonic becomes unstable at critical flow velocity of 45 km s^{-1} and wavelength $\lambda_{\mathrm{KH}} = 12\,\mathrm{Mm}$ with a linear growth rate $\gamma_{\mathrm{KH}} = 0.0075\,\mathrm{s}^{-1}$ at the magnetic field twist parameter $\varepsilon = 0.4$. We note that by exploring the KH instability (Zhelyazkov *et al.*, 2015a) in a high-temperature solar surge at coronal plasma densities, like that observed by Kayshap *et al.* (2013), for the same mode at the same position on the $k_z a$-axis and the same value of ε, one obtains a mode growth rate of $0.033\,\mathrm{s}^{-1}$ being exactly equal to the growth rate of the imaged KH instability in a coronal mass ejecta in the lower corona by Foullon *et al.* (2013). On the other hand, the aforementioned $\gamma_{\mathrm{KH}} = 0.0075\,\mathrm{s}^{-1}$ is of the same order as the growth rate of $0.003\,\mathrm{s}^{-1}$ of vortex-shaped features along the interface between an erupting (dimming) region and the surrounding corona imaged by the *SDO*/AIA as reported by Ofman and Thompson (2011). All these comparisons allow us to believe that the KH instability might be imaged in surges, too. The second magnetic field configuration (the right flux tube in Fig. 7.1), reveals some new aspects of the KH instability, notably for a fixed pair of magnetic fields twist parameters, in our case equal to $\varepsilon_{\mathrm{i}} = 0.3$ and $\varepsilon_{\mathrm{e}} = 0.01$, for a given high-harmonic mode we obtain two instability windows on the $k_z a$-axis, next to each other. Thus, the range of KH instability is extended. True, in the second instability windows the growth rates are much lower than in the standard instability windows, but nevertheless these weak/slow instabilities can occur. Moreover, the critical flow velocity for emerging KH instability might be relatively lower. For example, the $m = -4$ harmonic with wavelength of 11 Mm can become unstable at jet speed of only $\cong 27$ km s^{-1} and its growth rate is equal to $0.00094\,\mathrm{s}^{-1}$. Solar cool surges are generally small-scale eruptive events and their contributions to solar coronal heating due to triggered by the KH instability wave turbulence is modest. There are some cases, for instance as surges are detected in

UV and EUV spectral lines, when the developing KH instability can bring forth a noticeable contribution to the solar corona energy budget.

7.2   Kelvin–Helmholtz instability in dark mottles

Chromospheric fine-scale structures such as limb spicules, on-disk mottles, and dynamic fibrils are among the most popular objects for study in solar physics today. These jet-like plasma features, formed near the network boundaries, can protrude into the transition region and low corona (see, e.g., Beckers, 1968, 1972; Sterling, 2000; De Pontieu and Erdélyi, 2006) and act as conduits for channeling energy and mass from the solar photosphere into the upper solar atmosphere and the solar wind (see De Pontieu *et al.*, 2004; De Pontieu and Erdélyi, 2006; Morton, 2012).

Dark mottles are commonly observed in the quiescent solar chromosphere. They are small-scale jet-like features of relatively cool and dense material located at the boundaries of supergranular cells, ejected from the lower chromosphere at speeds of about $10–30\,\mathrm{km\,s^{-1}}$ (see Tsiropoula and Tziotziou, 2004; Rouppe van der Voort *et al.*, 2007). Mottles are often considered as the disk representation/counterpart of chromospheric spicules (see, e.g., Hansteen *et al.*, 2006). Mottles form into two kind of groups, namely rosettes and chains. They have lengths between $5''$ and $10''$, widths smaller than $1''$ and lifetimes of the order of $13–14\,\mathrm{min}$ (Bratsolis *et al.*, 1993). Dark mottles have temperatures between $7000\,\mathrm{K}$ and $15\,000\,\mathrm{K}$, electron densities of $4 \times 10^{10}–1 \times 10^{11}\,\mathrm{cm^{-3}}$ and gas pressure of the order of $0.2\,\mathrm{dyn\,cm^{-2}}$. According to Beckers (1963), the total number of mottles on the solar surface is estimated to be equal to 4×10^5. The exact nature of mottles remains the subject of an ongoing debate, with the majority of solar researchers agreeing that mottles and spicules are related, in the sense that they have similar temperatures, density profiles, widths, lengths, and lifetimes (Tsiropoula and Schmieder, 1997; Zachariadis *et al.*, 2001). Along with the widely accepted conception that the magnetic reconnection is the main driving mechanism of these structures, there exists an alternative suggestion that at least part of the quiet-Sun mottles are driven by magnetoacoustic shocks (Rouppe van der Voort *et al.*, 2007). Chromospheric small-scale jet-like structures, such as mottles, spicules, and fibrils, play an important role in the mass balance of the solar atmosphere. It is estimated that only a small percentage of the mass outflow provided by mottles is sufficient to compensate for the coronal mass loss due to the solar wind [see Tsiropoula and Tziotziou (2004)].

Physical parameters of mottles were studied using various instrumentations over past four decades (see, e.g., Bray, 1973; Tsiropoula and Schmieder, 1997;

Tsiropoula *et al.*, 1999; Tziotziou *et al.*, 2003; Tsiropoula and Tziotziou, 2004; Bostancı and Erdoğan, 2016; Bostancı, 2011). The reader can find an overview of the most important studies on the origin of mottles and their physical parameters in the review paper of Tsiropoula *et al.* (2012).

Recent ground-based and space-borne observations have shown a plethora of waves and oscillations in spicules, mottles, and fibrils (see Kukhianidze *et al.*, 2006; Zaqarashvili *et al.*, 2007; de Pontieu *et al.*, 2007; He *et al.*, 2009; Zaqarashvili and Erdélyi, 2009; Okamoto and De Pontieu, 2011; Kuridze *et al.*, 2012; Morton, 2012; Mathioudakis *et al.*, 2013). These oscillations are usually observed as periodic transverse displacements (see, for example, Zaqarashvili and Erdélyi, 2009; Okamoto and De Pontieu, 2011; Pietarila *et al.*, 2011; Kuridze *et al.*, 2012, 2013; Morton, 2012; Bostancı *et al.*, 2014). The observations support the idea that the chromospheric fine structures can be modeled as thin, overdense magnetic flux tubes that are waveguides for the transverse oscillations with periods that have an observational upper bound limited by their finite visible lifetime. This is also supported by three-dimensional numerical modeling of the chromosphere (Leenaarts *et al.*, 2012). In this regard, the observed transverse oscillations have been interpreted as fast kink MHD waves (see, e.g., Spruit, 1982; Erdélyi and Fedun, 2007). Spicules and mottles are typically short-lived features, with lifetimes of the order of typical wave periods. The observed waves also appear to propagate with phase velocities close to the local Alfvén speed. Mottles and spicules have typical lengths of 4–7.5 Mm, so waves propagate along the mottle in only 60–100 s.

One important issue associated with chromospheric jets like mottles, spicules, and dynamic fibrils is that the MHD transverse waves propagating along a jet can become unstable and the instability that arise is of the Kelvin–Helmholtz (KH) kind. The primary aim now is to establish whether MHD waves traveling along dark mottles' jets can become unstable within their velocity range of 10–30 km s^{-1}.

7.2.1 *Mottles models, basic parameters, and governing equations*

We explore the dark mottles in the chromospheric network observed by Bostancı (2011) using two-dimensional spectroscopic observations in Hα obtained with the Göttingen Fabry–Pérot Spectrometer in the Vacuum Tower Telescope at the Observatory del Teide, Tenerife. Cloud modeling was applied by her to measure the mottles' optical thickness, source function, Doppler width, and line of sight velocity. Using these measurements, the number density of hydrogen atoms in levels 1 and 2, total particle density, electron density, temperature, gas pressure, and mass density parameters were determined with the method of Tsiropoula and Schmieder (1997). We note that similar values of those quantities were obtained by Tsiropoula and Tziotziou (2004) observing a $150'' \times 120''$ solar region in the Hα line with the Multichannel Subtractive Double Pass spectrograph mounted on the French–Italian telescope THEMIS (Tenerife, Canary Islands) from 10:07 UT to 11:02 UT on May 14, 2000 with a time cadence of $\sim$40.5 s. Bostancı and Erdoğan (2016) presented chromospheric cloud modeling on the basis of Hα profile-sampling images (over a sizable two-dimensional field of view encompassing quiet-Sun network) taken by Cauzzi *et al.* (2009) with the Interferometric Bidimensional Spectrometer at the Dunn Solar Telescope. The values of mottles' parameters derived by Bostancı and Erdoğan (2016) are very close to those of Tsiropoula and Tziotziou (2004).

We need for our modeling at least the electron density and temperature inside a mottle's jet—those values taken from Table 2 in Bostancı, 2011 are $n_i = 7.0 \times 10^{10}\,\mathrm{cm}^{-3}$ and $T_i = 12\,000\,\mathrm{K}$, respectively. Bearing in mind that the main body of the mottle is positioned at the TR/quiet lower corona region, we assume that its (mottle's) magnetic field is $B_i = 4\,\mathrm{G}$ [Tsiropoula and Tziotziou (2004) suggest 4.1 G]; our choice for the environment's electron density and temperature are correspondingly $n_e = 5.0 \times 10^9\,\mathrm{cm}^{-3}$ and $T_e = 630\,000\,\mathrm{K}$. Then Alfvén and sound speeds inside the jets are $v_{Ai} \cong 33\,\mathrm{km\,s}^{-1}$ (more exactly, $32.96\,\mathrm{km\,s}^{-1}$) and $c_{si} \cong 13\,\mathrm{km\,s}^{-1}$ (recall that the mottles are dense and cool chromospheric plasma flows). The sound speed of surrounding plasma is $c_{se} \cong 93\,\mathrm{km\,s}^{-1}$. At a density contrast $\eta = \rho_e/\rho_i = 0.0714$, the pressure balance equation (equal sums of thermal and magnetic pressures in both media) yields an Alfvén speed $v_{Ae} = 87\,\mathrm{km\,s}^{-1}$, or equivalently a magnetic field $B_e \cong 2.8\,\mathrm{G}$. Thus we have a mottle with electron density of $\sim 10^{11}\,\mathrm{cm}^{-3}$ in a weaker (or null) magnetic field environment as is usually anticipated for the chromospheric dark mottles.

We model solar dark mottles as vertically moving magnetic flux tubes because they are usually observed as well-defined on-disk features (see the left panel in Fig. 7.2 of Kuridze *et al.*, 2013). Up to now, spicules, mottles, and dynamic fibrils were considered as untwisted flux tubes, but the very recent high-resolution (0.33-arc second) observations performed by De Pontieu *et al.* (2014) with NASA's *Interface Region Imaging Spectrograph* (*IRIS*; De Pontieu *et al.*, 2014a) reveal a chromosphere and TR that are replete with twist or torsional motions on sub-arc second scales, occurring in active regions, quiet Sun regions, and coronal holes alike. They coordinated observations with the Swedish 1-meter Solar Telescope (SST) to quantify these twisting motions and their association with rapid heating to at least TR temperatures. Thus, alongside untwisted magnetic flux tubes we can also depict the mottles as weakly twisted tubes. The equilibrium magnetic field and plasma density inside the tube of radius a (equal to the half-width $\Delta\ell/2$ of the mottle) are $\boldsymbol{B}_i$ and ρ_i, while $\boldsymbol{B}_e$ and ρ_e are the corresponding quantities in the mottle's environment. Our frame of reference is attached to the TR/coronal plasma which implies that $\boldsymbol{v}_0$ is the relative jet velocity with respect to its environment. As seen, we have the same magnetic field topology as in the surge case (left column in Fig. 7.1). Thus, we will not repeat the derivation of the wave dispersion relation of the MHD modes propagating along the incompressible moving jet. We will simply use Eq. (7.16). In the case of untwisted moving magnetic flux tube of compressible plasma, the dispersion relation of the MHD modes is that of Edwin and Roberts (1983) given by Eq. (4.30) in Chapter 4:

$$\frac{\rho_e}{\rho_i}(\omega^2 - k_z^2 v_{Ae}^2)\kappa_i \frac{I'_m(\kappa_i a)}{I_m(\kappa_i a)} - \left[(\omega - \boldsymbol{k}\cdot\boldsymbol{v}_0)^2 - k_z^2 v_{Ai}^2\right]\kappa_e \frac{K'_m(\kappa_e a)}{K_m(\kappa_e a)} = 0.$$

$$(7.25)$$

Recall that the squared attenuation coefficients κ_i and κ_e collectively are given by

$$\kappa^2 = -\frac{\left[(\omega - \boldsymbol{k}\cdot\boldsymbol{v}_0)^2 - k_z^2 c_s^2\right]\left[(\omega - \boldsymbol{k}\cdot\boldsymbol{v}_0)^2 - k_z^2 v_A^2\right]}{(c_s^2 + v_A^2)\left[(\omega - \boldsymbol{k}\cdot\boldsymbol{v}_0)^2 - k_z^2 c_T^2\right]},$$

where

$$c_T = \frac{c_s v_A}{(c_s^2 + v_A^2)^{1/2}}$$

is the tube speed.

7.2.2 *Numerical calculations and results*

As in the case of surges, we start with finding the solutions to the wave dispersion relation (7.25) governing the propagation of MHD modes along the compressible untwisted dark mottle. The input parameters are: the density contrast $\eta = 0.0714$, the reduced plasma betas $\tilde{\beta}_i = 0.1521$, $\tilde{\beta}_e = 1.1447$, $b = 0.706$, and the running Alfvén Mach number M_A. The numerical calculations show that the threshold Alfvén Mach number at which the kink mode ($m = 1$) becomes unstable is equal to $M_A = 3.79$ and the marginal dispersion and dimensionless growth rate curves are shown in Fig. 7.7 in red color. There one can see three additional set of curves illustrating the changes in the normalized wave phase velocity and the growth rate with increasing M_A. The critical flow velocity for the onset of KH instability, $v_0^{cr} = 3.79 \times 33 = 125\,\text{km s}^{-1}$, is, however, far beyond jet's speeds accessible for dark mottles/spicules in solar atmosphere. In the case of a weakly twisted flux tube (with twist parameter $\varepsilon = 0.025$), the threshold Alfvén Mach number is equal to 5.38417 that implies a much higher critical jet's velocity of $\cong 177\,\text{km s}^{-1}$. We

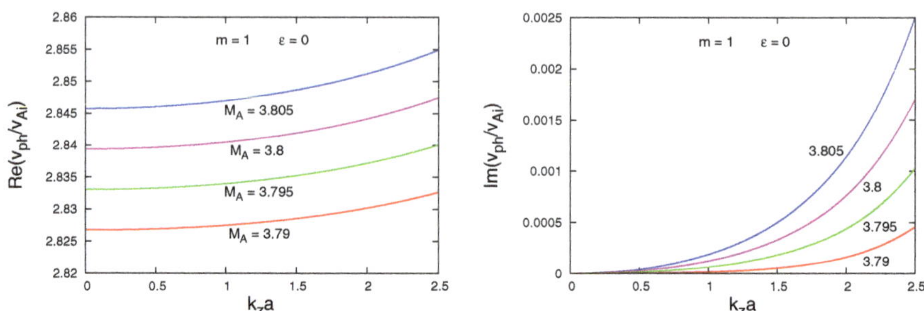

Fig. 7.7. (*Left panel*) Dispersion curves of unstable kink ($m = 1$) MHD mode propagating in an untwisted compressible plasma tube with $\eta = 0.0714$ and $b = 0.706$ for four values of the Alfvén Mach number $M_A = 3.79$, 3.795, 3.8, and 3.805. (*Right panel*) The normalized growth rates for the same values of M_A. Red curves in both plots correspond to the onset of KH instability.

note that this requirement applies for KH instability occurring in the longitudinal direction (subject to longitudinal shear flow) and it is less strict and far more plausible for that occurring in the transfer direction (subject to transverse shear flow). Therefore, in our case of longitudinal shear flow the only possible way for emerging KH instability in mottles is the excitation of higher MHD harmonics that can become unstable at sub-Alfvénic flow velocities in twisted tubes (Zaqarashvili *et al.*, 2010).

Numerical solutions to Eq. (7.16) for positive azimuthal mode numbers $m = 2, 3, 4$ yield (with $\varepsilon = 0.025$) threshold Alfvén Mach numbers not very different to the aforementioned M_A for the kink ($m = 1$) mode, that is, the instability's occurrence requires very high flow speeds inaccessible for small-scale chromospheric jets like mottles and spicules. The situation, however, distinctly changes when we perform the calculations for negative values of the mode number m.

For the fluting-like $m = -2$ mode, we obtain three instability windows (see Fig. 7.8) on the $k_z a$-axis whose position and width depend upon the magnetic field twist parameter ε. One observes that at various dimensionless wavenumbers, $k_z a$, we can have different mode growth rates depending on in which window the chosen $k_z a$ lies. If we take the mottle's width $\Delta \ell$ to be equal to $750\,\mathrm{km}$, the instability wavelength that corresponds to a given $k_z a$ is $\lambda_{\mathrm{KH}} = \pi \Delta \ell / k_z a$. It is clear that for very small normalized wavenumbers that wavelength would be much longer than the mottle's height. We assume that a $\lambda_{\mathrm{KH}} = 2.5\,\mathrm{Mm}$ is a realistic choice and this wavelength is "presented" in Fig. 7.8 by $k_z a = 0.9425$. At this position, the vertical purple line cuts off a normalized growth rate $\mathrm{Im}(v_{\mathrm{ph}}/v_{\mathrm{Ai}}) = 0.3151$ that yields a mode growth rate $\gamma_{\mathrm{KH}} = 0.025\,\mathrm{s}^{-1}$ (see the left panel in Fig. 7.8). The critical jet's velocity for the instability onset is $v_0^{\mathrm{cr}} = 27.2\,\mathrm{km\,s}^{-1}$ (see the

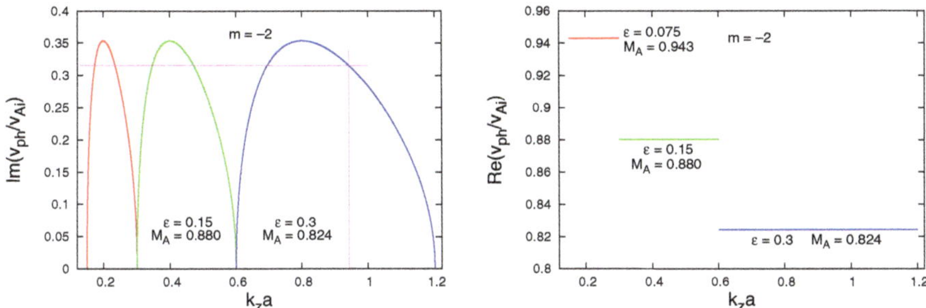

Fig. 7.8. (*Left panel*) Growth rates of the unstable $m = -2$ MHD mode propagating on twisted incompressible jets with $\varepsilon = 0.075$, 0.15, and 0.3 at the corresponding threshold M_A numbers. The growth rate of the unstable $m = -2$ harmonic at $\lambda_{KH} = 2.5\,\text{Mm}$ ($k_z a = 0.9425$) is $\gamma_{KH} = 0.025\,\text{s}^{-1}$. (*Right panel*) Marginal dispersion curves of the unstable $m = -2$ MHD mode for the critical M_A numbers as functions of the magnetic field twist parameter ε. At $k_z a = 0.9425$, the critical jet velocity is $v_0^{cr} = 27.2\,\text{km}\,\text{s}^{-1}$.

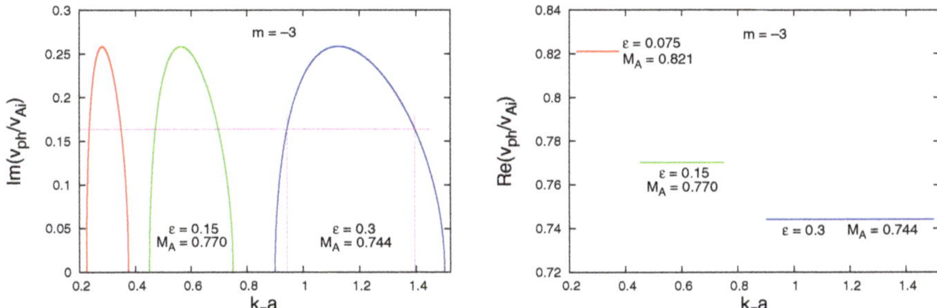

Fig. 7.9. (*Left panel*) Similar as Fig. 7.3 but for the case $m = -3$. The growth rate of the unstable $m = -3$ harmonic at $\lambda_{KH} = 2.5\,\text{Mm}$ ($k_z a = 0.9425$) is $\gamma_{KH} = 0.013\,\text{s}^{-1}$ whereas at $\lambda_{KH} \cong 1.7\,\text{Mm}$ ($k_z a = 1.3952$) is $\gamma_{KH} = 0.019\,\text{s}^{-1}$. The maximal growth rate $\gamma_{KH}^{max} = 0.024\,\text{s}^{-1}$ occurs at $\lambda_{KH} \cong 2.1\,\text{Mm}$ ($k_z a = 1.125$). (*Right panel*) Marginal dispersion curves of the unstable $m = -3$ MHD mode for the threshold M_A numbers as functions of the magnetic field twist parameter ε. For all unstable wavelengths, the critical jet velocity is $v_0^{cr} = 24.6\,\text{km}\,\text{s}^{-1}$.

right panel in the same figure). The critical velocities in the first and the second windows are equal correspondingly to $31.1\,\text{km}\,\text{s}^{-1}$ and $29.0\,\text{km}\,\text{s}^{-1}$.

Figures 7.9 and 7.10 show the results for the $m = -3$ and $m = -4$ MHD modes. The first observation, comparing Figs. 7.8–7.10, is that with increasing the modulus of the MHD harmonic, the range of wave propagation (for a set of three magnetic filed twist parameters equal to 0.075, 0.15, and 0.3) becomes wider. On the other hand, the instability windows do not recover all dimensionless wavenumber's values—there appear gaps between the windows. That is well-pronounced

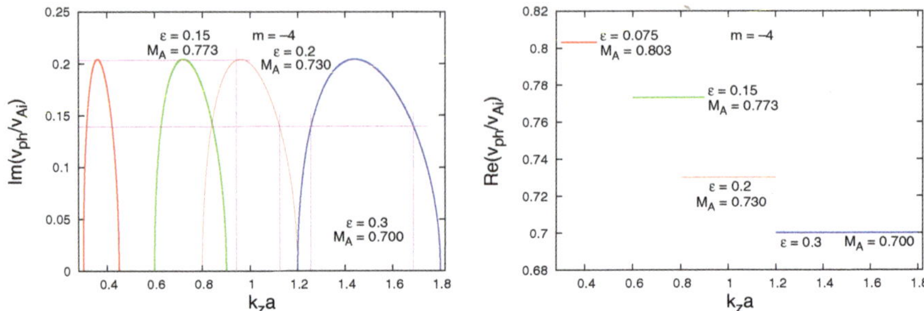

Fig. 7.10. (*Left panel*) Growth rates of the unstable $m = -4$ MHD mode propagating on twisted incompressible jets with $\varepsilon = 0.075$, 0.15, 0.2, and 0.3 at the corresponding threshold M_A numbers. The growth rate of the unstable $m = -4$ harmonic at $\lambda_{KH} = 2.5\,\mathrm{Mm}$ ($k_z a = 0.9425$) (in the $\varepsilon = 0.2$ instability window) is $\gamma_{KH} = 0.0165\,\mathrm{s}^{-1}$, whereas at $\lambda_{KH} = 2.1\,\mathrm{Mm}$ ($k_z a = 1.125$) in the same window it is $\gamma_{KH} = 0.0135\,\mathrm{s}^{-1}$. The growth rates at $\lambda_{KH} = 1.9\,\mathrm{Mm}$ ($k_z a = 1.2563$) and $\lambda_{KH} = 1.4\,\mathrm{Mm}$ ($k_z a = 1.6866$) in the the fourth window ($\varepsilon = 0.3$) are equal correspondingly to $0.015\,\mathrm{s}^{-1}$ and $0.020\,\mathrm{s}^{-1}$. (*Right panel*) Marginal dispersion curves of the unstable $m = -4$ MHD mode for the threshold M_A numbers as functions of the magnetic field twist parameter ε. The critical jet velocities in the third and fourth instability windows are equal to $24.1\,\mathrm{km\,s}^{-1}$ and $23.1\,\mathrm{km\,s}^{-1}$, respectively.

for the $m = -4$ MHD mode. Those gaps in Figs. 7.9 and 7.10 can be filled in by performing the calculation for some other values of the magnetic field twist parameter, ε, say with $\varepsilon = 0.2$ for the $m = -4$ harmonic (thin orange curves in Fig. 7.10). The second observation is that with increasing the modulus of the MHD mode the maximal values of the normalized phase velocity growth rate decrease. Common in all cases is the circumstance that one can really have unstable MHD harmonics (with reasonable wavelengths) only in the third window for $\varepsilon = 0.3$. Particularly, for the $m = -4$ MHD mode, one could also have reliable unstable wavelengths $\lambda_{KH} = 2.5\,\mathrm{Mm}$ and $\lambda_{KH} = 2.1\,\mathrm{Mm}$ at $\varepsilon = 0.2$.

Let us now give some numbers for critical velocities, unstable wavelengths and corresponding mode growth rates of the $m = -3$ and $m = -4$ MHD modes. For the $m = -3$ harmonic (see Fig. 7.9), the critical jet velocities for emergence of KH instability correspondingly are $27.1\,\mathrm{km\,s}^{-1}$, $25.4\,\mathrm{km\,s}^{-1}$, and $24.6\,\mathrm{km\,s}^{-1}$ as the lowest value is obtained in the third instability window with $\varepsilon = 0.3$. If we pick out a $k_z a = 0.9425$, that is, an unstable $\lambda_{KH} = 2.5\,\mathrm{Mm}$, the corresponding dimensionless wave phase velocity growth rate is equal to 0.1633 which implies a mode growth rate $\gamma_{KH} = 0.013\,\mathrm{s}^{-1}$. Note that we have the same normalized growth rate of 0.1633 at $k_z a = 1.3952$, that is, at $\lambda_{KH} = 1.7\,\mathrm{Mm}$, but the corresponding mode growth rate is higher than the previous one, namely being equal to $\gamma_{KH} = 0.019\,\mathrm{s}^{-1}$. The explanation of that difference is very simple: the nominator of the converting formula of going from dimensionless to actual growth rates contains the $k_z a$-value as a multiplier—that is why, the growth rate corresponding to the shorter wavelength is greater than that associated with the longer wavelength. It is curious to see how big is the growth rate for $\text{Im}(v_{\text{ph}}/v_{\text{Ai}})_{\text{max}} = 0.2582$ in the third window: this maximum occurs at $k_z a = 1.125$; that is, for $\lambda_{\text{KH}} = 2.1\,\text{Mm}$, the corresponding mode growth rate is $\gamma_{\text{KH}} = 0.025\,\text{s}^{-1}$. The critical flow velocity for occurring of these unstable $m = -3$ MHD modes (see the right panel in Fig. 7.9) is $v_0^{\text{cr}} = 24.6\,\text{km s}^{-1}$.

With exciting the $m = -4$ harmonic, the picture becomes more intriguing: first, one can have relatively shorter unstable wavelengths as well as lower critical flow velocities, and second, if we want to know how big is the growth rate at $\lambda_{\text{KH}} = 2.5\,\text{Mm}$ $(k_z a = 0.9425)$ we must fill the gap between the second (green) window and the third (blue) one by computing for some intermediate value of the magnetic field twist parameter: $\varepsilon = 0.2$ accommodates almost perfectly $k_z a = 0.9425$ (see the thin orange curve in the left panel of Fig. 7.10): the vertical purple line cuts across the orange curve very close to its maximum—$\text{Im}(v_{\text{ph}}/v_{\text{Ai}}) = 0.2032$ that yields $\gamma_{\text{KH}} = 0.0165\,\text{s}^{-1}$. Note that in the same instability window we can find out the growth rate associated with $\lambda_{\text{KH}} = 2.1\,\text{Mm}$—its value calculated from $\text{Im}(v_{\text{ph}}/v_{\text{Ai}}) = 0.1393$ is $\gamma_{\text{KH}} = 0.0135\,\text{s}^{-1}$. In the fourth (blue) window, as in the previous case of $m = -3$, for $\text{Im}(v_{\text{ph}}/v_{\text{Ai}}) = 0.1393$ we can have two unstable wavelengths of $1.9\,\text{Mm}$ (at $k_z a = 1.2563$) and $1.4\,\text{Mm}$ (at $k_z a = 1.6866$) with mode growth rates equal to $0.015\,\text{s}^{-1}$ and $0.020\,\text{s}^{-1}$, respectively. The highest dimensionless growth rate in that window is $\text{Im}(v_{\text{ph}}/v_{\text{Ai}})_{\text{max}} = 0.2041$ that yields $\gamma_{\text{KH}} = 0.025\,\text{s}^{-1}$ at $\lambda_{\text{KH}} \cong 1.6\,\text{Mm}$. The critical jet's velocities, in descending order, are equal to $26.5\,\text{km s}^{-1}$, $25.5\,\text{km s}^{-1}$, $24.1\,\text{km s}^{-1}$, and $23.1\,\text{km s}^{-1}$. The critical velocities that discussed in the orange and blue instability windows wavelengths are correspondingly equal to $24.1\,\text{km s}^{-1}$ and $23.1\,\text{km s}^{-1}$ (see the right panel in Fig. 7.10). All these examples illustrate the flexibility of our approach in studying the KH instability of higher MHD modes in chromospheric/TR jets.

7.2.3 *Discussion and conclusion*

In this chapter, we have studied the condition under which MHD modes traveling on cool solar dark mottles can become unstable against the Kelvin–Helmholtz instability. Our model for a dark mottle is a vertically moving untwisted or twisted magnetic cylindrical flux tube that is surrounded by plasma embedded in homogeneous magnetic field. In the second case, the twist of internal magnetic field is characterized by the ratio of azimuthal magnetic field component at the inner surface of the tube to its longitudinal component, ε. In that case, the tube becomes unstable to the kink instability when $\varepsilon > 2$. The sausage (pinch) instability needs even higher values. Here, we only consider the weak azimuthal magnetic field limit taking $\varepsilon \ll 1$, therefore the configuration is stable for the magnetic instabilities.

Our computations show that the kink ($m = 1$) MHD mode propagating on a moving untwisted flux tube can become unstable if the jet's speeds is higher than $125\,\mathrm{km\,s^{-1}}$. When the moving tube is considered to be twisted (with $\varepsilon = 0.025$), the critical velocity for an instability onset is even higher—then it is equal to $\cong 177\,\mathrm{km\,s^{-1}}$. Both velocities are inaccessible for dark mottles or spicules in the solar atmosphere. Hence, the transverse waves detected in dark mottles by Kuridze *et al.* (2012) were stable against KH instability. The authors used the formula for the kink speed (6.15) from Chapter 6 to compare with measured phase velocities of the kink mode. We note that used by them expression of c_{k} [Eq. (2) in Kuridze *et al.* (2012)] is applicable only in the case of an isolated flux tube ($B_{\mathrm{e}} = 0$). Generally that speed is higher than the Alfvén speed inside the tube, v_{Ai}, that is easy to see by presenting formula (6.15) from Chapter 6 in the form

$$c_{\mathrm{k}} = \left(\frac{1 + B_{\mathrm{e}}^2/B_{\mathrm{i}}^2}{1 + \rho_{\mathrm{e}}/\rho_{\mathrm{i}}} \right)^{1/2} v_{\mathrm{Ai}} = \left(\frac{1 + b^2}{1 + \eta} \right)^{1/2} v_{\mathrm{Ai}}.$$

In particular, for our dark mottle, modeled as an untwisted magnetic flux tube, the kink speed is equal to $\cong 39\,\mathrm{km\,s^{-1}}$ or in normalized form, $c_{\mathrm{k}}/v_{\mathrm{Ai}} = 1.182613$. The computations yield 1.182590. It is important to note that this expression of the kink speed is valid only in the case of static plasmas. When we have a flowing flux tube, the flow shifts upwards the kink-speed dispersion curve and also splits it into two separate curves, which for small Alfvén Mach numbers travel with normalized velocities $M_{\mathrm{A}} \mp c_{\mathrm{k}}/v_{\mathrm{Ai}}$ (Zhelyazkov, 2012). This implies that one should be very cautious when uses expression (6.15) from Chapter 6 for evaluating the actual phase velocity of the kink mode in dark mottles or spicules. The evolution of the pair of kink waves with increasing the flow speed/Alfvén Mach number essentially depends on the jet's and its environment's parameters [see the discussion in Zhelyazkov (2012)]. At the KH instability onset that kink mode normalized phase velocity is markedly larger than the kink speed in static plasma; for example in our case (see the left panel in Fig. 7.7), in the long wavelength limit, its magnitude is equal to 2.826711 or equivalently to $\cong 93\,\mathrm{km\,s^{-1}}$.

In their second article on transverse waves in solar mottles, Kuridze *et al.* (2013), in addition, used the tools of so-called solar magnetoseismology (see, e.g., Verth *et al.*, 2011; De Moortel and Nakariakov, 2012, and the references therein) to infer estimates of local plasma parameters that are otherwise difficult to measure directly. These seismological tools rely on accurate identification of MHD wave-modes and their properties. In their article, the authors succeeded in obtaining the flow velocity in mottles—their estimation yields $\approx 11\,\mathrm{km\,s}^{-1}$ only, once again confirming that it is unbelievable to detect KH instability of the kink mode in small-scale chromospheric jets like mottles and spicules. The solar magnetoseismology temperature diagnostic suggested that the particular dark mottle analyzed by Kuridze *et al.* (2013) was at the lower end of previous temperature range estimates of 7100–13 000 K by Tsiropoula *et al.* (1993).

Our computations have shown that the only possible way to emerge KH instability of MHD waves in solar dark mottles is to excite high harmonics with negative azimuthal mode numbers m. This can happen if the magnetic field in the moving flux tube is weakly twisted (magnetic field twist parameter $\varepsilon \ll 1$). For each mode number $m = -2, -3, -4$, one can obtain instability windows on the $k_z a$-axis of the dispersion curve/growth rate plots whose (windows's) position and width depend upon the value of ε. A specific feature of the dispersion curve of unstable MHD modes is that their normalized phase velocities coincide with the threshold Alfvén Mach numbers, that is the actual mode phase velocity equals the critical flow speed for rising the KH instability. Therefore, unstable perturbations are frozen in the flow, and consequently, they are vortices rather than waves. This is firmly based on physics because the KH instability in hydrodynamics deals with unstable vortices. All threshold Alfvén Mach numbers found in this paper yield acceptable critical speeds of the mottles that ensure the occurrence of KH instability—these speeds lie in the region of 21–$31\,\mathrm{km\,s}^{-1}$ and are accessible for chromospheric dark mottles. The obtained mode growth rates between $0.013\,\mathrm{s}^{-1}$ and $0.025\,\mathrm{s}^{-1}$ generally depend on the unstable wavelength (λ_{KH} in our notation), that is, on its position on the $k_z a$-axis. Bearing in mind that the length/heights of dark mottles are relatively short (only a few thousands kilometers) we searched for KH instability of MHD modes with wavelengths in the interval of 1400–2500 km and have obtained mode growth rates of several dozens inverse milliseconds. It is interesting to note that the found mode growth rates of $0.025\,\mathrm{s}^{-1}$ (for $\varepsilon = 0.3$ at $\lambda_{\mathrm{KH}} = 2.5\,\mathrm{Mm}$ for the $m = -2$ harmonic and at $\lambda_{\mathrm{KH}} = 1.6\,\mathrm{Mm}$ for the $m = -4$ harmonic) are of the same order as the growth rates of the $m = -3$ MHD mode in high-temperature solar surges (Zhelyazkov *et al.*, 2015a). We note that our approach of exploring the KH instability in solar jets is rather flexible; if, for instance, given excited unstable MHD mode falls over a $k_z a$-position on our plots that is outside a computed instability window, we have, by changing the magnetic field twist parameter, to accommodate that $k_z a$-position on another instability window. It was demonstrated in searching the KH instability of the $m = -4$ harmonic at $\lambda_{\mathrm{KH}} = 2.5\,\mathrm{Mm}$ (see the left panel in Fig. 7.10 and the corresponding discussion). If we have observational data such as jet's width, critical flow velocity and instability wavelength, there is no problem to compute the mode growth rate and if necessary one can tune the procedure to reach a very good agreement between computed growth rate and that deduced from observations. The chromospheric dark mottle is a small-scale jet as compared to EUV and X-ray jets or coronal mass ejections, however, the KH instability emergence in a collective of few dozens thousands mottles can yield a modest but nevertheless noticeable contribution to coronal heating or to bringing forth the solar wind.

Chapter 8

Kelvin–Helmholtz Instability in EUV Solar Jets

8.1 Observations, nature, and physical parameters of EUV jets

Solar jets are small-scale eruptions observed at different heights of the solar atmosphere. They are observed in the different parts of the solar surface such as: in coronal hole regions (Madjarska *et al.*, 2013; Zhelyazkov *et al.*, 2018), active regions (Schmieder *et al.*, 2013; Chandra *et al.*, 2015; Sterling *et al.*, 2017), quiet regions (for example Panesar *et al.*, 2016). Since their first observations are in Hα (Roy, 1973a), Ca II H (Shibata *et al.*, 2007; Chifor *et al.*, 2008a; Nishizuka *et al.*, 2008), EUV (Alexander and Fletcher, 1999), and soft X-ray (Shibata *et al.*, 1992), a very remarkable progress has been made to understand the physics of the solar jets including the high resolution data of *Solar Dynamics Observatory* (*SDO*) (Moschou *et al.*, 2013; Innes *et al.*, 2016), as well as by the *Interface Region Imaging Spectrograph* (*IRIS*) (Chen and Innes, 2016).

The solar jets can be divided into two categories, that is, "standard" and "blowout" jets (Moore *et al.*, 2013, 2015; Raouafi *et al.*, 2016; Chandra *et al.*, 2017). Initially, this classification was based on their morphology. In the case of standard jets their spires are thin and narrow during their entire lifetime and their bases remained relatively dim, except for the commonly observed compact jet bright point (JBP) on one side of the jet's base. In contrary to this, the blowout jets can be defined by broad spires. A blowout jet initiated as standard jet starts as a emerging bipole reconnecting with ambient open field. But sometime during this reconnection process, the emerging bipole is triggered unstable and erupts outward and it becomes the "blowout" jet. The "blowout" jet is associated with a small flux rope eruption. Sometime this erupting flux rope can become coronal mass ejection (CME) (see, for example, Liu *et al.*, 2015; Chandra *et al.*, 2017). The interpretation

125

of "blowout" jets was first proposed by Moore *et al.* (2010, 2013). Later on, these were reported and confirmed in several observations (Sterling *et al.*, 2016; Panesar *et al.*, 2016; Sterling *et al.*, 2017).

The accepted mechanism for the generation of solar jets is the magnetic reconnection. Conditions for the magnetic reconnection can be magnetic flux emergence (Heyvaerts *et al.*, 1997; Shibata *et al.*, 1992; Guo *et al.*, 2013), also flux cancellation (Priest *et al.*, 1994; Longcope, 1998; Innes *et al.*, 2010; Adams *et al.*, 2014), or can be both (Young and Muglach, 2014a; Chandra *et al.*, 2015). Several numerical simulations have been done in this direction. Most of these includes the magnetic flux emergence. Moreover, in case of Pariat and co-workers (Pariat *et al.*, 2017) one makes the condition for magnetic reconnection imposing horizontal photospheric twisting motions. Recently, it is proposed that the jets are small-scale phenomena and this physical mechanism is similar to the large-scale eruptions (Wyper *et al.*, 2017). Therefore to understand these small-scale phenomena is very crucial to explain the large-scale solar eruptions.

Concerning the physical parameters of various EUV jets, observed with various instruments, one can say that the electron number densities are in the range of 10^8–10^{10} cm^{-3}, and the temperatures are of 0.05–2.0 MK. Kim *et al.* (2007) have observed an active-region jet using various spectral lines formed at different temperatures as observed by *Hinode* EIS. They have found simultaneous blue-shift (upflows) up to maximum speed of -64 km s^{-1} and red-shift (downflows) up to 20 km s^{-1} at the base of the coronal jet as observed by *Hinode*/EIS. Moreover, Chifor *et al.* (2008b) have observed a blue-shift of the active region jet plasma with the apparent velocity of 150 km s^{-1} and comparatively a weak red-shifted plasma motion at the base of the jet. Chifor *et al.* (2008a) have reported that the EUV jet was co-spatial and consists of the similar kinematical features as of its soft X-ray counterpart. A similar conclusion has been drawn by Yang *et al.* (2011) who have presented simultaneous observations of three recurring jets in EUV and soft X-ray emissions that occurred in an active region on 2007 June 5. On comparing their morphological and kinematic characteristics in these two different wavelengths and related emissions, they found that EUV and soft X-ray jets have co-spatial triggering site, similar direction, size, and propulsion speeds. Yang *et al.* (2011) have also analyzed jet's spectral properties by using multiple spectral lines as observed by *Hinode* EIS. They have found that these jets have temperature in the range of 0.05–2.0 MK, while maximum electron densities have been reported as 6.6×10^9–3.4×10^{10} cm^{-3}. For each of these observed jets, a blue-shifted component of the plasma and a red-shifted base are simultaneously observed. These observed jets as reported by Yang *et al.* (2011) have maximum Doppler velocities ranging from 25 to 121 km s^{-1} for the blue-shifted plasma component, while it is observed in the range of 115–232 km s^{-1} for the red-shifted plasma component.

Table 8.1. Jets physical parameters at background magnetic field $B_e = 7\,\text{G}$.

Space craft	Jet speed (km s^{-1})	Temperature (MK)		Electron density ($\times 10^{10}$ cm^{-3})		Plasma beta	
		T_i	T_e	n_i	n_e	β_i	β_e
Hinode	150	1.20	2.00	10.00	5.980	8.76	8.48
SDO/AIA	332	1.65	1.81	0.945	0.638	1.55	0.82

These observational results are found to be consistent with the magnetic reconnection model of the coronal jets (Yokoyama and Shibata, 1995, 1996).

Recently, Chandra *et al.* (2015) have presented and discussed the multi-wavelength observations of five homologous recurrent EUV solar jets triggered in NOAA AR 11133 on December 11, 2010, which were observed by the Atmospheric Imaging Assembly (AIA; Lemen *et al.*, 2012) on board the *Solar Dynamic Observatory*. The speed of these jets were ranging between 86 and 267 km s^{-1}. They also found that all these EUV jets were triggered in the direction of open field lines as shown by a potential-field source-surface (PFSS) (Schatten *et al.*, 1969; Altschuler and Newkirk, Jr., 1969; Wang and Sheeley, Jr., 1992). Cheung *et al.* (2015) have presented the evolution of four homologous helical recurrent jets triggered from NOAA AR 11793 in the transition region and corona on July 21, 2013 by using the AIA/*SDO*. These jets have also been observed by the *Interface Region Imaging Spectrograph* (IRIS; De Pontieu *et al.*, 2014a) and the Doppler velocities in these jets have been estimated using the spectral data. They found that one edge of each of these jets was blue-shifted while the opposite edge was red-shifted simultaneously, which indicates that these jets have helical structure with the same sign of the helicity.

For studying Kelvin–Helmholtz instability in solar EUV jets we have chosen two events, namely an AR jet observed by the EUV Imaging Spectrometer (EIS) on board *Hinode* on 2007 January 15/16 (west of NOAA AR 10938) (Chifor *et al.*, 2008b) and one jet (the J$'_6$ one) from several EUV jets occurring during 2014 April 15–16 in the active region NOAA AR 12035 and observed by the Atmospheric Imaging Assembly (AIA) on board *SDO* satellite (Joshi *et al.*, 2017). Details of the observations of these two jets will be given in the corresponding subsections— here we only summarize their basic physical parameters listed in Table 8.1.

8.2 Jets geometry and the governing magnetohydrodynamic equations

As in the previous four chapters, we model each jet as a moving cylindrical magnetic flux tube of radius a surrounded by a static coronal plasma. We assume that

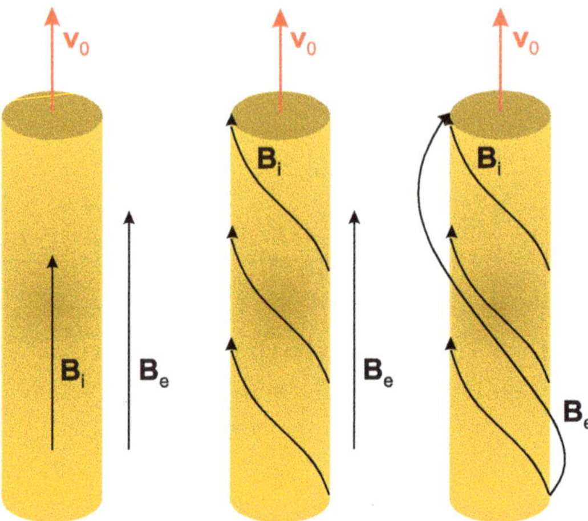

Fig. 8.1. Magnetic fields and flow velocity configurations in axially moving solar jet flux tubes. *Reprinted with permission.*

the jet and ambient electron number densities, n_i and n_e, respectively, are homogeneous. Concerning magnetic field topology we consider three cases illustrated in Fig. 8.1. In the first case (the left column in the picture), the magnetic fields inside and outside the tube are homogeneous accordingly with magnitudes B_i and B_e. In the second case (the middle column in Fig. 8.1), we assume that the magnetic field inside the jet/magnetic flux tube, $\boldsymbol{B}_i$, is slightly twisted, while the one of the environment, $\boldsymbol{B}_e$, is still homogeneous. In the latter case (the right column in the picture) both magnetic fields are twisted. The jet velocity $\boldsymbol{v}_0$ and wavevector $\boldsymbol{k}$ are directed along the central line of each flux tube, which (the line) is the z-axis of a cylindrical coordinate system that will be used in deriving the wave dispersion relations of propagating normal MHD modes in the jet. The primary reason for considering three types of magnetic flux tubes is to see, first, how the magnetic field twist of the internal magnetic field $\boldsymbol{B}_i$ will change/affect the condition for emerging of KHI of given MHD mode, and second, how an additionally involved twisted external magnetic field $\boldsymbol{B}_e$ will change the picture. In cylindrical coordinates, all perturbations associated with the wave propagation are proportional to $\exp\left[i(-\omega t + m\phi + k_z z)\right]$, where ω is the angular wave frequency, m is the azimuthal mode number, and k_z is the axial wavenumber, that is, $\boldsymbol{k} = (0, 0, k_z)$. Among the various MHD modes the kink one with $m = 1$ turns out to be the most unstable against the KHI—that is why we will focus our study on this mode only. The KHI occurs as we know when the jet velocity $\boldsymbol{v}_0$ exceeds some critical value, v_0^{cr}. That critical flow velocity depends upon two important parameters of the jet–environment system, notably the density contrast between the two media and the ratio of external to internal magnetic fields (more correctly the ratio of their axial components) (Zaqarashvili *et al.*, 2014a). The density contrast, η, recall is defined as the ratio n_e/n_i. For the magnetic fields ratio, b, we have, according to Fig. 8.1, three definitions of b, namely equal to B_e/B_i for a untwisted magnetic flux tube, B_e/B_{iz} for a slightly twisted tube, and B_{ez}/B_{iz} for the case when both magnetic field are twisted.

Under these conditions, for studying the propagation of MHD modes, we need altogether five different dispersion equations: one for the case of untwisted magnetic flux tube, two for the case of twisted internal magnetic field (depending on whether the environment is incompressible or cool plasma), and similarly two for the case of twisted internal and external magnetic fields. For untwisted moving magnetic flux tube, we can use Eq. (4.30) from Chapter 4

$$\frac{\rho_e}{\rho_i}(\omega^2 - k_z^2 v_{Ae}^2)\kappa_i \frac{I'_m(\kappa_i a)}{I_m(\kappa_i a)} - \left[(\omega - \boldsymbol{k}\cdot\boldsymbol{v}_0)^2 - k_z^2 v_{Ai}^2\right]\kappa_e \frac{K'_m(\kappa_e a)}{K_m(\kappa_e a)} = 0, \quad (8.1)$$

in which the squared wave attenuation coefficients $\kappa_{i,e}$ collectively are given by the expression

$$\kappa^2 = -\frac{\left[(\omega - \boldsymbol{k}\cdot\boldsymbol{v}_0)^2 - k_z^2 c_s^2\right]\left[(\omega - \boldsymbol{k}\cdot\boldsymbol{v}_0)^2 - k_z^2 v_A^2\right]}{(c_s^2 + v_A^2)\left[(\omega - \boldsymbol{k}\cdot\boldsymbol{v}_0)^2 - k_z^2 c_T^2\right]}, \quad (8.2)$$

where the tube speed (different inside and outside the jet) is $c_T = c_s v_A/\sqrt{c_s^2 + v_A^2}$.

We note that in the limit of incompressible plasmas ($c_{si,e} \to \infty$) the wave attenuation coefficients in both media become equal to the axial wavenumber k_z and the wave dispersion relation (8.1) takes the simple form of a quadratic equation, that provides solutions for the real and imaginary part of of the wave phase velocity in closed forms:

$$v_{ph} \equiv \frac{\omega}{k_z} = \frac{-M_A B \pm \sqrt{D}}{\eta A - B} v_{Ai}, \quad (8.3)$$

where $M_A = v_0/v_{Ai}$ is the Alfvén Mach number,

$$A = I'_m(k_z a)/I_m(k_z a), \quad B = K'_m(k_z a)/K_m(k_z a),$$

and the discriminant D is

$$D = M_A^2 B^2 - (\eta A - B)\left[(1 - M_A^2)B - Ab^2\right].$$

Obviously, if $D \geqslant 0$, then

$$\text{Re}(v_{ph}) = \frac{-M_A B \pm \sqrt{D}}{\eta A - B} v_{Ai}, \quad \text{Im}(v_{ph}) = 0,$$

else

$$\mathrm{Re}(v_{\mathrm{ph}}) = \frac{-M_A B}{\eta A - B} v_{\mathrm{Ai}}, \quad \mathrm{Im}(v_{\mathrm{ph}}) = \frac{\sqrt{-D}}{\eta A - B} v_{\mathrm{Ai}}.$$

We point out that our choice of the sign of $\sqrt{-D}$ in the expression for $\mathrm{Im}(v_{\mathrm{ph}})$ is plus although, in principle, it might also be minus—in that case, owing to arising instability the wave's energy is transferred to the jet.

For the case of twisted internal magnetic field, we will use Eq. (6.13) from Chapter 6

$$\frac{\left[(\omega - \boldsymbol{k} \cdot \boldsymbol{v}_0)^2 - \omega_{\mathrm{Ai}}^2\right] F_m(\kappa_i a) - 2m A \omega_{\mathrm{Ai}} / \sqrt{\mu \rho_i}}{\left[(\omega - \boldsymbol{k} \cdot \boldsymbol{v}_0)^2 - \omega_{\mathrm{Ai}}^2\right]^2 - 4\omega_{\mathrm{Ai}}^2 / \mu \rho_i}$$

$$= \frac{P_m(k_z a)}{\frac{\rho_e}{\rho_i}(\omega^2 - \omega_{\mathrm{Ae}}^2) + A^2 P_m(k_z a) / \mu \rho_i}, \tag{8.4}$$

where

$$F_m(\kappa_i a) = \frac{\kappa_i a I'_m(\kappa_i a)}{I_m(\kappa_i a)} \quad \text{and} \quad P_m(k_z a) = \frac{k_z a K'_m(k_z a)}{K_m(k_z a)}.$$

The wave attenuation coefficient inside the tube is given by

$$\kappa_i = k_z \left(1 - \frac{4A^2 \omega_{\mathrm{Ai}}^2}{\mu \rho_i \left[(\omega - \boldsymbol{k} \cdot \boldsymbol{v}_0)^2 - \omega_{\mathrm{Ai}}^2\right]^2} \right)^{1/2}, \tag{8.5}$$

where

$$\omega_{\mathrm{Ai}} = \frac{1}{\sqrt{\mu \rho_i}}(m A + k_z B_{iz})$$

is the local Alfvén frequency. If the environment is a cool medium, the argument $k_z a$ of the modified Bessel function K_m should be replaced by

$$\kappa_e a = k_z a \left[1 - (\omega / \omega_{\mathrm{Ae}})^2 \right]^{1/2},$$

where $\omega_{\mathrm{Ae}} = k_z B_e / \sqrt{\mu \rho_e}$.

In the most complicated case of both twisted magnetic fields (the right column in Fig. 8.1) we can use Eq. (7.24) from Chapter 7, applicable when the surrounding plasma is considered as incompressible medium. It is more instructive to re-derive that equation starting from the basic equations of ideal magnetohydrodynamics and obtain its modification when the environment is treated as a cool medium. Fortunately, the governing equations for space and time evolution of fluid velocity $\boldsymbol{v}$, magnetic field $\boldsymbol{B}$, and pressure p perturbations for incompressible and cool plasmas have practically the same form:

$$\rho\frac{\partial}{\partial t}\boldsymbol{v}_1 + \rho(\boldsymbol{v}_0\cdot\nabla)\boldsymbol{v}_1 + \nabla\left(p_1 + \frac{\boldsymbol{B}_0\cdot\boldsymbol{B}_1}{\mu}\right)$$
$$-\frac{1}{\mu}(\boldsymbol{B}_0\cdot\nabla)\boldsymbol{B}_1 - \frac{1}{\mu}(\boldsymbol{B}_1\cdot\nabla)\boldsymbol{B}_0 = 0, \tag{8.6}$$

$$\frac{\partial}{\partial t}\boldsymbol{B}_1 - \nabla(\boldsymbol{v}_1\times\boldsymbol{B}_0) - \nabla(\boldsymbol{v}_0\times\boldsymbol{B}_1) = 0, \tag{8.7}$$

$$\nabla\cdot\boldsymbol{v}_1 = 0, \tag{8.8}$$

$$\nabla\cdot\boldsymbol{B}_1 = 0. \tag{8.9}$$

We note that in the case of cool plasma, the thermal pressure p and its perturbation p_1 are equal to zero, and the fluid velocity perturbation, $\boldsymbol{v}_1$, in cylindrical coordinates, is presented as $\boldsymbol{v}_1 = (v_{1r}, v_{1\phi}, 0)$, while the magnetic field perturbation has its three non-zero components, notably $\boldsymbol{B}_1 = (B_{1r}, B_{1\phi}, B_{1z})$. Notice also that for a cool plasma the total pressure perturbation $p_{\text{tot}} = p_1 + \boldsymbol{B}_0\cdot\boldsymbol{B}_1/\mu$ reduces to magnetic pressure perturbation only.

The equilibrium magnetic field inside the moving flux tube is the same as in Chapters 5–7, namely $\boldsymbol{B}_i = (0, Ar, B_{iz})$, where, recall, A and B_{iz} are constants. The jet's flow velocity has only z component, equal to v_0. As unperturbed parameters depend on the r coordinate only, all the perturbations can be Fourier analyzed with $\exp\left[i(-\omega t + m\phi + k_z z)\right]$ and the governing MHD Eqs. (8.6) and (8.7) are:

$$-i(\omega - \boldsymbol{k}\cdot\boldsymbol{v}_0)v_{1r} + \frac{1}{\rho_i}\frac{d}{dr}p_{\text{tot}} - i\frac{1}{\sqrt{\mu\rho_i}}\omega_{Ai}B_{1r} + \frac{2A}{\mu\rho_i}B_{1\phi} = 0, \tag{8.10}$$

$$-i(\omega - \boldsymbol{k}\cdot\boldsymbol{v}_0)v_{1\phi} + i\frac{1}{\rho_i}\frac{m}{r}p_{\text{tot}} - i\frac{1}{\sqrt{\mu\rho_i}}\omega_{Ai}B_{1\phi} - \frac{2A}{\mu\rho_i}B_{1r} = 0, \tag{8.11}$$

$$-(\omega - \boldsymbol{k}\cdot\boldsymbol{v}_0)v_{1z} + \frac{1}{\rho_i}k_z p_{\text{tot}} - \frac{1}{\sqrt{\mu\rho_i}}\omega_{Ai}B_{1z} = 0, \tag{8.12}$$

$$(\omega - \boldsymbol{k}\cdot\boldsymbol{v}_0)B_{1r} + (mA + k_z B_{iz})v_{1r} = 0, \tag{8.13}$$

$$(\omega - \boldsymbol{k}\cdot\boldsymbol{v}_0)B_{1\phi} + (mA + k_z B_{iz})v_{1\phi} = 0, \tag{8.14}$$

$$(\omega - \boldsymbol{k}\cdot\boldsymbol{v}_0)B_{1z} + (mA + k_z B_{iz})v_{1z} = 0, \tag{8.15}$$

where

$$\omega_{Ai} = \frac{1}{\sqrt{\mu\rho_i}}(mA + k_z B_{iz}) \tag{8.16}$$

is the local Alfvén frequency inside the moving flux tube. From the above equations, one can obtain expressions for the fluid velocity components in terms of the total pressure perturbation, p_{tot}, and its derivative with respect to r, namely

$$v_{1r} = -\mathrm{i}\frac{1}{Y_{\mathrm{i}}}\frac{1}{\Omega_{\mathrm{i}}\rho_{\mathrm{i}}}\left(\frac{\mathrm{d}}{\mathrm{d}r} - Z_{\mathrm{i}}\frac{m}{r}\right)p_{\text{tot}}, \tag{8.17}$$

$$v_{1\phi} = \frac{1}{Y_{\mathrm{i}}}\frac{1}{\Omega_{\mathrm{i}}\rho_{\mathrm{i}}}\left(\frac{m}{r} - Z_{\mathrm{i}}\frac{\mathrm{d}}{\mathrm{d}r}\right)p_{\text{tot}}, \tag{8.18}$$

$$v_{1z} = \frac{1}{\Omega_{\mathrm{i}}\rho_{\mathrm{i}}}k_z\,p_{\text{tot}}. \tag{8.19}$$

Here,

$$\frac{1}{\Omega_{\mathrm{i}}} \equiv \frac{\omega - \boldsymbol{k}\cdot\boldsymbol{v}_0}{(\omega - \boldsymbol{k}\cdot\boldsymbol{v}_0)^2 - \omega_{\mathrm{Ai}}^2},$$

$$Z_{\mathrm{i}} \equiv \frac{2A\omega_{\mathrm{Ai}}}{\sqrt{\mu\rho_{\mathrm{i}}}\left[(\omega-\boldsymbol{k}\cdot\boldsymbol{v}_0)^2-\omega_{\mathrm{Ai}}^2\right]}, \quad \text{and} \quad Y_{\mathrm{i}} = 1 - Z_{\mathrm{i}}^2.$$

On using these expressions for fluid velocity perturbation in the constraint Eq. (8.8), after some algebra we get that the total pressure perturbation obeys the equation

$$\left[\frac{\mathrm{d}^2}{\mathrm{d}r^2} + \frac{1}{r}\frac{\mathrm{d}}{\mathrm{d}r} - \left(\frac{m^2}{r^2} + \kappa_{\mathrm{i}}^2\right)\right]p_{\text{tot}} = 0, \tag{8.20}$$

in which κ_{i} is

$$\kappa_{\mathrm{i}}^2 = k_z^2\left[1 - \frac{4A^2\omega_{\mathrm{Ai}}^2}{\mu\rho_{\mathrm{i}}(\Omega^2 - \omega_{\mathrm{Ai}}^2)}\right], \quad \text{where } \Omega \equiv \omega - \boldsymbol{k}\cdot\boldsymbol{v}_0.$$

Equation (8.20) is the Bessel equation whose solution bounded at the tube axis is

$$p_{\mathrm{i\,tot}} = \alpha_{\mathrm{i}}I_m(\kappa_{\mathrm{i}}r). \tag{8.21}$$

Here, I_m is the modified Bessel function of order m and α_{i} is a constant. Transverse displacement, $\xi_{\mathrm{i}r}$, can be obtained from expression (8.17) and has the form:

$$\xi_{\mathrm{i}r} = \frac{\alpha_{\mathrm{i}}}{r}\frac{(\Omega^2 - \omega_{\mathrm{Ai}}^2)\kappa_{\mathrm{i}}r\,I_m'(\kappa_{\mathrm{i}}r) - 2mA\omega_{\mathrm{Ai}}I_m(\kappa_{\mathrm{i}}r)/\sqrt{\mu\rho_{\mathrm{i}}}}{\rho_{\mathrm{i}}(\Omega^2 - \omega_{\mathrm{Ai}}^2)^2 - 4A^2\omega_{\mathrm{Ai}}^2/\mu}, \tag{8.22}$$

where the prime, $'$, implies a differentiation by the Bessel function argument.

Let us now go to the environment and find similar expressions for $p_{\mathrm{e\,tot}}$ and $\xi_{\mathrm{e}r}$, respectively. We consider the external magnetic field of the form

$$\boldsymbol{B}_{\mathrm{e}} = \left(0,\, B_{\mathrm{e}\phi}(a)\frac{a}{r},\, B_{\mathrm{e}z}(a)\frac{a^2}{r^2}\right) \tag{8.23}$$

(where for convenience we denote $B_{e\phi}(a) \equiv B_\phi$ and $B_{ez}(a) \equiv B_z$) and the plasma density in the form $\rho_0 = \rho_e(a/r)^4$ so that the Alfvén frequency

$$\omega_{\mathrm{Ae}} = \frac{1}{\sqrt{\mu\rho_0(r)}}\left[\frac{m}{r}B_{e\phi}(r) + k_z B_{ez}(r)\right] = \frac{r^2}{\sqrt{\mu\rho_e a^4}}\left(\frac{ma B_\phi}{r^2} + \frac{k_z a^2 B_z}{r^2}\right)$$

$$= \frac{m B_\phi + k_z a B_z}{\sqrt{\mu\rho_e a^2}} \tag{8.24}$$

is constant. This circumstance allows us to find analytical solution to the governing equations. Now momentum and Faraday equations are displayed in the form:

$$-\mathrm{i}\omega v_{1r} + \frac{1}{\rho_e}\frac{\mathrm{d}}{\mathrm{d}r}p_{\mathrm{tot}} - \mathrm{i}\frac{1}{\mu\rho_e}\left(\frac{ma B_\phi}{r^2} + \frac{k_z a^2 B_z}{r^2}\right)B_{1r} + 2\frac{a}{r^2}\frac{B_\phi}{\mu\rho_e}B_{1\phi} = 0, \tag{8.25}$$

$$-\mathrm{i}\omega v_{1\phi} + \frac{1}{\rho_e}\frac{m}{r}p_{\mathrm{tot}} - \mathrm{i}\frac{1}{\mu\rho_e}\left(\frac{ma B_\phi}{r^2} + \frac{k_z a^2 B_z}{r^2}\right)B_{1\phi} = 0, \tag{8.26}$$

$$-\mathrm{i}\omega v_{1z} + \frac{1}{\rho_e}k_z p_{\mathrm{tot}} - \mathrm{i}\frac{1}{\mu\rho_e}\left(\frac{ma B_\phi}{r^2} + \frac{k_z a^2 B_z}{r^2}\right)B_{1z} + 2\frac{a^2}{r^3}\frac{B_z}{\mu\rho_e}B_{1r} = 0, \tag{8.27}$$

$$\omega B_{1r} + \left(\frac{ma B_\phi}{r^2} + \frac{k_z a^2 B_z}{r^2}\right)v_{1r} = 0, \tag{8.28}$$

$$\mathrm{i}\omega B_{1\phi} + \mathrm{i}\left(\frac{ma B_\phi}{r^2} + \frac{k_z a^2 B_z}{r^2}\right)v_{1\phi} + 2\frac{a}{r^2}B_\phi v_{1r} = 0, \tag{8.29}$$

$$\mathrm{i}\omega B_{1z} + \mathrm{i}\left(\frac{ma B_\phi}{r^2} + \frac{k_z a^2 B_z}{r^2}\right)v_{1z} + 2\frac{a^2}{r^3}B_z v_{1r} = 0. \tag{8.30}$$

By using expression (8.24), from above equations, after a lengthy algebra, one obtains that

$$v_{1r} = -\mathrm{i}\frac{1}{Y_e}\frac{1}{\Omega_e\rho_e}\left(\frac{\mathrm{d}}{\mathrm{d}r} - Z_e\frac{m}{r}\right)p_{\mathrm{tot}}, \tag{8.31}$$

$$v_{1\phi} = \frac{1}{\Omega_e\rho_e}\left[\left(1 + \frac{Z_e^2}{Y_e}\right)\frac{m}{r} - \frac{Z_e}{Y_e}\frac{\mathrm{d}}{\mathrm{d}r}\right]p_{\mathrm{tot}}, \tag{8.32}$$

$$v_{1z} = \frac{1}{\Omega_e\rho_e}k_z p_{\mathrm{tot}}, \tag{8.33}$$

where now

$$\frac{1}{\Omega_e} \equiv \frac{\omega}{\omega^2 - \omega_{\mathrm{Ae}}^2}, \quad Z_e \equiv \frac{2B_\phi\omega_{\mathrm{Ae}}}{\sqrt{\mu\rho_e a^2}(\omega^2 - \omega_{\mathrm{Ae}}^2)}, \quad \text{and} \quad Y_e = 1 - \frac{\omega^2}{\omega_{\mathrm{Ae}}^2}Z_e^2.$$

With these expressions for fluid velocity perturbations, on using again the equation $\nabla \cdot \boldsymbol{v}_1 = 0$, we obtain that the total pressure perturbation obeys the equation

$$\left[\frac{d^2}{dr^2} + \frac{5}{r}\frac{d}{dr} - \left(\frac{n^2}{r^2} + \kappa_e^2 \right) \right] p_{\text{tot}} = 0, \tag{8.34}$$

which is an equation for Bessel function with complex order $\nu = \sqrt{4 + n^2}$, with a solution bounded at infinity in the form

$$p_{e\,\text{tot}} = \alpha_e \frac{a^2}{r^2} K_\nu(\kappa_e r), \tag{8.35}$$

where α_e is a constant, the wave attenuation coefficient is equal to

$$\kappa_e = k_z \left(1 - \frac{4 B_\phi^2 \omega^2}{\mu \rho_e (\omega^2 - \omega_{Ae}^2)^2 a^2} \right)^{1/2}, \tag{8.36}$$

and the term n^2 is given by the expression

$$n^2 = m^2 - \frac{4m^2 B_\phi^2}{\mu \rho_e a^2 (\omega^2 - \omega_{Ae}^2)} + \frac{8m B_\phi \omega_{Ae}}{\sqrt{\mu \rho_e} a (\omega^2 - \omega_{Ae}^2)}.$$

With the help of the relation $\xi_r = i v_{1r}/\omega$, on using Eq. (8.31), one obtains that

$$\xi_{er} = \alpha_e \frac{r(\omega^2 - \omega_{Ae}^2)\kappa_e r K_\nu'(\kappa_e r)}{a^2 \rho_e (\omega^2 - \omega_{Ae}^2)^2 - 4 B_\phi^2 \omega^2/\mu}$$

$$- \alpha_e \frac{r}{a} \left(\frac{2a(\omega^2 - \omega_{Ae}^2) + 2m B_\phi \omega_{Ae}/\sqrt{\mu \rho_e}}{a^2 \rho_e (\omega^2 - \omega_{Ae}^2)^2 - 4 B_\phi^2 \omega^2/\mu} \right) K_\nu(\kappa_e r). \tag{8.37}$$

Merging the solutions for p_{tot} and ξ_r in both media at the tube surface, $r = a$, one obtains the dispersion equation of the normal MHD modes propagating on the moving magnetic flux tube. As we already said, the boundary conditions at the tube surface are the continuity of the Lagrangian displacement

$$\xi_{ir}|_{r=a} = \xi_{er}|_{r=a} \tag{8.38}$$

[where ξ_{ir} and ξ_{er} are given by Eqs. (8.22) and (8.37)], and the total Lagrangian pressure (Bennett *et al.*, 1999)

$$p_{i\,\text{tot}} - \left. \frac{B_{i\phi}^2}{\mu a}\xi_{ir} \right|_{r=a} = p_{e\,\text{tot}} - \left. \frac{B_{e\phi}^2}{\mu a}\xi_{er} \right|_{r=a}, \tag{8.39}$$

where the total pressure perturbations are given by Eqs. (8.21) and (8.35), respectively. Using these boundary conditions, we recover the dispersion equation governing the oscillations in moving twisted flux tube surrounded by twisted incompressible magnetized plasma derived in Zaqarashvili *et al.* (2014a):

$$
\frac{([\omega - \mathbf{k} \cdot \mathbf{v}_0]^2 - \omega_{\mathrm{Ai}}^2) F_m(\kappa_i a) - 2m A \omega_{\mathrm{Ai}}/\sqrt{\mu \rho_i}}{\rho_i([\omega - \mathbf{k} \cdot \mathbf{v}_0]^2 - \omega_{\mathrm{Ai}}^2)^2 - 4A^2 \omega_{\mathrm{Ai}}^2/\mu}
$$
$$
= \frac{a^2(\omega^2 - \omega_{\mathrm{Ae}}^2) Q_\nu(\kappa_e a) - G}{L - H\left[a^2(\omega^2 - \omega_{\mathrm{Ae}}^2) Q_\nu(\kappa_e a) - G\right]},
\tag{8.40}
$$

where

$$
F_m(\kappa_i a) = \frac{\kappa_i a I'_m(\kappa_i a)}{I_m(\kappa_i a)}, \quad Q_\nu(\kappa_e a) = \frac{\kappa_e a K'_\nu(\kappa_e a)}{K_\nu(\kappa_e a)},
$$

$$
L = a^2 \rho_e(\omega^2 - \omega_{\mathrm{Ae}}^2)^2 - 4B_{e\phi}^2 \omega^2/\mu,
$$

$$
H = B_{e\phi}^2/\mu a^2 - A^2/\mu, \quad G = 2a^2(\omega^2 - \omega_{\mathrm{Ae}}^2) + 2ma B_{e\phi}\omega_{\mathrm{Ae}}/\sqrt{\mu \rho_e}.
$$

When surrounding plasma is treated as a cool medium, then, recall, the plasma pressure perturbation $p_1 = 0$ and the total pressure perturbation, p_{tot}, reduce to magnetic pressure perturbation only. Moreover, the axial component of fluid velocity perturbation also is zero, that is, $v_{1z} = 0$. Under these circumstances, among the six governing MHD Eqs. (8.25)–(8.30), only Eqs. (8.27) and (8.30) are slightly changed; we will denote the magnetic pressure perturbation by p_{mag1}. By using Eq. (8.9), we express B_{1z} in terms of B_{1r} and $B_{1\phi}$ and obtain

$$
B_{1z} = \mathrm{i}\frac{1}{k_z}\left(\frac{\mathrm{d}}{\mathrm{d}r} + \frac{1}{r}\right) B_{1r} - \frac{1}{k_z}\frac{m}{r} B_{1\phi}.
$$

Inserting this B_{1z} alongside with B_{1r} and $B_{1\phi}$ expressed via v_{1r} and $v_{1\phi}$ [by using Eqs. (8.28) and (8.29)] into Eq. (8.27), we get

$$
-\mathrm{i}\frac{1}{\rho_i}k_z^2 p_{\mathrm{mag1}} - \frac{\omega_{\mathrm{Ae}}^2}{\omega}\left(\frac{\mathrm{d}}{\mathrm{d}r} + 3\frac{1}{r}\right) v_{1r} - \mathrm{i}\frac{\omega_{\mathrm{Ae}}^2}{\omega}\frac{m}{r} v_{1\phi} = 0.
$$

Now, on using Eqs. (8.31) and (8.32) giving us fluid velocity perturbation components in terms of total (that is, magnetic) pressure perturbation, above equation takes the form

$$
\left[\frac{\mathrm{d}^2}{\mathrm{d}r^2} + \frac{3}{r}\frac{\mathrm{d}}{\mathrm{d}r} - \left(\frac{n_c^2}{r^2} + \kappa_{\mathrm{ec}}^2\right)\right] p_{\mathrm{mag1}} = 0,
\tag{8.41}
$$

which like before is the equation for Bessel function with complex order $\nu_{\rm c} = \sqrt{2 + n_{\rm c}^2}$, where

$$n_{\rm c}^2 = m^2 - \frac{4m^2 B_\phi^2}{\mu \rho_{\rm e} a^2 (\omega^2 - \omega_{\rm Ae}^2)} + \frac{4m B_\phi \omega_{\rm Ae}}{\sqrt{\mu \rho_{\rm e}} a (\omega^2 - \omega_{\rm Ae}^2)}.$$

Here, the label "c" stamps for *cool*. Note that in the cool environment the wave attenuation coefficient is given by

$$\kappa_{\rm ec} = k_z \left(1 - \frac{4 B_\phi^2 \omega^2}{\mu \rho_{\rm e} \left(\omega^2 - \omega_{\rm Ae}^2 \right)^2 a^2} \right)^{1/2} \left(\frac{\omega^2}{\omega_{\rm Ae}^2} - 1 \right)^{1/2}. \tag{8.42}$$

Further on, following the standard steps for deriving the wave dispersion relation, we arrive at Eq. (8.40) in which have to replace ν by $\nu_{\rm c}$ and $\kappa_{\rm e}$ by $\kappa_{\rm ec}$, respectively.

Having at hand all necessary wave dispersion equations, we can now apply them in studying the propagation characteristics of the MHD modes running on the observed EUV jets, that will be done in the next sections.

8.3   Kelvin–Helmholtz instability in an EUV jet observed by *Hinode*

Chifor *et al.* (2008b) have presented a study of an AR jet observed by the EUV Imaging Spectrometer (EIS) on board *Hinode* on 2007 January 15/16 (west of NOAA AR 10938). EIS covers two wavelength bands: 170–211 Å and 246–292 Å referred to as the short wavelength (SW) and long wavelength (LW) bands, respectively. The jet was observed at temperatures between 5.4 and 6.4 in Log $T_{\rm e}$. In the Ca XVII λ192 window at the location of the up-flow jet component, the O V λ192.90 line was seen next to the λ192.83 line [believed to be Fe XI according to Young *et al.* (2007)]. Chifor *et al.* (2008b) have observationally captured that there was also O V emissions near the base (the red-shifted plasma component) of the jet, but that could be due to the red-shifted Fe XI line. A strong blue-shifted component and the signature of a weak red-shifted plasma component at the base of the jet was observed around Log $T_{\rm e} = 6.2$ by Chifor *et al.* (2008b). The up-flow velocities exceeded 150 km s^{-1}, which is very significant observation. The jet plasma was seen over a wide range of the temperatures between 5.4 and 6.4 in Log $T_{\rm e}$. Using Fe XII λ186 and λ195 line ratios, Chifor *et al.* (2008a) have measured the electron densities above Log $N_{\rm e} = 11$ for the high-velocity up-flowing plasma component of the observed coronal jet.

Our choice of jet's physical parameters (listed in Table 8.1) are: Log $T_{\rm i} = 6.08$ that corresponds to $T_{\rm i} = 1.2$ MK, and Log $n_{\rm i} = 11$ that yields electron number density $n_{\rm i} = 1.0 \times 10^{11}$ cm^{-3}. Bearing in mind that this EUV jet is positioned in the TR/lower corona (see Fig. 53(b) in Cheung *et al.*, 2015), we assume that AR plasma has a typical temperature $T_e = 2\,\text{MK}$ and ambient coronal density $\rho_e = 1.0 \times 10^{-13}\,\text{g cm}^{-3}$ (Kuridze *et al.*, 2011), or equivalently an electron number density $n_e = 5.98 \times 10^{10}\,\text{cm}^{-3}$. Our choice for the background homogeneous magnetic field in TR/lower corona is $B_e = 7\,\text{G}$ (Chae *et al.*, 2003). With a density contrast $\eta = \rho_e/\rho_i = 0.598$, the sound and Alfvén speeds in both media are: $c_{si} \cong 128\,\text{km s}^{-1}$, $v_{Ai} = 47.6\,\text{km s}^{-1}$ (more exactly $v_{Ai} = 47.567\,\text{km s}^{-1}$), and $c_{se} = 166\,\text{km s}^{-1}$, $v_{Ae} \cong 62\,\text{km s}^{-1}$, respectively. We note that the magnetic field inside the jet is $B_i = 6.899\,\text{G}$ that yields a ratio of both magnetic fields $b = B_e/B_i = 1.014$. The plasma betas in the jet and its environment are correspondingly $\beta_i = 8.76$ and $\beta_e = 8.48$. Thus, we can, in principle, consider both media as incompressible plasmas. We emphasize that for our piecewise uniform plasma density profile the singularities of the Alfvénic continuous spectrum that would exist if the density were continuously varying are all concentrated in the point $r = a$ (Goossens *et al.*, 2012). Note that this is true for all modes with $m \neq 0$. Moreover, since we are interested in unstable MHD waves, they must be discrete modes (see Chapter 8 and Fig. 8.4 in Jardin *et al.*, 2010).

8.3.1 *Kelvin–Helmholtz instability of the kink ($m = 1$) mode*

It is well known that the normal modes propagating along a bounded homogeneous magnetized cylindrical plasma column can be pure surface waves, pseudosurface (body) waves, or leaky waves (Cally, 1986). The type of the wave crucially depends on the ordering of the basic speeds in both media (the column and its surrounding plasma), more specifically sound and Alfvén speeds as well as corresponding tube speeds. Borrowing for a while Cally's notation (a stands for internal Alfvén speed, a_e for external Alfvén one, c for internal sound speed, and c_e for the external one) the speeds' ordering in our case is

$$a < a_e < c < c_e.$$

After normalizing all velocities with respect to the external sound speed, c_e, we obtain the dimensionless Alfvén, sound, and tube speeds as follows:

$$A = 0.2865, \quad C = 0.771, \quad A_e = 0.3735,$$

$$C_T = 0.2886, \quad \text{and} \quad C_{Te} = 0.35.$$

In calculating C_T and C_{Te} we have used $c_{Ti} = 44.59\,\text{km s}^{-1}$ and $c_{Te} = 58.08\,\text{km s}^{-1}$, respectively. Thus we have the following relations:

$$A < C \quad \text{and} \quad A < C_{Te}, \quad \text{as well as} \quad A < C_{Te} < C.$$

Now looking at Table I in Cally (1986), we conclude that the kink ($m = 1$) mode must be a non-leaky pure surface mode of type S_+^- that implies an externally slow internally fast surface wave. By "fast" and "slow" here, it is meant that the normalized wave phase velocity V is either greater than or less than the lesser of the sound and Alfvén speeds. For an S_+^--type wave V should lies between A and C: $A \leqslant V \leqslant C$, and also $V < C_{\mathrm{Te}}$. We note that the first inequalities' chain in Table I is wrong—the correct one is listed above. It is curious to see whether the numerical solving Eq. (8.1) will confirm these predictions.

Before starting the numerical task, we have to normalize all variables and to specify the input parameters. Our normalization differs from that of Cally (1986)— we find more convenient the wave phase velocity, v_{ph}, and the other speeds to be normalized with respect to the Alfvén speed inside the jet, v_{Ai}—Cally's normal- ization is inapplicable for the case when the surrounding plasma is considered as an incompressible medium. The wavelength, $\lambda = 2\pi/k_z$, is normalized to the tube radius, a, that is equivalent to introducing a dimensionless wavenumber $k_z a$. For normalizing the Alfvén speed in the ambient coronal plasma, v_{Ae}, we need the density contrast, η, and the ratio of the magnetic fields $b = B_{\mathrm{e}}/B_{\mathrm{i}}$, to get $v_{\mathrm{Ae}}/v_{\mathrm{Ai}} = b/\sqrt{\eta}$. The normalization of sound speeds in both media requires the specification of the reduced plasma betas, $\tilde{\beta}_{\mathrm{i,e}} = c_{\mathrm{si,e}}/v_{\mathrm{Ai,e}}$. In the dimen- sionless analysis the flow speed, v_0, will be presented by the Alfvén Mach num- ber $M_{\mathrm{A}} = v_0/v_{\mathrm{Ai}}$. The input parameters for solving Eq. (8.1) are as follows: $\eta = 0.589$, $b = 1.014$, $\tilde{\beta}_{\mathrm{i}} = 7.3$, and $\tilde{\beta}_{\mathrm{e}} = 7.07$; during the calculations we will vary the Alfvén Mach number, M_{A}, from zero (static plasma) to values at which we will obtain unstable solutions. At a static magnetic flux tube ($M_{\mathrm{A}} = 0$), the kink waves propagate with the kink speed, which in our case is equal to $53.6\,\mathrm{km\,s^{-1}}$, or in dimensionless units, 1.1269, i.e., the wave is slightly super-Alfvénic one and v_{ph} lies between the internal and external Alfvén speeds. This is not surprising because the kink speed is not sensitive to the sound speed—that is why it is more natural C in the inequalities' chain in Cally's (Cally, 1986) Table I to be replaced by A_{e}. Our numerical calculation confirm that $v_{\mathrm{ph}} < c_{\mathrm{Te}}$. In Cally's dimensionless vari- ables our normalized kink speed of 1.1269 transforms into 0.3228 and we have $0.2865 < 0.3228 < 0.3735$ along with $0.3228 < 0.35$, that is, in agreement with aforementioned predictions. One can also conclude that the way of normalization is not too decisive—even with our normalization we get the correct inequalities' chain. Calculated during the solving transcendental Eq. (8.1) wave attenuation coefficients $m_{\mathrm{0i,e}}$ were (for $M_{\mathrm{A}} = 0$) real quantities confirming that the principal kink mode is a pure surface wave.

The flow shifts upwards the kink-speed dispersion curve as well as splits it into two separate curves, which for small Alfvén Mach numbers travel with velocities

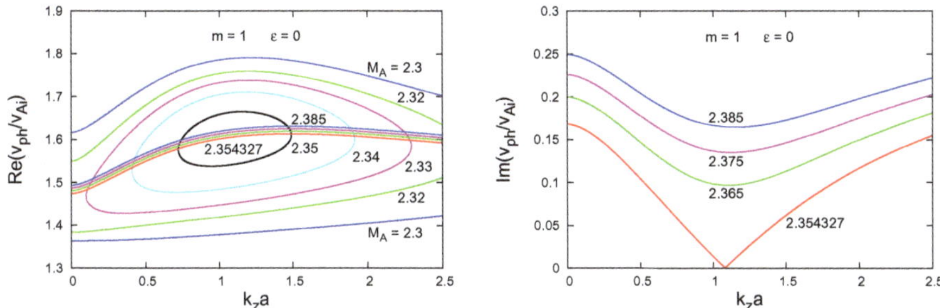

Fig. 8.2. (*Left panel*) Dispersion curves of stable and unstable kink ($m = 1$) MHD mode propagating on a moving untwisted flux tube of compressible plasma at $\eta = 0.598$ and $b = 0.36$. Unstable are the waves with dispersion curves located in the middle of the plot for four values of the Alfvén Mach number $M_A = 2.354327, 2.365, 2.375$, and 2.385. All other curves correspond to stable kink waves. (*Right panel*) The normalized growth rates of unstable waves for the same values of M_A. Red curves in both plots correspond to the onset of KH instability. *Reprinted with permission.*

$M_A \mp c_k/v_{Ai}$ (Zhelyazkov, 2012). (A similar duplication happens for the tube-speed dispersion curves, too.) At higher M_A, however, the behavior of each curve of the pair $M_A \mp c_k/v_{Ai}$ is completely different. As seen from the left panel of Fig. 8.2, for $M_A \geqslant 2.33$ both kink curves merge forming a closed dispersion curve. With increasing M_A the closed dispersion curves become smaller and narrower and the kink mode becomes unstable at $M_A^{cr} = 2.354327$. The growth rates of unstable kink modes are plotted in the right panel of Fig. 8.2. The red curves in all diagrams denote the marginal dispersion/growth rate curves: for $M_A < M_A^{cr}$ the kink waves are stable, otherwise they are unstable and the instability is of the KH-type. With $M_A^{cr} = 2.354327$ the kink mode will be unstable if the velocity of the moving flux tube exceeds $112\,\text{km s}^{-1}$—a speed, which is below the observationally evaluated jet speed of $150\,\text{km s}^{-1}$. It is important to underline that all the stable kink modes in our system are pure surface waves but the unstable ones are not—the latter become leaky waves (their external attenuation coefficients, m_{0e}s, are complex quantities with positive imaginary parts). This implies that wave energy is radiated outward in the surrounding medium. Thus, the KH instability plays a dual role: once in its nonlinear stage the instability can trigger wave turbulence and simultaneously the propagating KH-wave is radiating its energy outside.

The circumstance that the reduced plasma betas in both media are ~ 7, tempts us to consider the moving get and its surrounding magnetized plasma as incompressible media. Then the wave dispersion Eq. (8.1) becomes quadratic one giving us the wave phase velocity and its growth rate when the kink mode is unstable in closed forms [see Eq. (8.3)]. Calculated dispersion curves and growth rates are plotted in Fig. 8.3. One is immediately seen that the picture is very similar to that

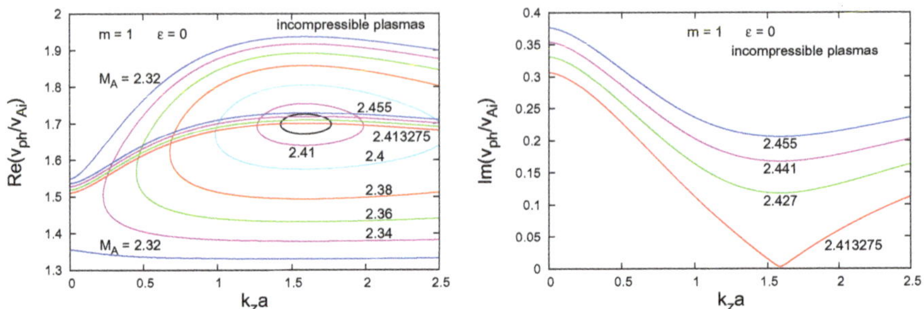

Fig. 8.3. (*Left panel*) The same as in Fig. 8.2, however, for incompressible plasma in both media. Unstable are the waves for four values of the Alfvén Mach number $M_\mathrm{A} = 2.413275$, 2.427, 2.441, and 2.455. The closed black curve corresponds to $M_\mathrm{A} = 2.4125$. All other curves correspond to stable kink waves. (*Right panel*) The normalized growth rates of unstable waves for the same values of M_A. Red curves in both plots correspond to the onset of KH instability. *Reprinted with permission.*

in Fig. 8.2. Note that both the dispersion curve of stable kink waves and the growth rate of the marginal unstable wave are shifted to the right. Moreover, the critical Alfvén Mach number now is a little bit higher, $M_\mathrm{A}^\mathrm{cr} = 2.413275$ that yields a critical flow speed of $114.8\,\mathrm{km\,s^{-1}}$. Unlike the unstable kink mode in compressible plasma, that in incompressible limit persists as a non-leaky surface wave, the two attenuation coefficients are real quantities.

When we model the EUV jet as a moving twisted magnetic flux tube two types of instabilities can develop in the jet, notably kink instability due to the twist of the magnetic field and KH instability owing to the tangential discontinuity of plasma flow at the tube boundary. A normal mode analysis (Dungey and Loughhead, 1954) and an energy consideration method (calculating the change in magnetic energy per unit length of the cylinder—see Lundquist, 1951) yield similar thresholds of the kink instability in twisted magnetic flux tubes in the form $B_{\mathrm{i}\phi}(a) > 2B_{\mathrm{i}z}$, or in our notation, as $\varepsilon > 2$. In the following, we will consider only weakly twisted tubes, $\varepsilon \ll 1$, and, therefore, only the KH instability may occur in the exploring jet configuration. In solving Eq. (8.4), we use the same method as in the case of untwisted tube. The new quantities for normalization that appear here are the local Alfvén frequencies $\omega_{\mathrm{Ai,e}}$. The natural way of normalizing ω_{Ai} is to multiply it by the tube radius, a, and after that divide by the Alfvén speed v_{Ai}, that is,

$$a\omega_{\mathrm{Ai}} = \frac{B_{\mathrm{i}z}}{\sqrt{\mu\rho_\mathrm{i}}} \left(m\frac{Aa}{B_{\mathrm{i}z}} + k_z a \right),$$

or

$$\frac{a\omega_{\mathrm{Ai}}}{v_{\mathrm{Ai}}} = m\frac{Aa}{B_{\mathrm{i}z}} + k_z a = m\varepsilon + k_z a.$$

Note that here the Alfvén speed is defined via the axial component of the twisted magnetic field, $v_{\text{Ai}} = B_{\text{iz}}/\sqrt{\mu\rho_{\text{i}}}$. Accordingly, the normalized local Alfvén frequency ω_{Ae} has the form

$$\frac{a\omega_{\text{Ae}}}{v_{\text{Ai}}} = k_z a \frac{B_e/B_{\text{iz}}}{\sqrt{\eta}} = k_z a \frac{b_{\text{twist}}}{\sqrt{\eta}},$$

where the parameter $b_{\text{twist}} = B_e/B_{\text{iz}} = b\sqrt{1+\varepsilon^2}$. For small values of the twist parameter ε, we take $b_{\text{twist}} \cong b$, and bearing in mind that b has a value close to 1, one can assume that $b = 1$. We begin our numerical calculations for the kink ($m = 1$) mode in the moving twisted tube with $\eta = 0.598$, $b_{\text{twist}} \cong b = 1$, and $\varepsilon = 0.025$. Dispersion curves and growth rates of unstable waves are plotted in Fig. 8.4. As seen, the curves are very similar to those shown in Figs. 8.2 and 8.3, of course, with the observation that the minimum of the marginal growth rate (red) curve of the kink mode in shifted to the right on the $k_z a$-axis as compared with the plot in the right panel of Fig. 8.2 for a similar curve of the $m = 1$ mode traveling on untwisted magnetic flux tube containing compressible plasma. On the other hand, the minima of the marginal growth rate curves in untwisted and twisted flux tubes in the limits of incompressible plasmas practically coincide (compare the right panels in Figs. 8.3 and 8.4). The critical flow velocity for emerging KH instability now is $v_0^{\text{cr}} = 114.4\,\text{km s}^{-1}$, calculated by using the "reduced" Alfvén speed $v_{\text{Ai}}/\sqrt{1+\varepsilon^2} = 47.522\,\text{km s}^{-1}$. It is worth pointing out that the incompressible plasma approximation in general yields slightly higher threshold Alfvén Mach numbers than the model of compressible media. But the value of $114.4\,\text{km s}^{-1}$ is an entirely acceptable and reasonable jet speed. It is instructive to note that the unstable kink ($m = 1$) mode is a non-leaky surface wave—its external

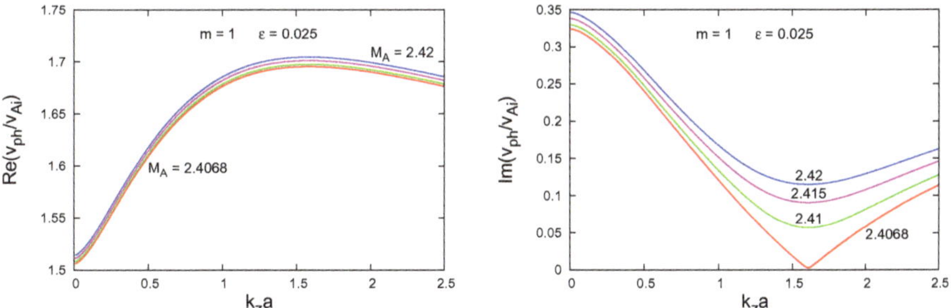

Fig. 8.4. (*Left panel*) Dispersion curves of unstable kink ($m = 1$) MHD mode propagating on a moving twisted flux tube of incompressible plasma at $\varepsilon = 0.025$, and for four values of the Alfvén Mach number $M_A = 2.4068, 2.41, 2.415$, and 2.42. (*Right panel*) The normalized growth rates for the same values of M_A. Red curves in both plots correspond to the onset of KH instability. *Reprinted with permission.*

attenuation coefficient, $\kappa_e = k_z$, is not changed by the instability, but the internal one, κ_i, becomes a complex quantity with positive real and imaginary parts. The same applies to the higher fluting-like ($m = 2$) mode and to the higher m modes with azimuthal mode numbers 3 and 4.

8.3.2 *Kelvin–Helmholtz instability of the $m = 2$, 3, and 4 modes*

It is interesting to see what kind of threshold Alfvén Mach numbers we will obtain as consider the excitations of higher MHD mode numbers. Next three figures show the results of numerical computations for the fluting-like $m = 2$ mode, the $m = 3$, and the $m = 4$ mode. Here we have a new phenomenon: the KH instabilities start at some critical $k_z a$-numbers along with the corresponding critical Alfvén Mach numbers. As one can see from Figs. 8.5–8.7, those critical wavenumbers have values of 0.676 trough 1.427 to 2.16099 for the $m = 2$, $m = 3$, and $m = 4$ MHD modes, respectively. If we assume that our EUV jet has a width $\Delta\ell = 4000\,\mathrm{km}$, then the corresponding critical wavelengths for a KH instability onset are $\lambda_{\mathrm{cr}}^{m=2} \cong 18.6\,\mathrm{Mm}$, $\lambda_{\mathrm{cr}}^{m=3} \cong 8.8\,\mathrm{Mm}$, and $\lambda_{\mathrm{cr}}^{m=3} \cong 5.8\,\mathrm{Mm}$, respectively. These critical wavelengths are of the order or less than the height of the EUV gets; for instance, Zhang and Ji (2014) observing three jets in the seven extreme-ultraviolet (EUV) filters of the AIA/*SDO*, have reported jets' heights between 12.8 and 26.8 Mm; the arithmetic mean diameter of those EUV jets was 3.7 Mm. One expects that the actual wavelengths of occurring KH instabilities of the higher m modes should be shorter than the discussed critical wavelengths. We pay attention to the fact that the threshold Alfvén Mach numbers for observing KH instability in the $m = 2$, $m = 3$, and $m = 4$ modes are very close, equal to 2.367, 2.365,

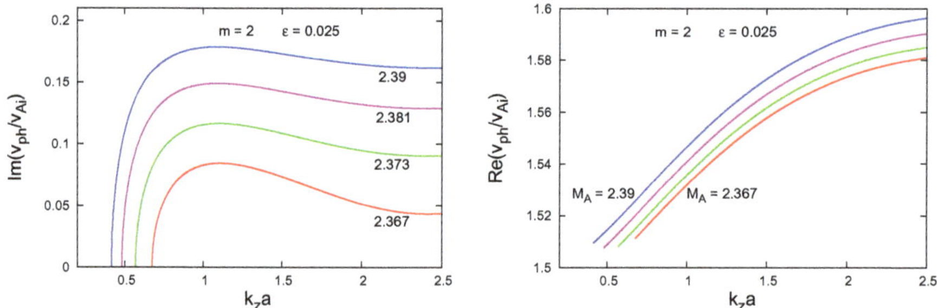

Fig. 8.5. (*Left panel*) Normalized growth rates of unstable fluting-like ($m = 2$) MHD mode propagating on a moving twisted flux tube of incompressible plasma at $\varepsilon = 0.025$, and for four values of the Alfvén Mach number $M_A = 2.367$, 2.373, 2.381, and 2.39. (*Right panel*) Wave dispersion curves for the same values of M_A. Red curves in both plots correspond to the onset of KH instability. *Reprinted with permission.*

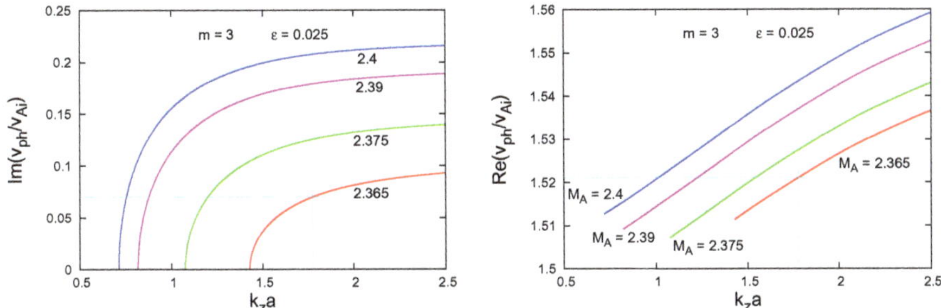

Fig. 8.6. (*Left panel*) Normalized growth rates of unstable $m = 3$ MHD mode propagating on a moving twisted flux tube of incompressible plasma at $\varepsilon = 0.025$, and for four values of the Alfvén Mach number $M_A = 2.365, 2.375, 2.39$, and 2.4. (*Right panel*) Wave dispersion curves for the same values of M_A. Red curves in both plots correspond to the onset of KH instability. *Reprinted with permission.*

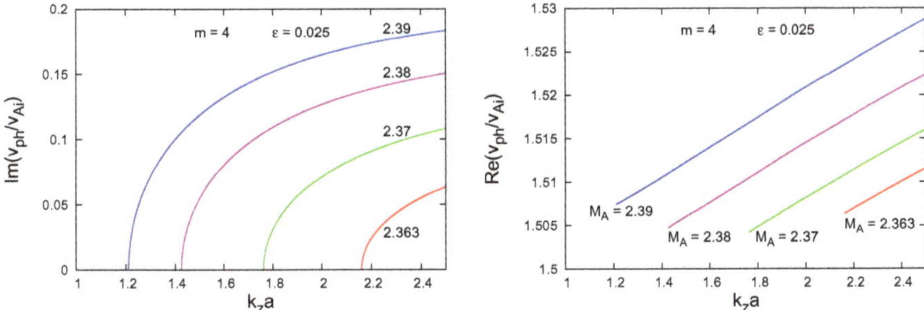

Fig. 8.7. (*Left panel*) Normalized growth rates of unstable $m = 4$ MHD mode propagating on a moving twisted flux tube of incompressible plasma at $\varepsilon = 0.025$, and for four values of the Alfvén Mach number $M_A = 2.363, 2.37, 2.38$, and 2.39. (*Right panel*) Wave dispersion curves for the same values of M_A. Red curves in both plots correspond to the onset of KH instability. *Reprinted with permission.*

and 2.363, respectively, and they yield critical flow velocities of 112.5, 112.4, and 112.3 km s^{-1}. Thus, we can conclude that all studied MHD modes propagating on the EUV jet in equal measure can become unstable against the KH instability—the required flow velocity shear lies between 112 and 114.4 km s^{-1}. We note also that the phase velocities of unstable higher m modes are in the range of 72–75 km s^{-1}, while those of the kink, $m = 1$, mode lie between 70 and 76.7 km s^{-1} in the case of untwisted jet, and between 71.6 and 81.6 km s^{-1} for a weakly twisted EUV jet. An evaluation of the instability growth rate requires the magnitude of the observed/detected wavelength (or equivalently, the $k_z a$-value) and the jet width. For example, if we assume that the KH instability of the $m = 3$ MHD harmonic rises (see Fig. 8.6) at $k_z a = 1.775$ (or with $\lambda_{KH} \cong 7.1$ Mm, assuming that $\Delta \ell = 4$ Mm), the corresponding normalized wave phase velocity growth rate is equal to 0.07 that yields a wave growth rate $\mathrm{Im}(\omega_{\mathrm{KH}}) \cong 3 \times 10^{-3}\,\mathrm{s}^{-1}$. Such growth rates, of few inverse milliseconds, were evaluated for the KH instability of high harmonic MHD modes (however, with negative signs of their azimuthal wave mode numbers) in cool surges (see, for instance, Subsection 7.3.1 or Zhelyazkov *et al.*, 2015b). The wave phase velocity of the considered unstable $m = 3$ mode (see the right panel in Fig. 8.6) is equal to $\approx 72.3\,\mathrm{km\,s}^{-1}$.

The question that immediately arises is how any change in physical parameters of the jet or its environment will influence the condition for KH instability occurrence. If we decrease, for instance, the electron density of the ambient coronal plasma to $5 \times 10^{10}\,\mathrm{cm}^{-3}$, making the density contrast η equal to 0.5 (that corresponds to a 2-times denser jet than its environment), the pressure balance equation cannot be satisfied for $B_{\mathrm{e}} = 7\,\mathrm{G}$. If we still want to keep that value, one must reduce the temperature inside the jet—$T_{\mathrm{i}} = 900\,000\,\mathrm{K}$ is an appropriate choice and we have a new value for the reference Alfvén speed, notably $v_{\mathrm{Ai}} = 63.09\,\mathrm{km\,s}^{-1}$. Performing all the computations with this new $\eta = 0.5$ and $b = 0.765$ (for the untwisted tube we need also of $\tilde{\beta}_{\mathrm{i}} = 3.11$ and $\tilde{\beta}_{\mathrm{e}} = 5.91$) we obtain that the kink, $m = 1$, mode (in both configurations) and harmonic MHD modes in a twisted magnetic tube with $\varepsilon = 0.025$ can become unstable at flow speeds of $140.3\,\mathrm{km\,s}^{-1}$ (at $M_{\mathrm{A}}^{\mathrm{cr}} = 2.224058$) for the compressible untwisted tube, and 147.5, 143, 141.8, and $141.6\,\mathrm{km\,s}^{-1}$ correspondingly for the $m = 1, 2, 3, 4$ modes in the incompressible tube with also incompressible environment. As seen, all these critical speeds are still accessible for our EUV jet. On the other hand, if we choose the background magnetic field to be equal to $10\,\mathrm{G}$ (at $\eta = 0.589$), that would imply that the reference Alfvén speed will increase to $68.457\,\mathrm{km\,s}^{-1}$, the critical jet speed for triggering a KH instability of the kink, $m = 1$, mode in an untwisted jet at $M_{\mathrm{A}}^{\mathrm{cr}} = 2.31995$ will become equal to $158.8\,\mathrm{km\,s}^{-1}$—a speed that is inaccessible for our EUV jet. The critical velocities for all modes in the twisted tube turn out to be even higher: they are equal to 164.7, 162, 161.8, and $161.5\,\mathrm{km\,s}^{-1}$ for the $m = 1, 2, 3, 4$ MHD modes, respectively. This conclusion confirms the famous Chandrasekhar's statement (Chandrasekhar, 1961) that a longitudinal magnetic field can, in general, stabilize the jet against the KH instability.

8.4   Kelvin–Helmholtz instability in an EUV jet observed by *SDO*/AIA

During 2014 April 15–16, the active region NOAA AR 12035 produced several jets. These jets are well observed by the Atmospheric Imaging Assembly (AIA) on board *SDO* satellite. The AIA observe the Sun in different UV and EUV wavelengths.

The pixel size and temporal evolution of the AIA data is 0.6 arcsec and 12 s, respectively. The dynamics and the kinematics of these jets were described by Joshi *et al.* (2017). They found two jet sites close to each other and noticed the slippage of jets from one site to other site. Using the Heliospheric Magnetic Imager (HMI, Schou *et al.*, 2012) photospheric magnetic field data, they found that the jets site were associated with the flux emergence as well the flux cancellation and presence of several null points (see their Fig. 10).

Here, we have selected one jet, notably J'_6 from Joshi *et al.*, 2017, Table 1) and model it for the Kelvin–Helmholtz instability. Our selection criterion is based on its high speed, large size, longer lifetime and visibility in all EUV channels relative to other jets. In addition to this criterion, this jet indicates eruption of blobs type structures at its boundary. We believe that these blobs like structures could be the result of KHI. Therefore, the chosen jet is a good candidate for KHI modeling. The blobs like structures were previously observed inside the jets in the past (Zhang and Ji, 2014). Zhang and Ji (2014) have reported the presence of blobs in the recurrent jets that occurred at the western edge of the NOAA active region 11259 on 2011 July 22. They have suggested that these blobs are plasmoids created by the magnetic reconnection as a result of tearing-mode instability. However, in our case the blobs type structures are observed at the boundary of the jets (not inside the jets) and could be due to the velocity shear. That is the reason, we interpret them as the indicator of KHI.

The jet starts $\approx$14:47 UT and reached its maximum length $\approx$14:57 UT and finally finish around 15:00 UT. The jet's maximum speed was $\approx$343 km s^{-1} at AIA 211 Å and the average speed of all the EUV channels was $\approx$332 km s^{-1}.

For a better image quality for *SDO*/AIA data, we have used the Multi-Gaussian Normalization (MGN) techniques (Morgan and Druckmüller, 2014) to enhanced the image quality. The evolution of the jet in AIA/EUV 171 and 193 Å is displayed in Fig. 8.8. During jet's evolution, we have noticed the propagation of plasma blobs along the jet boundary in all AIA EUV channels. To investigate these structures in more detail, we have enlarged the jet image in smaller field-of-view and this is shown in Fig. 8.9 at AIA 171, 193, 211, and 304 Å. An inspection of these images evidenced the movement of the plasma blobs at the jet's boundary. These structures are shown by the white arrows. These blob structures evidenced the KHI.

To model the jet for the KHI, we have computed the temperature and the density inside and outside of the jet. For this calculation, we have used the technique established by Aschwanden *et al.* (2013). This techniques uses the observations of six AIA EUV channels, i.e., 94, 131, 171, 193, 211 and 335 Å. The estimated values with errors for the jet are given in Table 8.1. An example of temperature

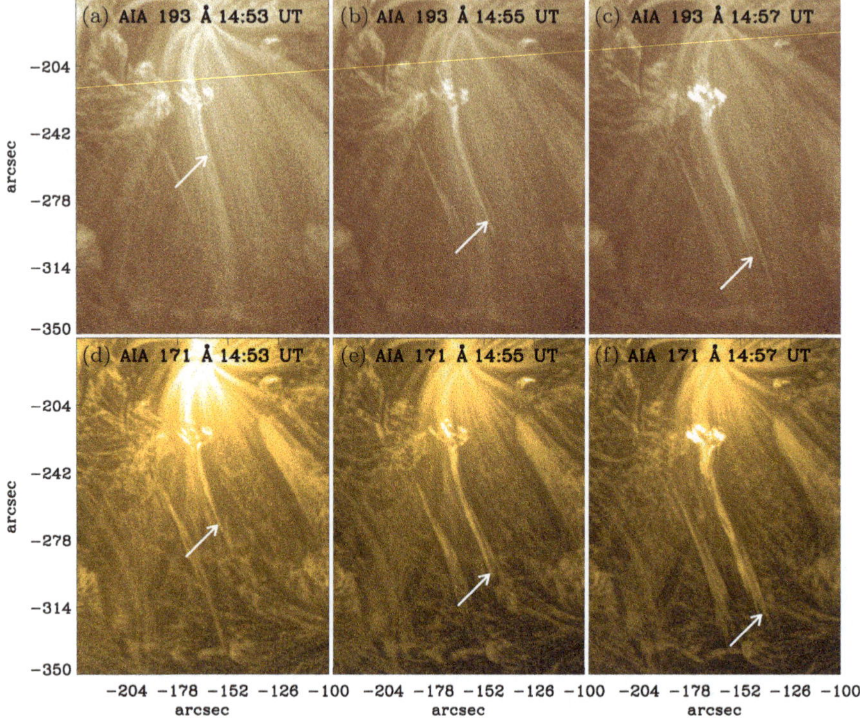

Fig. 8.8. Evolution of the jet in AIA 171 and 193 Å. *Reprinted with permission.*

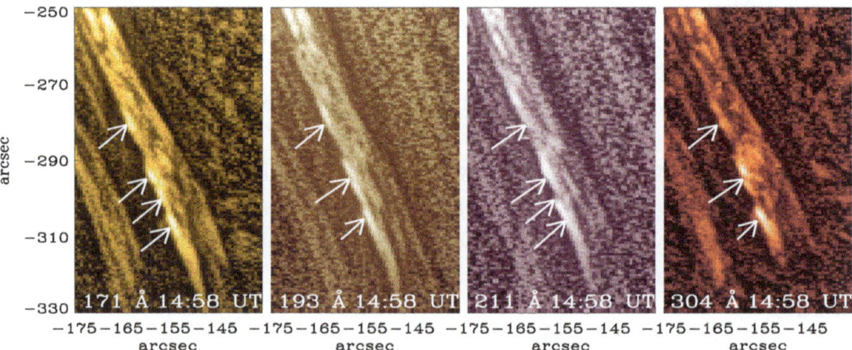

Fig. 8.9. AIA 171, 193, 211, and 304 Å images of the jet. The white arrows indicate different visible structures of blobs, which could be due to KHI. *Reprinted with permission.*

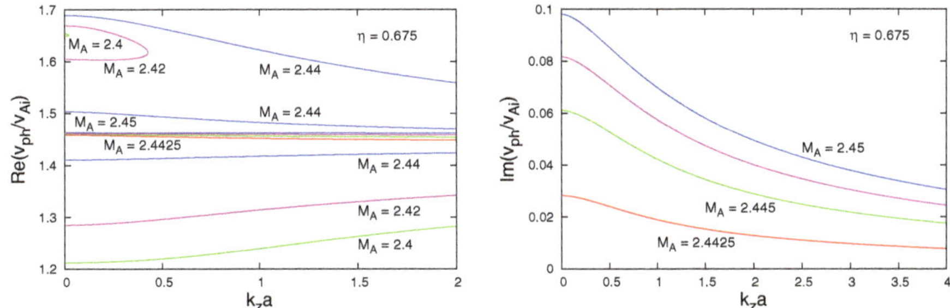

Fig. 8.10. (*Left panel*) Dispersion curves of stable and unstable kink ($m = 1$) mode propagating in a moving untwisted magnetic flux tube at $\eta = 0.675$, $b = 1.184$, and at various values of Alfvén Mach number M_A. The threshold Alfvén Mach number for KHI occurring is equal to 2.4425 (red curve). (*Right panel*) The normalized growth rates of the unstable kink mode for the same values of the input parameters. The purple curve has been calculated at $M_\text{A} = 2.4475$. *Reprinted with permission.*

and emission measure maps are presented in Fig. 8.10 during the peak phase of the jet.

According to Table 8.1, the density contrast η is equal to 0.675. The estimated temperatures inside and outside the jet define the following sound speeds in both media: $c_\text{si} = 151\,\text{km s}^{-1}$ and $c_\text{se} = 158\,\text{km s}^{-1}$, respectively. Regarding the coronal magnetic field value many people tried to find it through observations. The methods include Helioseismology, Radio observations as well as magnetic field modeling. According to very recent study done by Luna *et al.* (2017) the coronal magnetic field value ranges from 7 to 30 G. They have estimated these values using the Helioseismology of the filament and magnetic field modeling and the result from both these methods are consistent. Keeping these observational facts in mind, we assume that the background magnetic field is $B_\text{e} = 7\,\text{G}$, then the Alfvén speed in the environment, calculated from the standard formula $v_\text{A} = B/\sqrt{\mu\rho}$, is $v_\text{Ae} = 191\,\text{km s}^{-1}$. The Alfvén speed inside the flux tube can be derived from the total pressure balance equation (the sum of thermal and magnetic pressure to be constant, that is, to have the same values in both media) and it yields $v_\text{Ai} = 132.6\,\text{km s}^{-1}$. This implies an internal magnetic field $B_\text{i} = 5.9\,\text{G}$; thus the magnetic fields ratio is $b = 1.18$.

With aforementioned density contrasts, sound and Alfvén speeds, as well as magnetic fields ratios, prior to starting numerical calculation of wave dispersion relations of the kink mode ($m = 1$), we can make some predictions first for the nature of the propagating normal mode (be it pure surface, pseudosurface/body, or leaky wave) that depends on the speeds ordering (Cally, 1986); second, to evaluate the so-called *kink speed* c_k in a static magnetic flux tube expressed in terms of plasma densities ρ_i and ρ_e, and Alfvén speeds in both media:

$$c_k = \left(\frac{\rho_i v_{Ai}^2 + \rho_e v_{Ae}^2}{\rho_i + \rho_e} \right)^{1/2} = \left(\frac{1 + b^2}{1 + \eta} \right)^{1/2} v_{Ai}, \qquad (8.43)$$

which, as we know, is independent of sound speeds and characterizes the propagation of transverse perturbations; and third, to find the expected threshold/critical Alfvén Mach number M_A at which KHI would start—the later is determined by the inequality (Zaqarashvili *et al.*, 2014a):

$$M_A^2 > (1 + 1/\eta)(b^2 + 1). \qquad (8.44)$$

Here, the Alfvén Mach number is defined as v_0/v_{Ai}. These two formulas are written down for the kink mode propagating on a untwisted magnetic flux tube. When the magnetic field B_i is twisted, we should use modified values for the ratio b, and also for v_{Ai}—for computational reasons, the Alfvén speed inside the jet should be defined as $B_{iz}/\sqrt{\mu\rho_i}$.

Since in the numerical task we are working with dimensionless variables, one needs the normalization of several quantities. For example, the normalization of sound, c_s and tube, c_T, speeds in the attenuation coefficients κ_i and κ_e [see Eq. (8.2)], contained in dispersion Eq. (8.1), requires the values of both the reduced plasma betas $\tilde{\beta}_{i,e} = c_{i,e}^2/v_{Ai,e}^2$ and the magnetic fields ratio $b = B_e/B_i$. In the case of a twisted magnetic flux tube, for normalizing the local Alfvén frequency (8.16), along with the fixed input parameters $\eta = n_e/n_i$ and $b_{twist} = B_e/B_{iz}$, one needs to define the parameter $\varepsilon \equiv B_{i\phi}(a)/B_{iz}(a) = Aa/B_{iz}$ that characterizes the twisted magnetic field inside the tube. When the external magnetic field is twisted too, its normalization requires an additional ε, equal to $B_{e\phi}(a)/B_{ez}(a) = B_\phi/B_z$, which will have a subscript "2", while the similar parameter characterizing the twist of the internal magnetic field will possess the subscript "1".

Like the previous subsection, we begin the numerical calculations with the simplest jet's model as an untwisted axially moving flux tube (the left column in Fig. 8.1). The ordering of basic speeds (sound, Alfvén, and tube one) in both media is as follows:

$$c_{Ti} < c_{Te} < v_{Ai} < c_{si} < c_{se} < v_{Ae}.$$

According to Cally (1986) (see Table I there), at this ordering the kink ($m = 1$) mode, propagating in a static ($v_0 = 0$) tube, would be a pseudosurface/body wave of B_-^--type. In addition to $\eta = 0.675$, the other input parameters for the numerical task are: $\tilde{\beta}_i = 1.2946$, $\tilde{\beta}_e = 0.6836$, and $b = 1.184$. At Alfvén Mach number $M_A = 0$ (static flux tube), the anticipated normalized kink speed (8.43) has a magnitude of 1.1977. With these input parameters, the numerical computations of dispersion relation (8.1) confirm that the kink mode traveling on the tube is a pseudosurface/body wave possessing, at $k_z a \ll 1$, a normalized phase velocity equal to the normalized kink speed within four places after the decimal point. In moving flux tube, at relatively small M_As, the kink speed splits into a pair of phase velocities (Zhelyazkov, 2012), whose dispersion curves initially go almost parallel to each other, but for higher Alfvén Mach numbers, when one reaches the region of expected $M_A > 2.44$ for occurring of KHI according to the criterion (8.44), their behavior becomes completely different. Look, for example, at the green and purple curves labeled by $M_A = 2.4$ and $M_A = 2.42$ in the left panel of Fig. 8.10. The low-speed curves have more or less normal move while the high-speed ones turn over at some $k_z a$-values forming semi-closed dispersion curves. The instability arises at the threshold Alfvén Mach number equal to 2.4425—indeed rather close to the predicted value. This value of M_A tells us that the KHI should emerge at a critical speed of the jet equal to $323.9\,\mathrm{km\,s^{-1}}$, which is less than the average jet speed of $332\,\mathrm{km\,s^{-1}}$ (see Table 8.1). The marginal red curves divides the 1D M_A-space into two regions: for all $M_A < 2.4425$ the kink wave is a stable MHD mode, while in the opposite case it becomes unstable; alongside the marginal red curve in the left panel of Fig. 8.10, one can see other three dispersion curves (in green, purple, and blue colors) presenting the unstable kink mode at the corresponding Alfvén Mach numbers. The normalized growth rates of the same clutch of unstable kink waves are presented in the right panel of Fig. 8.10.

For the computation of unstable kink ($m = 1$) mode propagating in axially moving twisted flux tube (the middle column in Fig. 8.1), our choice for the magnetic field twist parameter is $\varepsilon = 0.025$. The other two input parameters are $\eta = 0.675$ and $b = 1$. The dispersion curves and corresponding normalized growth rates of the unstable mode for four values of the Alfvén Mach number are displayed in Fig. 8.11. There is an unexpected peculiarity—the threshold Alfvén Mach number for instability arising turns out to be lower than that in untwisted flux tube, notably equal to 2.30839, which yields a critical flow speed of $\approx 306\,\mathrm{km\,s^{-1}}$— $26\,\mathrm{km\,s^{-1}}$ less than the average jet speed. An extensive study of KHI in an EUV jet situated on the west side of NOAA AR 10938 and observed on board *Hinode* on 2007 January 15/16 showed just the opposite inequality: 2.4068 vs. 2.354327, or $114.4\,\mathrm{km\,s^{-1}}$ vs. $112\,\mathrm{km\,s^{-1}}$ (Zhelyazkov *et al.*, 2016).

It is curious to see what will be the wave growth rate, γ_{KH}, instability developing time, $\tau_{\mathrm{KH}} = 2\pi/\gamma_{\mathrm{KH}}$, and the kink wave phase velocity, v_{ph}, for a given wavelength. Bearing in mind that our jet has a width of 4 Mm and height of 152 Mm, if we chose $\lambda_{\mathrm{KH}} = 4\,\mathrm{Mm}$ to be a reasonable wavelength of the unstable mode, then the aforementioned instability parameters, determined by the crossed points

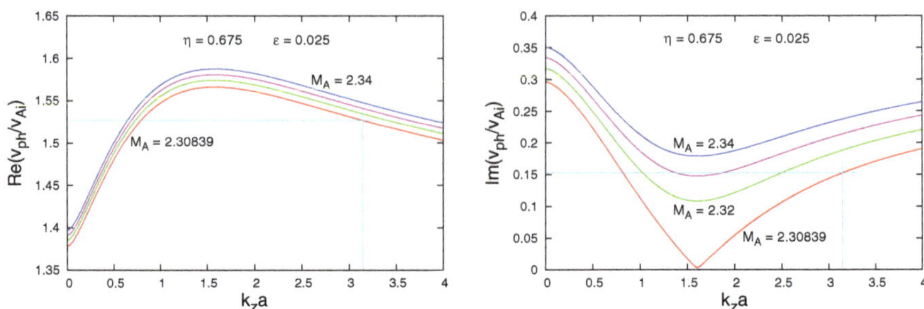

Fig. 8.11. (*Left panel*) Dispersion curves of unstable kink ($m = 1$) mode propagating in a moving twisted magnetic flux tube at $\eta = 0.675$, $b = 1$, and $\varepsilon = 0.025$ and at four various values of Alfvén Mach number M_A. The threshold Alfvén Mach number for KHI occurring is equal to 2.30839 (red curve). (*Right panel*) The normalized growth rates of the unstable kink mode for the same values of the input parameters. The purple curve has been calculated at $M_A = 2.33$. *Reprinted with permission.*

of cyan lines and red marginally dispersion and growth rate curves at $k_z a = \pi$ in Fig. 8.11, are:

$$\gamma_{\mathrm{KH}} = 31.7 \times 10^{-3}\,\mathrm{s}^{-1}, \quad \tau_{\mathrm{KH}} = 198.4\,\mathrm{s} = 3.3\,\mathrm{min},$$

$$\text{and} \quad v_{\mathrm{ph}} = 202.3\,\mathrm{km\,s}^{-1},$$

respectively. As seen, the KHI evolution time of 3.3 min is much less than the jet lifetime of 16 min (see Table 1 in Joshi *et al.*, 2017). Note that the two cross points in Fig. 8.11 can be considered as a "phase portrait" of KHI in the dimensionless phase velocity–wavenumber-plane.

In the most complicated case when the external magnetic field is also twisted (the right column in Fig. 8.1), as we have mentioned, it is necessary to introduce two magnetic field twist parameters, ε_1 for the internal field, and ε_2 for the external one, respectively. We take ε_1, as in the previous case, to be equal to 0.025; our choice for ε_2 is 0.001. Thus, with $\eta = 0.675$, $b = 1$, $\varepsilon_1 = 0.025$, and $\varepsilon_2 = 0.001$, the solutions to the wave dispersion relation (8.40) of the kink ($m = 1$) mode at five values of the Alfvén Mach number are graphically presented in Fig. 8.12. Here, we are faced with a distinctly different issue, namely at relatively low Alfvén Mach numbers, in the very long wavelength limit, $k_z a \ll 1$, one appears two branches of the dispersion curve separated by a gap (see the cyan curves in Fig. 8.12). With increasing the magnitude of the Alfvén Mach number, that gap becomes narrower and at some M_A the two branches merge forming a continuous dispersion curve—this is the marginal dispersion curve and the corresponding Alfvén Mach number is the threshold one for appearance of the KHI—in our case its value is 3.6075. We would like to notice that the red normalized growth rate curve has

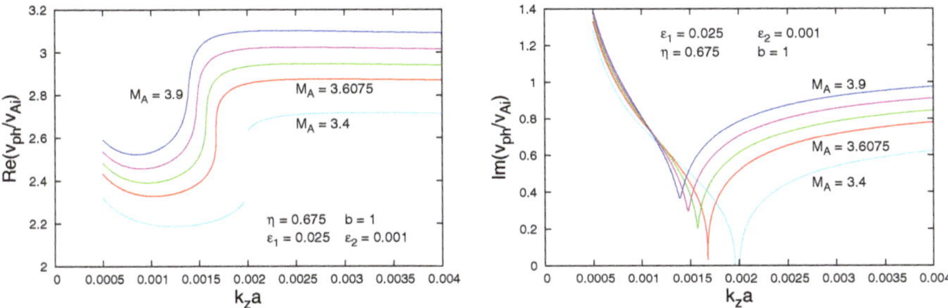

Fig. 8.12. (*Left panel*) Dispersion curves of unstable kink ($m = 1$) mode propagating in a moving twisted magnetic flux tube surrounded by a twisted magnetized plasma at $\eta = 0.675, b = 1, \varepsilon_1 = 0.025$, $\varepsilon_2 = 0.001$ and at five various values of Alfvén Mach number M_A. The two cyan curves labeled by $M_A = 3.4$ are associated with the propagation of a spurious unstable kink ($m = 1$) mode. The threshold Alfvén Mach number for KHI occurring is equal to 3.6075 (red curve). (*Right panel*) The normalized growth rates of the unstable and spurious unstable kink mode for the same values of the input parameters. The green and purple curves have been calculated at $M_A = 3.7$ and 3.8, respectively. *Reprinted with permission.*

no cusp at $k_z a = 0.00168$—it is a normal smooth curve. For larger values of $k_z a$ both the dispersion and growth rate curves go with gradually increasing magnitudes. But at such high threshold Alfvén Mach number $M_A = 3.6075$ and Alfvén speed of $132.52\,\mathrm{km\,s^{-1}}$, the required flow speed for instability onset is equal to $478\,\mathrm{km\,s^{-1}}$—a value which is inaccessible by the jet under consideration. Hence, even a very small twisted external magnetic field suppresses the KHI occurrence. We would like to note that an EUV jet with greater density contrast (smaller η) might change the picture in the sense to have observable Kelvin–Helmholtz instability even in this configuration. Our experience shows that the density contrast and in partially the values of temperatures control the occurrence of KH instability in any observable jet.

Chapter 9

Kelvin–Helmholtz Instability in X-ray Solar Jets

9.1 Observations and nature of the X-ray jets

X-ray jets were discovered by the Soft X-ray Telescope (SXT) on board *Yohkoh* (Tsuneta *et al.*, 1991), as transient X-ray energy release and enhancement with apparent collimated ballistic motions of the plasma associated with the flares in X-ray bright points, emerging flux regions, or active regions (for details, see Shibata *et al.*, 1992). As it has been pointed out by Shimojo *et al.* (1996), jets from X-ray bright points in active regions most likely appear at the western edge of preceding sunspots and exhibit a recurrent plasma propulsion in the solar atmosphere. These X-ray jets are confined plasma dynamics with typical morphological properties, e.g., $(1\text{–}40) \times 10^4$ km length, and the width of $5 \times 10^3\text{–}10^5$ km. Such jets possess apparent velocities of $10\text{–}1000 \, \text{km s}^{-1}$, and lifetime of $100\text{–}16\,000 \, \text{s}$ (Shimojo *et al.*, 1996). The electron densities of the X-ray jets are of the order of $(0.7\text{–}4) \times 10^9 \, \text{cm}^{-3}$. Their temperatures lie in the range of 3–8 MK with an average temperature of 5.6 MK (Shimojo and Shibata, 2000).

In terms of spatial location, jets can be classified as polar jets (Culhane *et al.*, 2007) and active region jets (Chifor *et al.*, 2008b). A study of polar jet parameters based on *Hinode* XRT observations was carried out by Savcheva *et al.* (2007) who showed that jets preferably occur inside the polar coronal holes. Culhane *et al.* (2007) have found from *Hinode*'s Extreme-ultraviolet Imaging Spectrometer (EIS) $40''$ slot observations of a polar coronal hole that jet temperature ranges from 0.4 to 5.0 MK. The jet velocities had typical values that are mostly less than the Sun's escape velocity $(618 \, \text{km s}^{-1})$, therefore, in consequence most of the jets fall back in the lower solar atmosphere after their triggering. Using the XTR on *Hinode*, Cirtain *et al.* (2007) concluded that X-ray jets in polar coronal holes have two distinct velocities: one near the Alfvén speed ($\sim$800 km s^{-1}), and another near the sound speed (200 km s^{-1}). Moreover, they were the first to give an evidence for the propagation of Alfvén waves in solar X-ray jets. Kim *et al.* (2007) presented the morphological and kinematic characteristics of three small-scale X-ray/EUV jets simultaneously observed by the *Hinode* XRT and the *Transition Region and Coronal Explorer* (*TRACE*). While observing the coronal jets, for two different wavelength bands, they obtain matching characteristics for their projected speed (90–310 km s^{-1}), lifetime (100–2000 s), and size (1.1–5 $\times$ 10^5 km). Yang *et al.* (2011) presented simultaneous observations of three recurring jets in EUV and soft X-ray (SXR), which occurred in an active region on 2007 June 5. On comparing their morphological and kinematic properties, the authors have found that EUV and SXR jets had similar on-set locations, directions, size and terminal velocities. The three observed jets were having maximum Doppler velocities ranging from 25 to 121 km s^{-1} in the Fe xii λ195 line and from 115 to 232 km s^{-1} in the He ii λ256 line. Extensive multi-instrument observations obtained simultaneously with the SUMER spectrometer on board the *Solar and Heliospheric Observatory* (*SOHO*), with EIS and XRT on board *Hinode*, and with the Extreme-ultraviolet imagers (EUVI) of the Sun–Earth Connection Coronal and Heliospheric Investigation (SECCHI) instrument suite on board the Ahead and Behind *STEREO* spacecrafts were performed by Madjarska (2011). There have been derived in great detail the dynamic process of X-ray jet formation and evolution. In particular, for the first time there was found spectroscopically a temperature of 12 MK (Fe xxiii 263.76 Å) and density of 4 $\times$ 10^{10} cm^{-3} in the quiet Sun. The author has clearly identified two types of up-flows in which the first one was the collimated up-low along the open magnetic fields, and the second was the formation of a plasma cloud from the expelled bright point small-scale loops. Chandrashekhar *et al.* (2014a) studied the dynamics of two jets seen in a polar coronal hole with a combination of EIS and XRT/*Hinode* data. They found no evidence of helical motions in these events, but detected a significant shift of the jet position in a direction normal to the jet axis, with a drift velocity of about 27 and 7 km s^{-1}, respectively.

The launch of the *Solar Dynamics Observatory* (*SDO*) with the Atmospheric Imaging Assembly (AIA) opens a new page in observing the solar jets. Moschou *et al.* (2013) have reported high cadence observations of solar coronal jets observed in the Extreme Ultraviolet (EUV) 304 Å using Atmospheric Imaging Assembly (AIA) instrument on board *SDO*. They registered, in fact coronal hole jets, with speeds of 94 to 760 km s^{-1}, and lifetimes of the order of several tens of minutes. A detailed description of the dynamical behavior of a jet in an on-disk coronal hole observed with AIA/*SDO* was presented by Chandrashekhar *et al.* (2014b). Their study reveals new evidence of plasma flows prior to the jet's initiation along the small-scale loops at the base of the jet. The authors have also found further evidence that flows along the jet consist of multiple, quasi-periodic small-scale plasma ejections. In addition, spectroscopic analysis estimates temperature as Log 5.89 ± 0.08 K and electron densities as Log 8.75 ± 0.05 cm^{-3} in the observed jet. Measured properties of the registered transverse wave have provided evidence that a strong damping of the wave occurred as it propagates along the jet with speeds of ~ 110 km s^{-1}. Using the magneto-seismological inversion, observed plasma and wave parameters, the jet's magnetic field is estimated as $B = 1.21 \pm 0.2$ G. Sterling *et al.* (2015) have reported high-resolution X-ray and extreme ultraviolet observations of 20 randomly selected X-ray jets that form in coronal holes at the solar polar caps. In each jet, converse to the widely accepted emerging magnetic flux model, a miniature version of the filament eruptions that initiated coronal mass ejections drove the jet producing reconnection process. Formation of a rotating jet during the filament eruption on 2013 April 10–11 on the base of multi-wavelength and multi-viewpoint observations with *STEREO*/SECCHI/EUVI and *SDO*/AIA was reported by Filippov *et al.* (2015). The confined eruption of the filament within a null-point topology, which is also known as an Eiffel tower magnetic field configuration, forms a twisted jet after magnetic reconnection near the null point. The sign of the helicity in the jet is observed same as that of the sign of the helicity in the filament. It is noteworthy that the untwisting motion of the reconnected magnetic field lines gives rise to the accelerating plasma along the jet.

It is well established that the magnetic reconnection at different heights in the solar atmosphere plays a key role in triggering jet-like events. The magnetic reconnection between open and closed fields (standard reconnection scenario) is one of the well-known processes of the jet's occurrence (Shimojo *et al.*, 2001; Kim *et al.*, 2007). The jets emerging by this kind of mechanism are known as *standard jets* (Shibata *et al.*, 1992, 2007; Kamio *et al.*, 2007; Moore *et al.*, 2010). Some other observational and simulation studies showed that the reconnection at the magnetic null in a fan-spine magnetic topology can also trigger jet-like events (Pariat *et al.*, 2009; Filippov *et al.*, 2009; Pariat *et al.*, 2010; Jiang *et al.*, 2015). Eruptions of small arches, filaments, and flux ropes from these type of magnetic field configurations can be responsible for reconnection and jets' occurring. These type of jets are known as *blowout jets* (Moore *et al.*, 2010; Sterling *et al.*, 2010; Pucci *et al.*, 2013). Moore *et al.* (2013) used the full-disk He ii 304 Å movies from the Atmospheric Imaging Assembly on *SDO* to study the cool ($T \sim 10^5$ K) component of X-ray jets observed in polar coronal holes by XRT. The AIA 304 Å movies revealed that most polar X-ray jets spin as they erupt. The authors examined 54 X-ray jets that were found in polar coronal holes in XRT movies sporadically taken during the first year of continuous operation of AIA (2010 May through

2011 April). These 54 jets were big and bright enough in the XRT images to be categorized as a standard jet or as a blowout jet. From the X-ray movies, 19 of the 54 jets appeared like standard jets, 32 appeared as blowout jets, and three were ambiguous and were not falling in any category. Moore *et al.* (2015) have studied 14 large-scale solar coronal jets observed in Sun's pole. In EUV movies from the *SDO*/AIA, each jet was very similar to most X-ray and EUV jets erupting in coronal holes. However, each was exceptional in that it went higher than most of the standard coronal jets. They were detected in the outer corona beyond 2.2 R_{Sun} in images as observed from the *Solar and Heliospheric Observatory/Large Angle Spectroscopic Coronagraph* (*LASCO*/C2 coronagraph, Brueckner *et al.*, 1995).

Schmieder *et al.* (2013) proposed a new model between the standard and blowout models, where magnetic reconnection occurs in the bald patches around some twisted field lines, one of whose foot points is open. Pariat *et al.* (2015) included the magnetic field inclination and photospheric field distribution and performed another 3D numerical MHD model for the two different types of jet events: standard and blowout jets. We note also that stereoscopic studies (multiple points of view) have been carried out by the EUVI/SECCHI imagers on board the twin *STEREO* spacecraft to estimate the expected speed, motion, and morphology of polar coronal jets (Patsourakos *et al.*, 2008). To sum up, the flux emergence (Gontikakis *et al.*, 2009; Chandra *et al.*, 2015) and flux cancellation (Chae *et al.*, 1998; Bellot Rubio and Beck, 2005) are the two main triggering processes that are known to be responsible for jets' occurrence. In few observational and primarily numerical studies, the wave-induced reconnection has also been suggested as a cause for the onset of jet-like events (Heggland *et al.*, 2009; Innes *et al.*, 2011; Srivastava and Murawski, 2011; Chandra *et al.*, 2015).

The first modeling of KH instability in X-ray jets was carried out by Vasheghani Farahani *et al.* (2009) who explored transverse wave propagation along the coronal hole soft X-ray jets detected by Cirtain *et al.* (2007). Vasheghani Farahani and colleagues analyzed analytically, in the limit of thin magnetic flux tube, the dispersion relation of the kink MHD mode and have obtained that this mode is unstable against the KH instability when the critical jet velocity is equal to $4.47v_A = 3576\,\mathrm{km\,s^{-1}}$ ($v_A = 800\,\mathrm{km\,s^{-1}}$ is the Alfvén speed inside the jet). A numerical solving of the same dispersion relation when considering the jet and its environment as cold magnetized plasmas, carried out by Zhelyazkov (2013), yielded a little bit lower critical flow speed for the instability onset, namely $4.31v_A = 3448\,\mathrm{km\,s^{-1}}$. The lowest critical jet speed of $4.025v_A = 3220\,\mathrm{km\,s^{-1}}$ was derived by numerically solving the wave dispersion relation without any approximations, that is, treating both media as compressible plasmas. But even the latter critical jet speed is still too high the KH instability to be detected/observed in coronal hole soft X-ray jets. The reason for obtaining such high critical speeds is the circumstance that

Vasheghani Farahani *et al.* (2009) assumed an electron number density of the order of $10^8\,\mathrm{cm}^{-3}$ and magnetic field strength of $10\,\mathrm{G}$.

In this chapter, we study the propagation of kink and higher MHD modes in standard active region soft X-ray jets, notably jets #'s 8, 11, and 16 of Shimojo and Shibata's set of sixteen observed flares and jets (Shimojo and Shibata, 2000), and have shown that with one order higher electron densities, $\sim\!10^9\,\mathrm{cm}^{-3}$, and moderate magnetic field, $\sim\!7\,\mathrm{G}$, MHD modes in high-speed jets, like these ones, can become unstable against the KH instability at accessible jets speeds, except for jet #16 which requires a higher flow velocity. In the following section, we list the basic physical parameters of jets, the topology of magnetic fields inside and outside the moving cylindrical flux tube modeling each jet, and also give the wave dispersion equations for both untwisted and twisted tubes.

9.2 Magnetic field topology, physical parameters, and MHD wave dispersion relations

As in the previous chapters, we consider the soft X-ray jet as a straight cylinder of radius a and density ρ_i embedded in a uniform field environment with density ρ_e. We study the propagation of MHD waves in two magnetic configurations, notably in untwisted and twisted flux tubes. For an untwisted tube, magnetic fields in both media are homogeneous and directed along the the z-axis of our cylindrical coordinate system (r, ϕ, z): $\boldsymbol{B}_\mathrm{i} = (0, 0, B_\mathrm{i})$ and $\boldsymbol{B}_\mathrm{e} = (0, 0, B_\mathrm{e})$, respectively (see Fig. 9.1).

The magnetic field inside the twisted tube is helicoidal, $\boldsymbol{B}_\mathrm{i} = (0, B_{\mathrm{i}\phi}(r), B_{\mathrm{i}z}(r))$, while outside the tube the magnetic field is uniform and directed along the

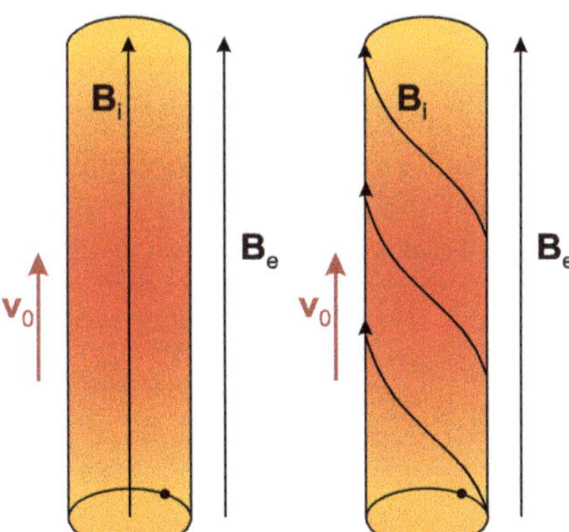

Fig. 9.1. Equilibrium magnetic fields of a soft X-ray solar jet in an untwisted flux tube (*Left panel*) and in a weakly twisted flux tube (*Right panel*). *Reprinted with permission.*

tube axis, $\boldsymbol{B}_e = (0, 0, B_e)$. Note that we assume a magnetic fields' equilibrium with uniform twist for which the magnetic field inside the tube is $\boldsymbol{B}_i = (0, Ar, B_{iz})$, where A and B_{iz} are constant. The parameter that characterizes the uniform magnetic field twist is the ratio $B_{i\phi}(a)/B_{iz} \equiv \varepsilon$, i.e., $\varepsilon = Aa/B_{iz}$. Our frame of reference for studying the wave propagation in the jet is attached to the surrounding magnetoplasma—thus $\boldsymbol{v}_0 = (0, 0, v_0)$ represents the relative jet velocity, if there is any flow in the environment. The jump of the tangential velocity at the tube boundary then initiates the magnetic KH instability onset when the jump exceeds a critical value.

Before dealing with MHD wave equations, it is necessary to specify what kind of plasma each medium is (the moving tube and its environment). As seen from the Events List in Shimojo and Shibata (2000), the physical parameters of the selected three jets are presented in Table 9.1.

It is seen from Table 9.1 that the interiors of the three jets can be considered as incompressible plasmas while their environments are cool media. Hence, for our study of MHD mode propagation at the two configurations (see Fig. 9.1), we need three different wave dispersion relations: one for compressible plasmas (untwisted flux tube), second for incompressible untwisted flux tube and cool environment, and third for the same two media but at twisted internal magnetic field. The first and second wave dispersion relations can be presented by one equation, for instance, Eq. (4.30) from Chapter 4

$$\frac{\rho_e}{\rho_i}(\omega^2 - k_z^2 v_{Ae}^2)\kappa_i \frac{I_m'(\kappa_i a)}{I_m(\kappa_i a)} - \left[(\omega - \boldsymbol{k} \cdot \boldsymbol{v}_0)^2 - k_z^2 v_{Ai}^2\right]\kappa_e \frac{K_m'(\kappa_e a)}{K_m(\kappa_e a)} = 0, \quad (9.1)$$

in which the squared wave attenuation coefficients $\kappa_{i,e}$ are defined by the expression

$$\kappa^2 = -\frac{[(\omega - \boldsymbol{k} \cdot \boldsymbol{v}_0)^2 - k_z^2 c_s^2][(\omega - \boldsymbol{k} \cdot \boldsymbol{v}_0)^2 - k_z^2 v_A^2]}{(c_s^2 + v_A^2)[(\omega - \boldsymbol{k} \cdot \boldsymbol{v}_0)^2 - k_z^2 c_T^2]}, \quad (9.2)$$

where the tube speed (different inside and outside the jet) is $c_T = c_s v_A/\sqrt{c_s^2 + v_A^2}$. In the case when the tube interior is incompressible medium ($c_{si} \to \infty$) and the

Table 9.1. Jets physical parameters at background magnetic field $B_e \cong 7\,\mathrm{G}$.

Jet number	Jet speed (km s^{-1})	Temperature (MK)		Electron density ($\times 10^9$ cm^{-3})		Plasma beta	
		T_i	T_e	n_i	n_e	β_i	β_e
8	385	5.2	2.0	4.10	3.95	31.5	0.560
11	437	5.5	2.0	2.90	2.60	7.37	0.403
16	532	7.4	2.0	1.00	0.95	1.00	0.147

surrounding plasma is a cool medium ($c_{se} \to 0$ and subsequently $c_{Te} \to 0$), the wave attenuation coefficients in the two media are: $\kappa_i = k_z$ and $\kappa_e = k_z\left[1 - (\omega/k_z v_{Ae})^2\right]^{1/2}$.

For the case of twisted internal magnetic field, we will use a modification of Eq. (6.13) from Chapter 6, namely

$$\frac{\left[(\omega - \boldsymbol{k} \cdot \boldsymbol{v}_0)^2 - \omega_{Ai}^2\right] F_m(\kappa_i a) - 2m A \omega_{Ai}/\sqrt{\mu \rho_i}}{\left[(\omega - \boldsymbol{k} \cdot \boldsymbol{v}_0)^2 - \omega_{Ai}^2\right]^2 - 4\omega_{Ai}^2/\mu \rho_i}$$
$$= \frac{P_m(\kappa_e a)}{\frac{\rho_e}{\rho_i}(\omega^2 - \omega_{Ae}^2) + A^2 P_m(\kappa_e a)/\mu \rho_i}, \tag{9.3}$$

where

$$F_m(\kappa_i a) = \frac{\kappa_i a I'_m(\kappa_i a)}{I_m(\kappa_i a)} \quad \text{and} \quad P_m(\kappa_e a) = \frac{\kappa_e a K'_m(\kappa_e a)}{K_m(\kappa_e a)}.$$

The wave attenuation coefficient inside the tube is given by

$$\kappa_i = k_z \left(1 - \frac{4 A^2 \omega_{Ai}^2}{\mu \rho_i \left[(\omega - \boldsymbol{k} \cdot \boldsymbol{v}_0)^2 - \omega_{Ai}^2\right]^2}\right)^{1/2}, \tag{9.4}$$

where

$$\omega_{Ai} = \frac{1}{\sqrt{\mu \rho_i}}(m A + k_z B_{iz}),$$

while in the environment it has the form

$$\kappa_e a = k_z a\left[1 - (\omega/\omega_{Ae})^2\right]^{1/2}, \tag{9.5}$$

in which $\omega_{Ae} = k_z B_e/\sqrt{\mu \rho_e}$.

For each jet firstly we shall study the dispersion characteristics of the kink ($m = 1$) mode in untwisted moving magnetic flux tube (in two approaches, notably (i) considering the jet and its surrounding plasma as compressible media, and (ii) treating the jet as an incompressible medium while its environment is assumed to be cool plasma), and later on, shall explore the same thing for the kink ($m = 1$) and higher ($m > 1$) MHD modes propagating in an incompressible twisted moving flux tube, surrounded by cool plasma.

9.2.1 *Kelvin–Helmholtz instability of MHD modes in untwisted flux tubes*

We begin with jet #11. From the physical parameters listed in Table 9.1, using the total pressure balance equation, at equilibrium magnetic field $B_e = 6.7\,\mathrm{G}$, one can obtain the sound and Alfvén speeds in both media, namely $c_{si} = 275\,\mathrm{km\,s^{-1}}$, $v_{Ai} = 111\,\mathrm{km\,s^{-1}}$, and $c_{se} = 166\,\mathrm{km\,s^{-1}}$, $v_{Ae} = 286\,\mathrm{km\,s^{-1}}$, respectively. The density contrast is $\eta = 0.896$ and the ratio of the axial external and internal magnetic fields is correspondingly $B_e/B_i \equiv b = 2.44$. (For the twisted tube, the second parameter has the form $b_{twist} = B_e/B_{iz}$.)

Among the various MHD wave spectra that exist in a static magnetic field flux tube of compressible plasma, surrounded by compressible medium, the most interesting for us is the kink-speed wave whose speed for jet #11 is equal to

$$c_k = \sqrt{(1 + b^2)/(1 + \eta)}\, v_{Ai} = 212.7\,\mathrm{km\,s^{-1}},$$

or in dimensionless form $c_k/v_{Ai} = 1.9166$.

As we know, the normal modes propagating in a static homogeneously magnetized flux tube can be pure surface waves, pseudosurface (body) waves, or leaky waves (see Cally, 1986). The type of the wave crucially depends on the ordering of the basic speeds in both media (the flux tube and its surrounding plasma), more specifically of sound and Alfvén speeds as well as corresponding tube speeds. In our case the ordering is

$$v_{Ai} < c_{se} < c_{si} < v_{Ae},$$

which, with $c_{Ti} \cong 103\,\mathrm{km\,s^{-1}}$ and $c_{Ti} \cong 143.6\,\mathrm{km\,s^{-1}}$ after normalizing them with respect to the external sound speed, c_{se}, yields (in Cally's notation)

$$A = 0.6687, \quad C = 1.6566, \quad A_e = 1.7229,$$

$$C_T = 0.62, \quad \text{and} \quad C_{Te} = 0.8649.$$

Since $A < C$ and $A < C_{Te}$ along with $A < C_{Te} < C$, according to Cally's classification, the kink mode in a rest flux tube must be pure surface mode type S_+^- (for detail, see Cally, 1986). For an S_+^--type wave V should lies between A and C: $A \leqslant V \leqslant C$, and also $V < C_{Te}$. In our normalization, the normalized wave velocity should be bracketed between 1 and $\sqrt{\tilde{\beta}_i} = 2.478$, i.e., $1 \leqslant \omega/(k_z v_{Ai}) \leqslant 2.478$. The typical wave dimensionless wave velocity $\omega/(k_z v_{Ai})$ is the normalized kink speed $c_k/v_{Ai} = 1.9166$, which, as expected, lies between 1 and 2.478. Concerning the relation between normalized wave phase velocity and external tube speed, in our case we have the opposite inequality, that is, $V > C_{Te}$, that in Cally's normalization reads as $1.2805 > 0.8649$, and in ours as $1.9166 > 1.2937$. The reason for that unexpected change in the mutual relations between two velocities is the relatively bigger value of parameter b ($=2.44$) that yields a larger magnitude of c_k, than, for example, in the case when b is close to 1—then the inequality $V < C_{Te}$ is satisfied (see, e.g., Zhelyazkov *et al.*, 2016).

A reasonable question is how the flow will change the dispersion characteristics of the kink ($m = 1$) mode. Calculations show that the flow shifts upwards the kink-speed dispersion curve and splits it into two separate curves (Zhelyazkov, 2012). (A similar duplication is observed for the tube-speed v_{Ti} dispersion curves, too.) The evolution of the pair of kink-speed dispersion curves can be seen in the left panel of Fig. 9.2—the input parameters for solving the dispersion equation (9.1) of the kink mode ($m = 1$) traveling in a moving flux tube of compressible plasma surrounded by compressible coronal medium are: $\eta = 0.896$, $b = 2.44$, $\tilde{\beta}_{\mathrm{i}} = 6.141$, and $\tilde{\beta}_{\mathrm{e}} = 0.336$. During calculations the Alfvén Mach number, M_{A}, was varied from zero (static plasma) to values at which we obtained unstable solutions. Let us first note that at $M_{\mathrm{A}} = 0$ we get the kink-speed dispersion curve which at a very small dimensionless wavenumber $k_z a$ ($=0.005$) yields the value of the normalized wave phase velocity equal to 1.9168—very close to the previously estimated magnitude of 1.9166. This observation implies that our code for solving the wave dispersion relation is correct. For small Alfvén Mach numbers, the pair of kink-speed modes travel with velocities $M_{\mathrm{A}} \mp c_{\mathrm{k}}/v_{\mathrm{Ai}}$ (Zhelyazkov, 2012) (in that case, their dispersion curves go practically parallel). At higher M_{A}, however, the behavior of each curve of the pair $M_{\mathrm{A}} \mp c_{\mathrm{k}}/v_{\mathrm{Ai}}$ turns out to be completely different. As seen from the left panel of Fig. 9.2, for $M_{\mathrm{A}} \geqslant 3.785$ both kink curves break and merge forming a family of semi-closed dispersion curves. A further increasing in M_{A} leads to a separation of those curves in opposite directions. This behavior of the kink ($m = 1$) mode dispersion curves signals us that we are in a range of the $\mathrm{Re}(v_{\mathrm{ph}}/v_{\mathrm{Ai}})$–$k_z a$-plane where one

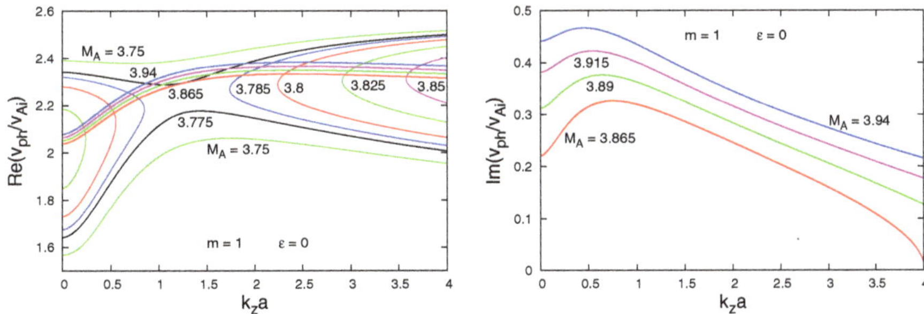

Fig. 9.2. (*Left panel*) Dispersion curves of stable and unstable kink ($m = 1$) MHD mode propagating in a moving untwisted magnetic flux tube of compressible plasma (modeling jet &11) at $\eta = 0.896$ and $b = 2.44$. Unstable dispersion curves located *above the middle of the plot* have been calculated for four values of the Alfvén Mach number $M_{\mathrm{A}} = 3.865$, 3.89, 3.915, and 3.94. (*Right panel*) The normalized growth rates of the unstable mode for the same values of M_{A}. Red curves in both plots correspond to the onset of KH instability. *Reprinted with permission.*

can expect the occurrence of KH instability. A prediction of the Alfvén Mach number M_A at which the instability will occur can be found from the inequality (Zaqarashvili *et al.*, 2014a)

$$|m|M_A^2 > (1 + 1/\eta)(|m|b^2 + 1), \tag{9.6}$$

which gives for the kink ($m = 1$) mode $M_A > 3.839$. In our case, the KH instability start at that M_A, for which the left-hand side semi-closed curve disappears—this happens at $M_A^{cr} = 3.865$. The growth rates of unstable kink modes are plotted in the right panel of Fig. 9.2. The red curves in all diagrams denote the marginal dispersion/growth rate curves: for values of the Alfvén Mach number smaller than M_A^{cr} the kink mode is stable, otherwise it becomes unstable and the instability is of the KH-type. With $M_A^{cr} = 3.865$ the kink mode will be unstable when the velocity of the moving flux tube is higher than $429 \, \mathrm{km \, s^{-1}}$—a speed, which is below than the observationally measured jet speed of $437 \, \mathrm{km \, s^{-1}}$. We would like to underline that all the stable kink modes in the untwisted ($\varepsilon = 0$) moving tube modeling jet #11 are pure surface modes while the unstable ones are not—the latter become partly surface and partly leaky modes (their external attenuation coefficients, m_{0e}s, are complex quantities with positive imaginary parts). This circumstance means that wave energy is radiated outward in the surrounding medium, which allows us to claim the KH instability plays a dual role: once in its nonlinear stage the instability can trigger wave turbulence and simultaneously the propagating KH-mode is radiating its energy outside.

When numerically solving Eq. (9.1) for the kink ($m = 1$) mode treating the jet's plasma as incompressible medium and its environment as cool plasma, we obtain dispersion curves' and growth rates' patterns very similar to those shown in Fig. 9.2 (see Fig. 9.3). Computations show, as expected, that the stable kink mode is a non-leaky surface mode, while the unstable one possesses a real internal attenuation coefficient, $m_{0i} = k_z$, not changed by the instability, but the external one, m_{0e}, becomes a complex quantity with positive real and negative imaginary parts. Now the threshold Alfvén Mach number is a little bit bigger and yields a critical jet speed for the instability onset of $\cong 442 \, \mathrm{km \, s^{-1}}$, being with $5 \, \mathrm{km \, s^{-1}}$ higher than the observationally measured $437 \, \mathrm{km \, s^{-1}}$. We note that, as a rule, jet's incompressible plasma approximation yields slightly higher threshold Alfvén Mach numbers than the model of compressible media. This ascertainment shows how sensitive to jet's and its environment's media treatment is the occurrence of KH instability of the kink mode in our jet—one can expect instability onset at an accessible jet speed only if the both media are considered as compressible magnetized plasmas (Zhelyazkov *et al.*, 2017).

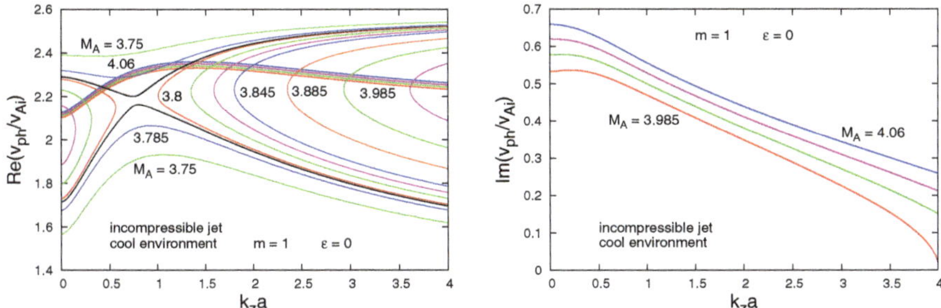

Fig. 9.3. (*Left panel*) Dispersion curves of stable and unstable kink ($m = 1$) MHD mode propagating in a moving untwisted magnetic flux tube of incompressible plasma surrounded by cool medium at the same input parameters as in Fig. 9.2. The dispersion curves of unstable modes located *above the middle of the plot* have been calculated for $M_A = 3.985, 4.01, 4.035$, and 4.06. Alfvén Mach numbers associated with the non-labeled black, green, and purple curves are equal to $3.796, 3.815$, and 3.83, respectively, while those of far right green and purple semi-closed curves have magnitudes of 3.925 and 3.965. (*Right panel*) Similar plots as in Fig. 9.2. *Reprinted with permission.*

The basic physical parameters of jet #8, namely $n_i = 4.1 \times 10^9\,\text{cm}^{-3}$ and $T_i = 5.2\,\text{MK}$ (see Table 9.1), for a low density contrast of $\eta = 0.963$ assuming as before that the temperature of surrounding plasma is $T_e = 2.0\,\text{MK}$ and the background magnetic field, B_e, equal to $7\,\text{G}$, yield the following sound and Alfvén speeds: $c_{se} = 166\,\text{km s}^{-1}$, $c_{si} = 267.5\,\text{km s}^{-1}$, $v_{Ae} = 242.8\,\text{km s}^{-1}$, and $v_{Ai} = 52.22\,\text{km s}^{-1}$. The speed ordering, $v_{Ai} < c_{se} < v_{Ae} < c_{si}$, according to Cally's (Cally, 1986) classification, tells us that in a rest magnetic flux tube the propagating wave must be a pure surface mode of type S_+^-. The numerical computations confirm this, as well as reproduce the normalized kink speed of 3.3342 within three places after the decimal point. With the flow inclusion, the behavior of the pair kink-speed dispersion curves in the region of the instability onset, as seen from the left panel of Fig. 9.4, is completely different; more specifically, the lower semi-closed kink-speed dispersion curves at $M_A = 6.625$ and 6.65 (the orange curve) correspond to pseudosurface (body) waves, while the higher kink-speed dispersion curves are associated with surface waves. Note that at $M_A = 6.668$ the surface wave dispersion curves brakes in two parts: the right-hand one merges with the lower kink-speed curve at $k_z a = 0.2337$ (see the blue line) while its left-hand side forms a narrow semi-closed dispersion curve of pure surface mode. The threshold Alfvén Mach number is equal to 6.6704 (the prediction one according to (9.6) is 6.67) which implies that the critical flow speed for KH instability onset is equal to $348.3\,\text{km s}^{-1}$—a value far below the jet speed of $385\,\text{km s}^{-1}$. This circumstance guarantees that even if one considers the jet medium as incompressible plasma and

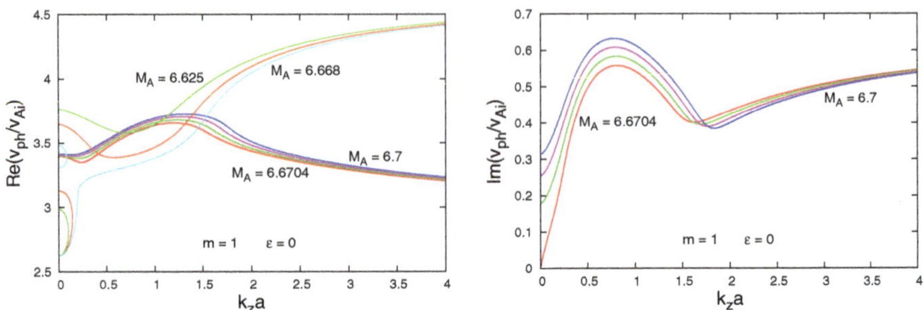

Fig. 9.4. (*Left panel*) Dispersion curves of stable and unstable kink ($m = 1$) MHD mode propagating in a moving untwisted magnetic flux tube of compressible plasma (modeling jet #8) at $\eta = 0.963$ and $b = 4.56$. Unstable mode's dispersion curves are pictured *in the middle of the plot* for $M_A = 6.6704$, 6.68, 6.69, and 6.7. (*Right panel*) The corresponding normalized growth rates of unstable modes for the same values of M_A. *Reprinted with permission.*

its environment as cool plasma we will get an accessible jet's speed for instability onset.

In a similar way, calculating the sound and Alfvén speeds for jet #16 and its environment (with $n_i = 1.0 \times 10^9 \, \text{cm}^{-3}$, $T_i = 7.4 \, \text{MK}$, $\eta = 0.95$, $T_e = 2.0 \, \text{MK}$, and $B_e = 6.7 \, \text{G}$) we get: $c_{se} = 166 \, \text{km s}^{-1}$, $c_{si} \cong 319 \, \text{km s}^{-1}$, $v_{Ae} \cong 474 \, \text{km s}^{-1}$, $v_{Ai} = 350 \, \text{km s}^{-1}$, and plasma betas $\beta_i \cong 1.0$ and $\beta_e = 0.14$. The magnetic fields ratio, b, is equal to 1.32 giving $B_i \cong 5.1 \, \text{G}$. In this case, the speed ordering is $c_{se} < c_{si} < v_{Ai} < v_{Ae}$ and the kink ($m = 1$) mode propagating in a rest magnetic flux tube is a pseudosurface (body) wave of type B_+^- (see Table I in Cally, 1986). The numerical calculations confirm this mode type and yield a normalized value of the kink speed very close to the predicted one of 1.1858. Now the pattern of stable kink-speed dispersion curves is more complicated (see the left panel of Fig. 9.5)—while the regular lower kink-speed curves correspond to pseudosurface (body) waves, the higher kink-speed dispersion curves have the form of narrow and long semi-closed up to $M_A = 2.3$ loops and are in fact bulk waves: both attenuation coefficients are purely imaginary numbers. As expected, the dispersion curves of unstable kink mode possess complex attenuation coefficients (with positive real and imaginary parts for m_{0i} and positive real and negative imaginary part for m_{0e}). The threshold Alfvén Mach number being equal to 2.359305 determines a critical flow velocity of $825.8 \, \text{km s}^{-1}$ for KH instability onset which is much higher than the jet #16 speed of $532 \, \text{km s}^{-1}$. This consideration shows that even a relatively high-speed observed soft X-ray jet cannot become unstable—the reason for that in this case is the low density of the jet's plasma.

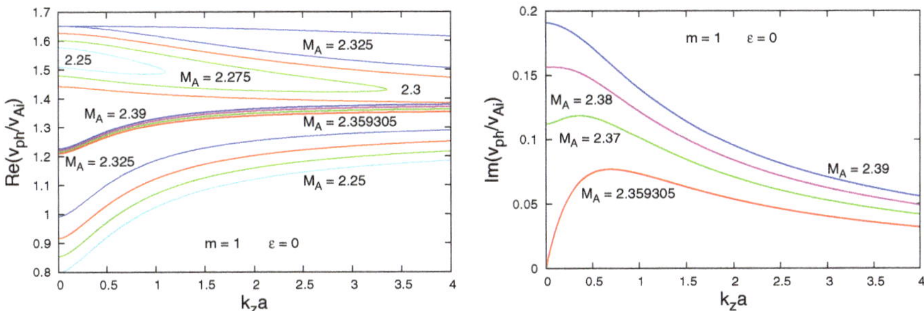

Fig. 9.5. (*Left panel*) Dispersion curves of stable and unstable kink ($m = 1$) MHD mode propagating in a moving untwisted magnetic flux tube of compressible plasma (modeling jet #16) at $\eta = 0.95$ and $b = 1.32$. Dispersion curves of unstable modes located *in the middle of the plot* have been calculated for $M_A = 2.359305$, 2.37, 2.38, and 2.39. (*Right panel*) Plots of corresponding normalized growth rates. *Reprinted with permission.*

9.2.2 *Kelvin–Helmholtz instability of MHD modes in twisted flux tubes*

When numerically solving wave dispersion equation (9.3) we need the normalization of the local Alfvén frequencies $\omega_{\mathrm{Ai,e}}$. The normalization of these frequencies is performed by multiplying each of them by the tube radius, a, and divide by the Alfvén speed $v_{\mathrm{Ai}} = B_{\mathrm{iz}}/\sqrt{\mu\rho_i}$ to get

$$\frac{a\omega_{\mathrm{Ai}}}{v_{\mathrm{Ai}}} = m\frac{Aa}{B_{\mathrm{iz}}} + k_z a = m\varepsilon + k_z a \quad \text{and} \quad \frac{a\omega_{\mathrm{Ae}}}{v_{\mathrm{Ai}}} = k_z a \frac{B_{\mathrm{e}}/B_{\mathrm{iz}}}{\sqrt{\eta}} = k_z a \frac{b_{\mathrm{twist}}}{\sqrt{\eta}},$$

where $b_{\mathrm{twist}} = B_{\mathrm{e}}/B_{\mathrm{iz}} = b\sqrt{1 + \varepsilon^2}$. For small values of ε, we will take $b_{\mathrm{twist}} \cong b$.

We begin our calculations for the kink ($m = 1$) mode in the moving twisted flux tube (modeling jet #11) with $\eta = 0.896$, $b_{\mathrm{twist}} \cong b = 2.44$, and $\varepsilon = 0.025$. The results of numerical task are presented in Fig. 9.6, from which one sees that the kink mode dispersion and growth rate curves are very similar to those shown in Fig. 9.3. The critical flow velocity for emerging KH instability now is $v_0^{\mathrm{cr}} = 442.6\,\mathrm{km\,s}^{-1}$, calculated by using the "reduced" Alfvén speed $v_{\mathrm{Ai}}/\sqrt{1 + \varepsilon^2} = 110.96\,\mathrm{km\,s}^{-1}$. KH instability of the ($m = -1$) mode will occur if the jet speed exceeds $442\,\mathrm{km\,s}^{-1}$, which value is beyond the observationally derived jet speed of $437\,\mathrm{km\,s}^{-1}$. We note that while the wave attenuation coefficient of the $m = 1$ mode propagating in untwisted moving magnetic flux tube of incompressible plasma surrounded by a cool medium is not changed by the flow, in the twisted flux tube (in the same approximations) that attenuation coefficient becomes a complex number with positive imaginary part, like the attenuation coefficient in the cool environment.

According to the instability criterion (9.6), the occurrence of KH instability of higher MHD modes would require lower threshold Alfvén Mach numbers (and

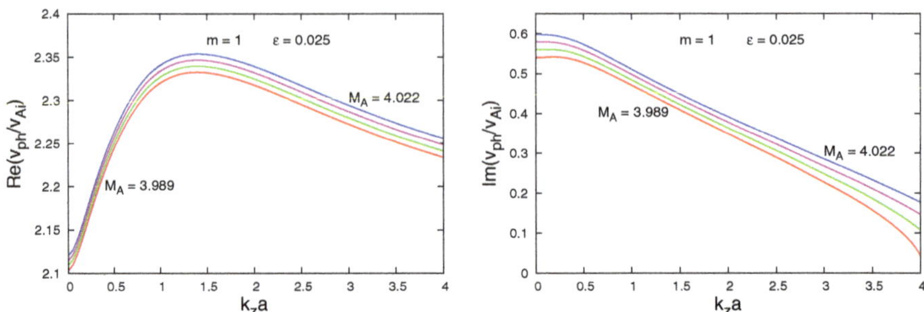

Fig. 9.6. (*Left panel*) Dispersion curves of unstable kink ($m = 1$) MHD mode propagating in a moving twisted flux tube of incompressible plasma (modeling jet #11) at $\eta = 0.896$ and $\varepsilon = 0.025$, and for $M_A = 3.989$, 4.0 (green curve), 4.011 (purple curve), and 4.022 (blue curve). (*Right panel*) The normalized growth rates for the same values of M_A. *Reprinted with permission.*

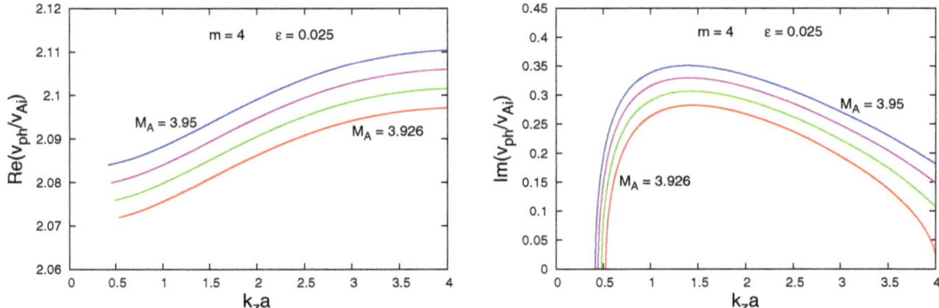

Fig. 9.7. (*Left panel*) Dispersion curves of unstable ($m = 4$) MHD mode propagating in a moving twisted flux tube of incompressible plasma at the same parameters as in Fig. 9.6 for $M_A = 3.926$, 3.934 (green curve), 3.942 (purple curve), and 3.95. (*Right panel*) The normalized growth rates for the same values of M_A. *Reprinted with permission.*

accordingly lower critical jet speeds) than those of the kink ($m = \pm 1$) mode. Our computations show, however, that with $\varepsilon = 0.025$, the threshold Alfvén Mach numbers are generally still high to initiate the KH instability. Their magnitudes for the $m = 2$, 3, and 4 modes are 3.965, 3.943, and 3.926, respectively. The corresponding critical velocities for instability onset are accordingly equal to 440.0, 437.5, and 435.6 km s^{-1}. As seen, only the $m = 4$ MHD mode can become unstable against KH instability. A clutch of unstable dispersion curves and normalized growth rates are shown in Fig. 9.7. Here we observe a new phenomenon: the KH instability starts at some critical $k_z a$-number along with the corresponding threshold Alfvén Mach number. That critical dimensionless wavenumber for the $m = 4$ MHD mode (look at Fig. 9.7) is equal to 0.528. If we assume that jet #11 has a width $\Delta \ell = 5 \times 10^3$ km, then the critical wavelength for a KH instability emergence is $\lambda_{\mathrm{cr}}^{m=4} \cong 29.8$ Mm.

A substantial decrease in the threshold Alfvén Mach number in moving twisted magnetic flux tubes can be obtained by increasing the magnetic field twist parameter, say taking, for instance, $\varepsilon = 0.4$. The dispersion curves and normalized wave growth rates of the kink ($m = \pm 1$) mode are plotted in Figs. 9.8 and 9.9. One can immediately see the rather complicated forms both of the dispersion curves and dimensionless growth rates of the $m = -1$ kink mode—such a complication was absent at $\varepsilon = 0.025$. We note that computations were performed at $b_{\mathrm{twist}} = b/\sqrt{1 + \varepsilon^2}$, and the reference "reduced" Alfvén speed inside the jet now is $v_{\mathrm{Ai}}/\sqrt{1 + \varepsilon^2} = 103$ km s^{-1}. With this Alfvén speed, the critical velocities

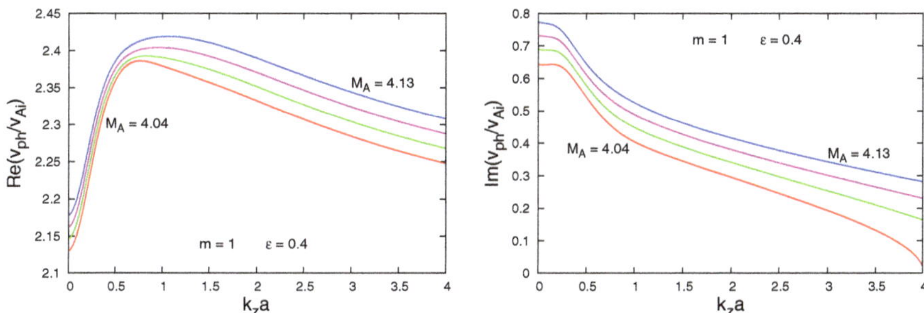

Fig. 9.8. (*Left panel*) Dispersion curves of unstable ($m = 1$) MHD mode propagating in a moving twisted flux tube of incompressible plasma (modeling jet #11) at $\eta = 0.896$ and $\varepsilon = 0.4$, and for $M_A = 4.04$, 4.07 (green curve), 4.1 (purple curve), and 4.13. (*Right panel*) The normalized growth rates for the same values of M_A. *Reprinted with permission.*

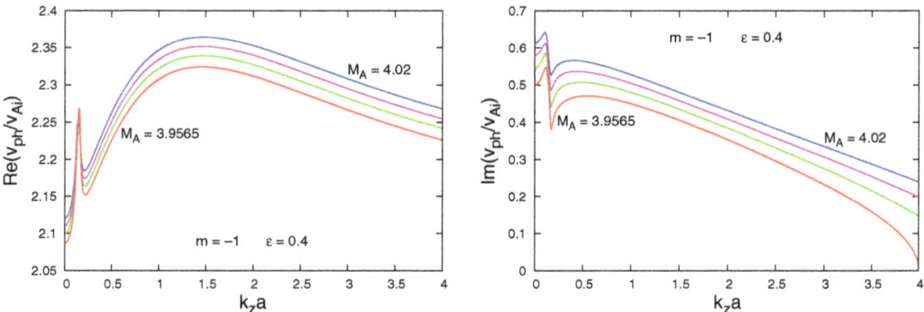

Fig. 9.9. (*Left panel*) Dispersion curves of unstable ($m = -1$) MHD mode propagating in a moving twisted flux tube of incompressible plasma (modeling jet #11) at the same parameters as in Fig. 9.8 and for $M_A = 3.9565$, 3.98 (green curve), 4.0 (purple curve), and 4.02. (*Right panel*) The normalized growth rates for the same values of M_A. *Reprinted with permission.*

for the instability occurrence are equal to $416\,\mathrm{km\,s^{-1}}$ (for the $m = 1$ mode) and $\cong 408\,\mathrm{km\,s^{-1}}$ for the $m = -1$ mode, respectively.

In solving dispersion equation (9.3) for the $m = 2$ MHD mode, we start with a little bit bigger Alfvén Mach number than that predicted by instability criterion (9.6) and have obtained three different kinds of dispersion curves and normalized growth rates (see Fig. 9.10), notably curves with distinctive change of the normalized wave phase velocity in a relatively narrow $k_z a$-interval (the orange and blue curves in the left panel of Fig. 9.10), two piece-wise dispersion curves (the purple and green curves in the same plot), and one almost linear dispersion curve (the red one which is in fact the marginal dispersion curve obtained for the threshold Alfvén Mach number equal to 4.062). No less interesting are the dimensionless growth rate curves corresponding to these three kinds of dispersion curves. For the threshold Alfvén Mach number of 4.062, we obtained one instability window and instability starts at the critical dimensionless wavenumber of 1.944, or equivalently at $\lambda_{\mathrm{cr}}^{m=2} \cong 8.0\,\mathrm{Mm}$. For the slightly superthreshold Alfvén Mach numbers of 4.075 and 4.1, we got two different instability windows (see the green and purple curves in the right panel of Fig. 9.10) and a further increase in M_A leads to merging of those two instability windows—see the blue and orange curves in the same plot. The critical flow velocity at which KH instability arises is equal to $\cong 418\,\mathrm{km\,s^{-1}}$.

We have skipped the complicated dispersion and growth rate curves pictured in Fig. 9.10 for the $m = 3$ and $m = 4$ MHD modes—we have calculated only a few curves for Alfvén Mach numbers close to the corresponding threshold ones (see Figs. 9.11 and 9.12). One can observe that for the magnetic field twist parameter $\varepsilon = 0.4$ the critical dimensionless wavenumbers are shifted far on the right—their

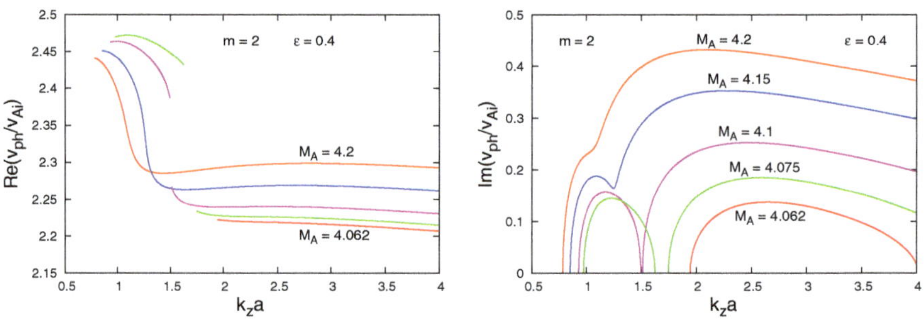

Fig. 9.10. (*Left panel*) Dispersion curves of unstable ($m = 2$) MHD mode propagating in a moving twisted flux tube of incompressible plasma (modeling jet #11) at the same parameters as in Fig. 9.8, and for $M_\mathrm{A} = 4.2, 4.15, 4.1, 4.075$, and 4.062. (*Right panel*) The normalized growth rates for the same values of M_A. *Reprinted with permission.*

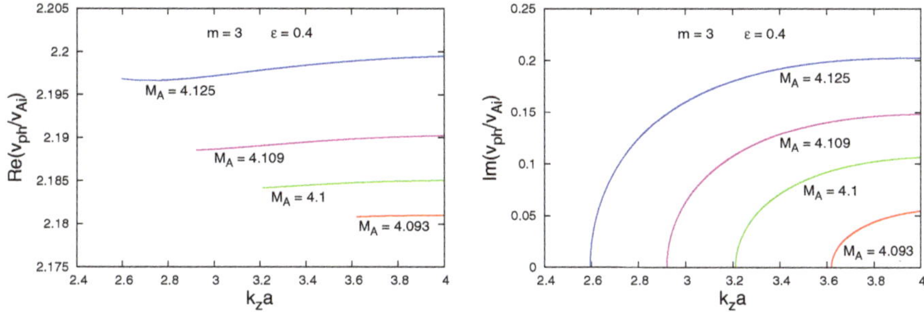

Fig. 9.11. (*Left panel*) Dispersion curves of unstable ($m = 3$) MHD mode propagating in a moving twisted flux tube of incompressible plasma (modeling jet #11) at the same parameters as in Fig. 9.8, and for $M_A = 4.093, 4.1, 4.109$, and 4.125. (*Right panel*) The normalized growth rates for the same values of M_A. *Reprinted with permission.*

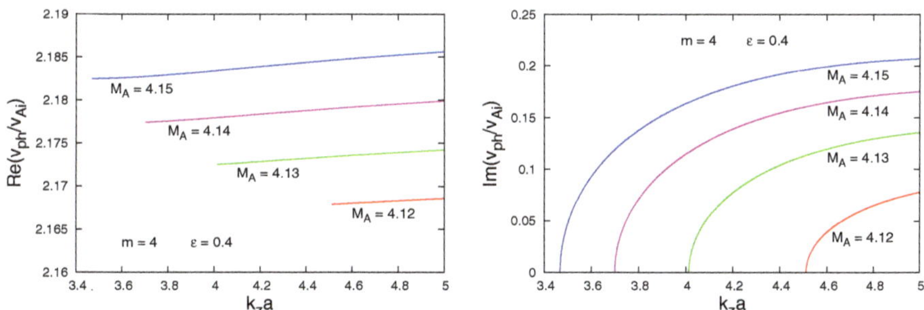

Fig. 9.12. (*Left panel*) Dispersion curves of unstable ($m = 4$) MHD mode propagating in a moving twisted flux tube of incompressible plasma (modeling jet #11) at the same parameters as in Fig. 9.8, and for $M_A = 4.12, 4.13, 4.14$, and 4.15. (*Right panel*) The normalized growth rates for the same values of M_A. *Reprinted with permission.*

values for both modes are equal to 3.619 and 4.512, respectively, that yield $\lambda_{cr}^{m=3} \cong 4.3\,\mathrm{Mm}$ and $\lambda_{cr}^{m=4} \cong 3.5\,\mathrm{Mm}$. The critical jet speeds at which KH instability emerges are equal correspondingly to $\cong 422$ and $424\,\mathrm{km\ s^{-1}}$. We note that all the threshold Alfvén Mach numbers of higher MHD modes ($m \geqslant 2$) traveling on the moving twisted magnetic flux tube are larger than the predicted ones, but while for $\varepsilon = 0.025$ they (Alfvén Mach numbers) are decreasing with increasing the mode number, at $\varepsilon = 0.4$ we have just the opposite ordering. We would like to notice that finding the marginal dispersion and growth rate curves of the higher modes at $\varepsilon = 0.4$ turns out to be a laborious computational task.

To finish our survey on KH instability of MHD modes in moving twisted magnetic flux tubes, we will briefly consider how a weak twist of the internal magnetic field, $\varepsilon = 0.025$, will change the critical jet speeds for the occurrence of KH instability in jet #8. We are not going to graphically present the dispersion curves and growth rates of unstable modes because simply there is nothing special in their shape—in the incompressible plasma jet approximation and cool environment at small magnetic field twist the curves look, more or less, similar. Recall that at density contrast of 0.963 and magnetic fields ratio $b = 4.56$, the threshold Alfvén Mach number for the kink ($m = 1$) mode is equal to 6.9047, which for the reference Alfvén speed of $52.22\,\mathrm{km\,s}^{-1}$ yields a critical flow velocity of $360.6\,\mathrm{km\,s}^{-1}$, that is lower than the jet speed of $385\,\mathrm{km\,s}^{-1}$. The KH instability of $m = 2, 3,$ and 4 MHD modes should occur at flow velocities of $\cong 259, 357,$ and $355.5\,\mathrm{km\,s}^{-1}$ and start at critical dimensionless wavenumbers $(k_z a)_{\mathrm{cr}}$ equal respectively to 0.054, 0.119, and 0.205.

9.3   Concluding remarks

Our study shows that the instability onset for a given MHD mode m is very sensitive to the main input parameters: density contrast, η, magnetic field ratio, b, the magnetic field twist ε, and especially to the background magnetic field B_{e}. If we assume that for jet #11 $B_{\mathrm{e}} = 10\,\mathrm{G}$, Alfvén speeds in the two media have new values, notably $v_{\mathrm{Ai}} = 320.38\,\mathrm{km\,s}^{-1}$ and $v_{\mathrm{Ae}} = 427.5\,\mathrm{km\,s}^{-1}$. With these new Alfvén speeds the input parameters for solving wave dispersion relations (9.1) and (9.3) have changed and more specifically $\tilde{\beta}_{\mathrm{i}} = 0.7376$, $\tilde{\beta}_{\mathrm{e}} = 0.1506$, and $b = 1.2636$. Now the critical jet velocity for the instability onset dramatically increases—it becomes $751\,\mathrm{km\,s}^{-1}$ for the kink mode in untwisted tube and $774\,\mathrm{km\,s}^{-1}$ in weakly twisted ($\varepsilon = 0.025$) flux tube.

It is intriguing to see what critical jet speeds we will obtain for a numerically modeled solar X-jet—a case in point is the study of Miyagoshi and Yokoyama (2003). These authors have presented MHD numerical simulations of solar X-ray jets based on magnetic reconnection model that includes chromospheric evaporation. Peculiar to their study is that total pressure balance equation excludes the magnetic pressure inside the jet saying that it is very weak, that is (in their notation),

$$p_{\mathrm{jet}} = p_{\mathrm{cor}} + \frac{B^2}{2\mu},$$

where $B \equiv B_{\mathrm{e}}$ is the background (coronal) magnetic field. Supposing that $n_{\mathrm{cor}} = 10^9\,\mathrm{cm}^{-3}$, $T_{\mathrm{cor}} = 10^6\,\mathrm{K}$, $B = 10\,\mathrm{G}$, and $n_{\mathrm{jet}} = 4.5 \times 10^9\,\mathrm{cm}^{-3}$, they obtain from their total balance equation the temperature of the jet, $T_{\mathrm{jet}} = 6.7 \times 10^6\,\mathrm{K}$—indeed a reasonable value. However, to satisfy the standard total pressure balance equation (assuming a very small, but finite value for the magnetic field in the jet, B_{i}), with the aforementioned values for both the temperature and jet density, we were forced to double the coronal density, i.e., to take $n_{\mathrm{cor}} \equiv n_{\mathrm{e}} = 2.0 \times 10^9 \ \mathrm{cm}^{-3}$. In such a case the sound and Alfvén speeds in the jet and its environment are: $c_{\mathrm{si}} \cong 304 \ \mathrm{km \ s}^{-1}$, $c_{\mathrm{se}} = 117 \ \mathrm{km \ s}^{-1}$, $v_{\mathrm{Ai}} = 47.67 \ \mathrm{km \ s}^{-1}$, and $v_{\mathrm{Ae}} = 487.5 \ \mathrm{km \ s}^{-1}$. The ordering of sound and Alfvén speeds is the same as for the #11 X-ray jet in the Shimojo and Shibata paper (Shimojo and Shibata, 2000) that implies the propagation of pure surface stable modes. The magnetic field inside the jet is $B_{\mathrm{i}} \cong 1.47 \, \mathrm{G}$— hence the magnetic fields ratio is $b = 6.817$. Thus, the input parameters for solving (9.1) and (9.3) are: $\eta = 0.444$, $\tilde{\beta}_{\mathrm{i}} = 40.5872$, $\tilde{\beta}_{\mathrm{e}} = 0.0695$, and as we already find, $b = 6.817$. Note that this relatively high value of b would suspect a rather big value of the threshold Alfvén Mach number—the instability criterion yields $M_{\mathrm{A}} > 12.42$. Numerically found threshold M_{A} for initiating KH instability of the kink ($m = 1$) mode in moving untwisted magnetic flux tube of compressible plasma is 12.435 that yields a critical velocity of $607 \ \mathrm{km \ s}^{-1}$. For the case of weakly twisted flux tube ($\varepsilon = 0.025$) in the approximation incompressible jet and cool plasma environment, we obtained for the same kink mode a critical velocity of $620 \ \mathrm{km \ s}^{-1}$. The critical speeds for the higher ($m = 2$–4) modes are accordingly of 615, 609, and $606 \ \mathrm{km \ s}^{-1}$.

The three examples of coronal X-jets embedded in a background magnetic field of $10 \, \mathrm{G}$ show that KH instability of basically the kink ($m = 1$) mode can occur if jets' speeds lie in the range of 500–$700 \ \mathrm{km \ s}^{-1}$. Such velocities are generally accessible for high-speed jets; lower critical speeds between 400 and $500 \ \mathrm{km \ s}^{-1}$ one can observe/detect at moderate external magnetic fields in the range of 6–$7 \, \mathrm{G}$. This requirement is in agreement with the Pucci *et al.* (2013) evaluation that the magnetic field inside a standard X-ray jet would be equal to $2.8 \, \mathrm{G}$, while for a blowout jet that value should be around $4.5 \, \mathrm{G}$. To be honest, we can say that the most of soft X-ray solar jets with speeds below of 400–$450 \ \mathrm{km \ s}^{-1}$ are stable against the KH instability—this instability can develop only in relatively dense high-speed jets. Note, however, that each soft X-ray jet is a unique event—it requires a separate careful exploration of the instability conditions. Arising KH instability in turn might trigger wave turbulence considered as an effective mechanism for coronal heating.

Chapter 10

Kelvin–Helmholtz Instability in Rotating Solar Jets

10.1 Observations and nature of the rotating solar jets

It was established that more of the solar jets continuously observed by the Extreme-ultraviolet Imaging Spectrometer (EIS) on board *Hinode* satellite, Atmospheric Imaging Assembly (AIA) on board the *Solar Dynamics Observatory* (*SDO*), as well as from the *Interface Region Imaging Spectrograph* (*IRIS*) alongside the Earth-based solar telescopes possess rotational motion. Such tornado-like jets, termed macrospicules, were firstly detected in the transition region by Pike and Mason (1998) using observations by the *Solar and Heliospheric Observatory* (*SOHO*). Rotational motion in macrospicules was also explored by Kamio *et al.* (2010), Curdt and Tian (2011), Bennett and Erdélyi (2015), and Kiss *et al.* (2017, 2018). Type II spicules, according to De Pontieu *et al.* (2012) and Martínez-Sykora *et al.* (2013), along with the coronal hole EUV jets (Nisticò *et al.*, 2009; Liu *et al.*, 2009b; Nisticò *et al.*, 2010; Chen *et al.*, 2012; Hong *et al.*, 2013; Young and Muglach, 2014a,b; Moore *et al.*, 2015), and X-ray jets (Moore *et al.*, 2013), can rotate, too. Rotating EUV jets emerging from a swirling flare (Zhang and Ji, 2014) or formed during a confined filament eruption (Filippov *et al.*, 2015) confirm once again the circumstance that the rotational motion is a common property of many kinds of jets in the solar atmosphere.

The first scenario for the numerical modeling of hot X-ray jets was reported by Heyvaerts *et al.* (1997) and the basic idea was that a bipolar magnetic structure emerges into a unipolar pre-existing magnetic field and reconnects to form hot and fast jets that are emitted from the interface between the fields into contact. Later on, by examining many X-ray jets in *Hinode*/X-Ray Telescope coronal X-ray movies of the polar coronal holes, Moore *et al.* (2010) found that there is a dichotomy of polar X-ray jets, namely "standard" and "blowout" jets exist. Fang *et al.* (2014) studied the formation of rotating coronal jets through numerical simulation of the emergence of a twisted magnetic flux rope into a pre-existing open magnetic field. Another scenario for the nature of solar jets was suggested by Sterling *et al.* (2015), according to which the X-ray jets are due to flux cancellation and/or "mini-eruptions" rather than emergence. An alternative model for solar polar jets due to an explosive release of energy via reconnection was reported by Pariat *et al.* (2009). Using three-dimensional MHD simulations, the authors demonstrated that this mechanism does produce massive, high-speed jets. In subsequent two articles (Pariat *et al.*, 2015, 2016), Pariat and co-authors presented several parametric studies of a three-dimensional numerical MHD model for straight and helical solar jets. On the other side, Panesar *et al.* (2016) have shown that the magnetic flux cancellation can trigger the solar quiet-region coronal jets and they claim that the coronal jets are driven by the eruption of a small-scale filament, called a "minifilament". The small-scale chromospheric jets, like microspicules, were first numerically modeled by Murawski *et al.* (2011). Using the FLASH code, they solved the two-dimensional ideal MHD equations to model a macrospicule, whose physical parameters match those of a solar spicule observed. Another mechanism for the origin of macrospicules was proposed by Kayshap *et al.* (2013), who numerically modeled the triggering of a macrospicule and a jet.

The first modeling of the KHI in a rotating cylindrical magnetized plasma jet was done by Bondenson *et al.* (1987). Later on, Bodo *et al.* (1989, 1996) carried out a study of the stability of flowing cylindrical jet immersed in constant magnetic field $\boldsymbol{B}_0$. The authors used the standard procedure for exploring the MHD wave propagation in cylindrical flows considering that all the perturbations of the plasma pressure p, fluid velocity $\boldsymbol{v}$, and magnetic field $\boldsymbol{B}$, are $\propto \exp\left[\mathrm{i}(-\omega t + kz + m\theta)\right]$. Here, ω is the angular wave frequency, k the propagating wavenumber, and m the azimuthal mode number. Using the basic equations of ideal magnetohydrodynamics, Bodo *et al.* (1989, 1996) derived a Bessel equation for the pressure perturbation and an expression for the radial component of the fluid velocity perturbation. The solutions found that in both media (the jet and its environment) are merged at the perturbed tube boundary through the conditions for continuity of the total (thermal plus magnetic) pressure and the Lagrangian displacement. The latter is defined as the ratio of radial velocity perturbation component and the angular frequency in the corresponding medium. The obtained dispersion relation is used for examining the stability conditions of both axisymmetric, $m = 0$ (Bodo *et al.*, 1989), and non-axisymmetric, $|m| \geqslant 1$ modes (Bodo *et al.*, 1996). In a recent article, Bodo *et al.* (2016) performed a linear stability analysis of magnetized rotating cylindrical jet flows in the approximation of zero thermal pressure. They focused their analysis on the effect of rotation on the current-driven mode and on the unstable modes introduced by rotation. In particular, they found that rotation has a stabilizing effect on the current-driven mode only for rotation velocities of the order on the Alfvén speed. The more general case, when both the magnetic field and jet flow velocity are twisted, was studied by Zaqarashvili *et al.* (2015) and Cheremnykh *et al.* (2018), whose dispersion equations for modes with $m \geqslant 2$, represented in different ways, yield practically identical results.

The main goal of this chapter is to suggest a way of using the wave dispersion relation derived in Zaqarashvili *et al.* (2015) to study the possibility for the rising and development of KHI in rotating twisted solar jets. Among the enormous large number of observational studies of rotating jets with different origin or nature, we chose those which provide the magnitudes of axial and rotational speeds, jet width and height alongside the typical plasma parameters like electron number densities and electron temperature of the spinning structure and its environment. Thus, the targets of our exploration are: (i) the spinning coronal hole jet of 2010 August 21 (Chen *et al.*, 2012); (ii) the rotating coronal hole jet of 2011 February 8 (Young and Muglach, 2014a); (iii) the twisted rotating jet emerging from a filament eruption on 2013 April 10–11 (Filippov *et al.*, 2015); and (iv) the rotating macrospicule observed by Pike and Mason (1998) on 1997 March 8.

10.2 The geometry, magnetic field, and physical parameters in a jet model

We model whichever jet as an axisymmetric cylindrical magnetic flux tube with radius a and electron number density n_i (or equivalently, homogeneous plasma density ρ_i) moving with velocity U (see Fig. 10.1). We consider that the jet environment is a rest plasma with homogeneous density ρ_e immersed in a homogeneous background magnetic field B_e. This field, in cylindrical coordinates (r, ϕ, z), possesses only an axial component, i.e., $B_e = (0, 0, B_e)$. The magnetic field inside the tube, B_i, and the jet velocity, U, we assume, are uniformly twisted and are given by the vectors

$$B_i = (0, B_{i\phi}(r), B_{iz}) \quad \text{and} \quad U = (0, U_\phi(r), U_z), \tag{10.1}$$

respectively. We note that B_{iz} and U_z are constant. Concerning the azimuthal magnetic and flow velocity components, we suppose that they are linear functions of the radial position r and evaluated at $r = a$ they correspondingly are equal to $B_{i\phi}(a) \equiv B_\phi = Aa$ and $U_\phi = \Omega a$, where A and Ω are constants. Here, Ω is the jet angular speed, deduced from the observations. Hence, in equilibrium,

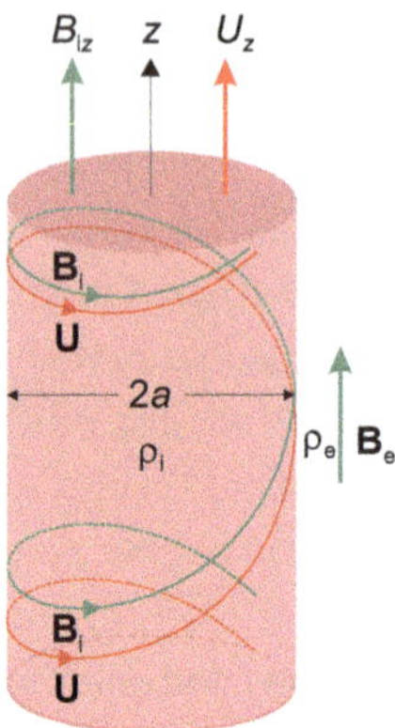

Fig. 10.1. Equilibrium magnetic fields of a rotating solar jet in a twisted flux tube.

the rigidly rotating plasma column, that models the jet, must satisfy the following force-balance equation (see, e.g., Chandrasekhar, 1961; Goossens *et al.*, 1992)

$$\frac{\mathrm{d}}{\mathrm{d}r}\left(p_i + \frac{B_i^2}{2\mu}\right) = \frac{\rho_i U_\phi^2}{r} - \frac{B_{i\phi}^2}{\mu r}, \tag{10.2}$$

where μ is the plasma permeability and $p_t = p_i + B_i^2/2\mu$ with $B_i^2 = B_{i\phi}^2(r) + B_{iz}^2$ is the total (thermal plus magnetic) pressure. According to Eq. (10.2), the radial gradient of the total pressure should balance the centrifugal force and the force owing to the magnetic tension. After integrating Eq. (10.2) from 0 to a, taking into account the linear dependence of U_ϕ and $B_{i\phi}$ on r, we obtain that

$$p_t(a) = p_t(0) + \frac{1}{2}\rho_i U_\phi^2(a) - \frac{B_{i\phi}^2(a)}{2\mu},$$

where $p_t(0) = p_1(0) + B_{iz}^2/2\mu$. (Integrating Eq. (10.2) from 0 to any r one can find the radial profile of p_t inside the tube. Such an expression of $p_t(r)$, obtained, however, from an integration of the momentum equation for the equilibrium variables, has been obtained in Zhelyazkov *et al.* (2018)—see Eq. (2) there.) It is clear from a physical point of view that the internal total pressure (evaluated at $r = a$) must be balanced by the total pressure of the surrounding plasma which implies that

$$p_1(0) + \frac{B_{iz}^2}{2\mu} - \frac{B_{i\phi}^2(a)}{2\mu} + \frac{1}{2}\rho_i U_\phi^2(a) = p_e + \frac{B_e^2}{2\mu}.$$

This equation can be presented in the form

$$p_1(0) + \frac{1}{2}\rho_i U_\phi^2(a) + \frac{B_{iz}^2}{2\mu}(1 - \varepsilon_1^2) = p_e + \frac{B_e^2}{2\mu}, \tag{10.3}$$

where $p_1(0)$ is the thermal pressure at the magnetic tube axis, and p_e denotes the thermal pressure in the environment. In the pressure balance Eq. (10.3), the number $\varepsilon_1 \equiv B_\phi/B_{iz} = Aa/B_{iz}$ represents the magnetic field twist parameter. Similarly, we define $\varepsilon_2 \equiv U_\phi/U_z$ as a characteristics of the jet velocity twist. We would like to underline that the choice of plasma and environment parameters must be such that the total pressure balance Eq. (10.3) is satisfied. In our case, the value of ε_2 is fixed by observationally measured rotational and axial velocities while the magnetic field twist, ε_1, has to be specified when using Eq. (10.3). We have to note that Eq. (10.3) is a corrected version of the pressure balance equation used in Zhelyazkov *et al.* (2018) and Zhelyazkov and Chandra (2018).

From measurements of n and T for similar coronal hole EUV jets (Nisticò *et al.*, 2009, 2010), we take n inside the jet to be $n_i = 1.0 \times 10^9 \, \text{cm}^{-3}$, and assume that the electron temperature is $T_i = 1.6\,\text{MK}$. The same quantities in the environment are, respectively, $n_e = 0.9 \times 10^9 \, \text{cm}^{-3}$ and $T_e = 1.0\,\text{MK}$. Note that the electron number density of the blowout jet observed by Young and Muglach (2014a) is in one order lower. The same applies for its environment. We consider that the background magnetic field for both hole coronal jets is $B_e = 3\,\text{G}$. The values of n and T of the rotating jet emerging from a filament eruption, observed by Filippov *et al.* (2015), were evaluated by us and they are $n_i = 4.65 \times 10^9 \, \text{cm}^{-3}$ and $T_i = 2.0\,\text{MK}$, respectively. From the same data set, we have obtained $n_e = 4.02 \times 10^9 \, \text{cm}^{-3}$ and $T_e = 2.14\,\text{MK}$. The background magnetic field, B_e, with which the pressure balance Eq. (10.3) is satisfied, is equal to $6\,\text{G}$. For the rotating macrospicule, we assume that $n_i = 1.0 \times 10^{10} \, \text{cm}^{-3}$ and $n_e = 1.0 \times 10^9 \, \text{cm}^{-3}$ to have at least one order denser jet with respect to the surrounding plasma. Our choice for macrospicule temperature is $T_i = 5.0 \times 10^5 \, \text{K}$, while that of its environment is supposed to be $T_e = 1.0 \times 10^6 \, \text{K}$. The external magnetic field, B_e, was taken as $5\,\text{G}$. All aforementioned physical parameters of the jets are summarized in Table 10.1. The plasma beta was calculated using $(6/5)c_s^2/v_A^2$, where $c_s = (\gamma k_B T/m_{\text{ion}})^{1/2}$ is the sound speed (in which $\gamma = 5/3$, k_B is the Boltzmann's constant, T the electron temperature, and m_{ion} the ion or proton mass), and $v_A = B/(\mu n_{\text{ion}} m_{\text{ion}})^{1/2}$ is the Alfvén speed, in which expression B is the full magnetic field $= (B_\phi^2 + B_z^2)^{1/2}$, and n_{ion} is the ion or proton number density.

Table 10.1. Jets physical parameters derived from observational data.

Kind of jet	B_e (G)	T_e (MK)	T_i (MK)	n_e ($\times 10^9\,\mathrm{cm}^{-3}$)	n_i ($\times 10^9\,\mathrm{cm}^{-3}$)	β_e	β_i
Standard coronal hole	3	1.00	1.6	0.90	1.00	0.348	2.079
Blowout coronal hole	3	2.00	1.7	0.55	0.17	0.116	0.115
Filament eruption	6	2.14	2.0	4.02	4.65	0.831	17.24
Macrospicule	5	1.00	0.5	0.19	1.00	0.139	2.248

10.3 Wave dispersion relation

A dispersion relation for the propagation of high-mode ($m \geqslant 2$) MHD waves in
a magnetized axially moving and rotating twisted jet was derived by Zaqarashvili
et al. (2015) and Cheremnykh *et al.* (2018). That equation was obtained, however,
under the assumption that both media (the jet and its environment) are incompressible plasmas. As seen from the last column in Table 10.1, plasma beta is greater
than 1 in the first, third, and forth jets which implies that the plasma of each of
the aforementioned jets can be considered as a nearly incompressible fluid (Zank
and Matthaeus, 1993). It is seen from the same table that the plasma beta of the
second jet is less than one as is in each of the jet environments and that is why
it is reasonable to treat them as cool media. Thus, the wave dispersion relation,
derived, for instance, in Zaqarashvili *et al.* (2015), has to be modified. In fact, we
need two modified versions: one for the incompressible jet—cool environment
configuration, and other for the cool jet—cool environment configuration. We are
not going to present in details the derivation of the modified dispersion equations
on the basis of the governing MHD equations, but will only sketch the essential steps in that procedure. The main philosophy in deriving the wave dispersion
equation is to find solutions for the total pressure perturbation, p_{tot}, and for the
radial component, ξ_r of the Lagrangian displacement, $\boldsymbol{\xi}$, and merge them at the
tube perturbed boundary through the boundary conditions for their (p_{tot} and ξ_r)
continuity (Chandrasekhar, 1961). In the case of the first configuration, we start
with the linearized ideal MHD equations, governing the incompressible dynamics
of the perturbations in the spinning jet

$$\frac{\partial}{\partial t}\boldsymbol{v} + (\boldsymbol{U} \cdot \nabla)\boldsymbol{v} + (\boldsymbol{v} \cdot \nabla)\boldsymbol{U} = -\frac{\nabla p_{\mathrm{tot}}}{\rho_i} + \frac{(\boldsymbol{B}_i \cdot \nabla)\boldsymbol{b}}{\rho_i \mu} + \frac{(\boldsymbol{b} \cdot \nabla)\boldsymbol{B}_i}{\rho_i \mu}, \tag{10.4}$$

$$\frac{\partial}{\partial t}\boldsymbol{b} - \nabla \times (\boldsymbol{v} \times \boldsymbol{B}_i) - \nabla \times (\boldsymbol{U} \times \boldsymbol{b}) = 0, \tag{10.5}$$

$$\nabla \cdot \boldsymbol{v} = 0, \tag{10.6}$$

$$\nabla \cdot \boldsymbol{b} = 0, \tag{10.7}$$

where $\boldsymbol{v} = (v_r, v_\phi, v_z)$ and $\boldsymbol{b} = (b_r, b_\phi, b_z)$ are the perturbations of fluid velocity and magnetic field, respectively, and p_{tot} is the perturbation of the total pressure, $p_{\text{t}} = p_{\text{i}} + B_{\text{i}}^2/2\mu$. The Lagrangian displacement, $\boldsymbol{\xi}$, can be found from the fluid velocity perturbation, $\boldsymbol{v}$, using the relation (Chandrasekhar, 1961)

$$\boldsymbol{v} = \frac{\partial \boldsymbol{\xi}}{\partial t} + (\boldsymbol{U} \cdot \nabla)\boldsymbol{\xi} - (\boldsymbol{\xi} \cdot \nabla)\boldsymbol{U}. \tag{10.8}$$

Further on, assuming that all perturbations are $\propto \exp\left[i(-\omega t + m\phi + k_z z)\right]$ and considering that the rotation and the magnetic field twists in the jet are uniform, that is,

$$U_\phi(r) = \Omega r \quad \text{and} \quad B_{\text{i}\phi}(r) = Ar, \tag{10.9}$$

where Ω and A are constants, from the above set of Eqs. (10.4)–(10.8), we obtain the following dispersion equation of the MHD wave with mode number m (for details, see Zaqarashvili *et al.*, 2015):

$$\frac{(\sigma^2 - \omega_{\text{Ai}}^2)F_m(\kappa_{\text{i}}a) - 2m(\sigma\Omega + A\omega_{\text{Ai}}/\sqrt{\mu\rho_{\text{i}}})}{\rho_{\text{i}}(\sigma^2 - \omega_{\text{Ai}}^2)^2 - 4\rho_{\text{i}}(\sigma\Omega + A\omega_{\text{Ai}}/\sqrt{\mu\rho_{\text{i}}})^2}$$
$$= \frac{P_m(\kappa_{\text{e}}a)}{\rho_{\text{e}}(\omega^2 - \omega_{\text{Ae}}^2) - (\rho_{\text{i}}\Omega^2 - A^2/\mu)P_m(\kappa_{\text{e}}a)}, \tag{10.10}$$

where

$$F_m(\kappa_{\text{i}}a) = \frac{\kappa_{\text{i}}aI_m'(\kappa_{\text{i}}a)}{I_m(\kappa_{\text{i}}a)} \quad \text{and} \quad P_m(\kappa_{\text{e}}a) = \frac{\kappa_{\text{e}}aK_m'(\kappa_{\text{e}}a)}{K_m(\kappa_{\text{e}}a)}.$$

In above expressions, the prime means differentiation of the Bessel functions with respect to their arguments,

$$\kappa_{\text{i}}^2 = k_z^2\left[1 - 4\left(\frac{\sigma\Omega + A\omega_{\text{Ai}}/\sqrt{\mu\rho_{\text{i}}}}{\sigma^2 - \omega_{\text{Ai}}^2}\right)^2\right] \quad \text{and} \quad \kappa_{\text{e}}^2 = k_z^2[1 - (\omega/\omega_{\text{Ae}})^2]$$

are the squared wave amplitude attenuation coefficients in the jet and its environment, in which

$$\omega_{\text{Ai}} = \left(\frac{m}{r}B_{\text{i}\phi} + k_z B_{\text{iz}}\right)\bigg/\sqrt{\mu\rho_{\text{i}}} \quad \text{and} \quad \omega_{\text{Ae}} = k_z B_{\text{e}}/\sqrt{\mu\rho_{\text{e}}}$$

are the local Alfvén frequencies in both media, and

$$\sigma = \omega - \frac{m}{r}U_\phi - k_z U_z$$

is the Doppler-shifted angular wave frequency in the jet. We note that in the case of incompressible coronal plasma (Zaqarashvili *et al.*, 2015), $\kappa_e = k_z$, because at an incompressible environment the argument of the modified Bessel function of second kind, K_m, and its derivative, K'_m, is $k_z a$.

The basic MHD equations for an ideal cool plasma are, generally, the same as the set of Eqs. (10.4)–(10.8) with Eq. (10.6) replaced by the continuity equation

$$\frac{\partial \rho_1}{\partial t} = -\nabla \cdot (\rho_0 \boldsymbol{v}_1 + \rho_1 \boldsymbol{U}) = 0.$$

Recall that for cold plasmas the total pressure reduces to the magnetic pressure only, that is $p_t = B_1^2/2\mu$, the z component of the velocity perturbation is zero, i.e., $\boldsymbol{v}_1 = (v_{1r}, v_{1\phi}, 0)$, while $\boldsymbol{B}_1 = (B_{1r}, B_{1\phi}, B_{1z})$. The above equation, which defines the density perturbation, is not used in the derivation of the wave dispersion relation because we are studying the propagation and stability of Alfvén-wave-like perturbations of the fluid velocity and magnetic field. Following the standard scenario for deriving the MHD wave dispersion relation (Zhelyazkov and Chandra, 2018), we finally arrive at

$$\frac{(\sigma^2 - \omega_{Ai}^2) F_m(\kappa_i^c a) - 2m(\sigma\Omega + A\omega_{Ai}/\sqrt{\mu\rho_i})}{\rho_i(\sigma^2 - \omega_{Ai}^2)^2 - 4\rho_i(\sigma\Omega + A\omega_{Ai}/\sqrt{\mu\rho_i})^2}$$
$$= \frac{P_m(\kappa_e^c a)}{\rho_e(\omega^2 - \omega_{Ae}^2) - (\rho_i\Omega^2 - A^2/\mu) P_m(\kappa_e^c a)}, \tag{10.11}$$

where

$$F_m(\kappa_i^c a) = \frac{\kappa_i^c a I'_m(\kappa_i^c a)}{I_m(\kappa_i^c a)} \quad \text{and} \quad P_m(\kappa_e^c a) = \frac{\kappa_e^c a K'_m(\kappa_e^c a)}{K_m(\kappa_e^c a)}.$$

Here, the wave attenuation coefficient in the internal medium has the form

$$\kappa_i^c = k_z \left\{ 1 - 4 \left(\frac{\sigma\Omega + A\omega_{Ai}/\sqrt{\mu\rho_i}}{\sigma^2 - \omega_{Ai}^2} \right)^2 \right\}^{1/2} \left(1 - \frac{\sigma^2}{\omega_{Ai}^2} \right)^{1/2},$$

while that in the environment, with $\Omega = 0$ and $A = 0$, is given by

$$\kappa_e^c = k_z \left(1 - \frac{\omega^2}{\omega_{Ae}^2} \right)^{1/2}.$$

Note that (i) both dispersion relations, (10.10) and (10.11), have similar forms—the difference is in the expressions for the wave attenuation coefficient inside the jet, namely $\kappa_i^c = \kappa_i\left[1 - \sigma^2/\omega_{Ai}^2\right]^{1/2}$; and (ii) the wave attenuation coefficients in the environments are not surprisingly the same, that is, $\kappa_e^c = \kappa_e \equiv k_z\left[1 - \omega^2/\omega_{Ae}^2\right]^{1/2}$.

10.4 Numerical solutions, wave dispersion, and growth rate diagrams

In studying at which conditions the high ($m \geqslant 2$) MHD modes in a jet—coronal plasma system become unstable, that is, all the perturbations to grow exponentially in time, we have to consider the wave angular frequency, ω, as a complex quantity: $\omega \equiv \mathrm{Re}(\omega) + \mathrm{i}\,\mathrm{Im}(\omega)$ in contrast to the wave mode number, m, and propagating wavenumber, k_z, which are real quantities. Recall that $\mathrm{Re}(\omega)$ is responsible for the wave dispersion while the $\mathrm{Im}(\omega)$ yields the wave growth rate. In the numerical task for finding the complex solutions to the wave dispersion relation (10.10) or (10.11), it is convenient, as we usually do, to normalize all velocities with respect to the Alfvén speed inside the jet, defined as $v_{Ai} = B_{iz}/\sqrt{\mu\rho_i}$, and the lengths with respect to a. Thus, we have to search the real and imaginary parts of the non-dimensional wave phase velocity, $v_{ph} = \omega/k_z$, that is, $\mathrm{Re}(v_{ph}/v_{Ai})$ and $\mathrm{Im}(v_{ph}/v_{Ai})$ as functions of the normalized wavenumber $k_z a$. The normalization of the other quantities like the local Alfvén and Doppler-shifted frequencies alongside the Alfvén speed in the environment, $v_{Ae} = B_e/\sqrt{\mu\rho_e}$, requires the usage of both twist parameters, ε_1 and ε_2, and also of the magnetic fields ratio, $b = B_e/B_{iz}$. The non-dimensional form of the jet axial velocity, U_z, is given by the Alfvén Mach number $M_A = U_z/v_{Ai}$. Another important non-dimensional parameter is the density contrast between the jet and its surrounding medium, $\eta = \rho_e/\rho_i$. Hence, the input parameters in the numerical task of finding the solutions to the transcendental Eqs. (10.10) or (10.11) (in complex variables) are: m, η, ε_1, ε_2, b, and M_A. Zaqarashvili *et al.* (2015) have established that KHI in an untwisted ($A = 0$) rotating flux tube with negligible longitudinal velocity can occur if

$$\frac{a^2\Omega^2}{v_{Ai}^2} > \frac{1+\eta}{1+|m|\eta}\frac{(k_z a)^2}{|m|-1}(1+b^2). \tag{10.12}$$

This inequality says that each MHD wave with mode number $m \geqslant 2$, propagating in a rotating jet can become unstable. This instability condition can be used also in the cases of slightly twisted spinning jets, provided that the magnetic field twist parameter, ε_1, is a number lying in the range of 0.001–0.005, simply because the numerical solutions, for example, to Eq. (10.10) show that practically there is no difference between the instability ranges at $\varepsilon_1 = 0$, and at 0.001 or 0.005. An important step in our study is the supposition that the deduced from observations jet axial velocity, U_z, is the threshold speed for the KHI occurrence. Then, for fixed values of m, η, $U_\phi = \Omega a$, v_{Ai}, and b, the inequality (10.12) can be rearranged to define the upper limit of the instability range on the $k_z a$-axis

$$(k_z a)_{\mathrm{rhs}} < \left\{ \left(\frac{U_\phi}{v_{\mathrm{Ai}}} \right)^2 \frac{1 + |m|\eta}{1 + \eta} \frac{|m| - 1}{1 + b^2} \right\}^{1/2}. \tag{10.13}$$

According to the above inequality, the KHI can occur for non-dimensional wavenumbers $k_z a$ less than $(k_z a)_{\mathrm{rhs}}$. On the other hand, one can talk for instability if the unstable wavelength, $\lambda_{\mathrm{KH}} = 2\pi / k_z$, is shorter than the height of the jet, H, which means that the lower limit of the instability region is given by

$$(k_z a)_{\mathrm{lhs}} > \frac{\pi \Delta \ell}{H}, \tag{10.14}$$

where $\Delta \ell$ is the jet width. Hence, the instability range in the $k_z a$-space is $(k_z a)_{\mathrm{lhs}} < k_z a < (k_z a)_{\mathrm{rhs}}$. Note that the lower limit, $(k_z a)_{\mathrm{lhs}}$, is fixed by the width and height of the jet, while the upper limit, $(k_z a)_{\mathrm{rhs}}$, depends on several jet–environment parameters. At fixed U_ϕ, v_{Ai}, η, and b, the $(k_z a)_{\mathrm{rhs}}$ is determined by the MHD wave mode number, $|m|$. As seen from inequality (10.13), with increasing the m, that limit shifts to the right, that is, the instability range becomes wider. The numerical solutions to the wave dispersion relation (10.10) confirm this and for given m one can obtain a series of unstable wavelengths, $\lambda_{\mathrm{KH}} = \pi \, \Delta \ell / k_z a$, as the shortest one takes place at $k_z a \approx (k_z a)_{\mathrm{rhs}}$. For relatively small mode numbers, when even the shortest unstable wavelengths turn out to be a few tens megameters, that could hardly be associated with the observed KH ones. As observations show, the KHI vortex-like structures running at the boundary of the jet, have the size of the width or radius of the flux tube (see, for instance, Fig. 1 in Zhelyazkov *et al.*, 2018). Therefore, we have to look for such an m, whose instability range would accommodate the expected unstable wavelength presented by its non-dimensional wavenumber, $k_z a = \pi \Delta \ell / \lambda_{\mathrm{KH}}$. An estimation of the required mode number for an $\varepsilon_1 = 0.005$-rotating flux tube can be obtained by presenting the instability criterion (10.12) in the form

$$\eta |m|^2 + (1 - \eta)|m| - 1 - \frac{(k_z a)^2 (1 + \eta)(1 + b^2)}{(U_\phi / v_{\mathrm{Ai}})^2} > 0. \tag{10.15}$$

We will use this inequality for obtaining the optimal m for each of the studied jets by specifying the value of that $k_z a$ (along with the other aforementioned input parameters) which corresponds to the expected unstable wavelength λ_{KH}.

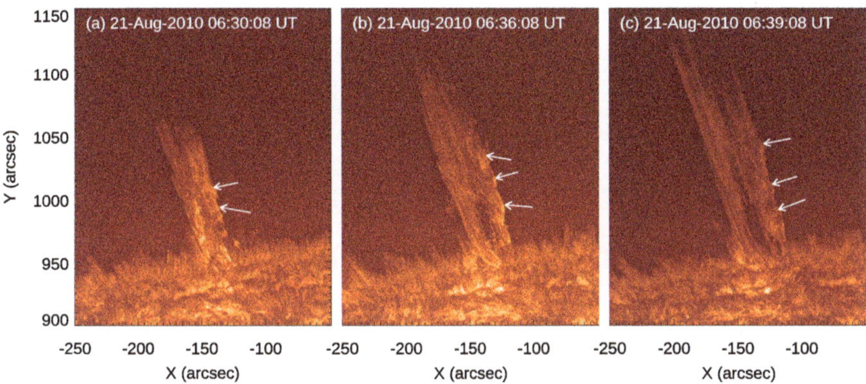

Fig. 10.2. AIA 304 Å images showing the detailed evolution of the jet observed on 2010 August 21. The small moving blobs on the right side boundary of the jet as indicated by white arrows, could be produced by a KHI. *Reprinted with permission.*

10.4.1 *Kelvin–Helmholtz instability in a standard polar coronal hole jet*

Chen *et al.* (2012) observationally studied the jet event of 2010 August 21, which occurred in the coronal hole region, close to the north pole of the Sun. Figure 10.2 presents the jet's evolution in AIA 304 Å. The jet started around 06:07 UT, reached its maximum height around 06:40 UT. During the evolution of the jet between 06:32 and 06:38 UT, small-scale moving blobs appeared on the right boundary. We interpret these blobs, shown by arrows in Fig. 10.2, as evidence of KHI. By tracking six identified moving features in the jet, Chen *et al.* (2012) found that the plasma moved at an approximately constant speed along the jet's axis. Inferred from linear and trigonometric fittings to the axial and transverse heights of the six tracks, the authors have found that the mean values of the axial velocity, U_z, transfer/rotational velocity, U_ϕ, angular speed, Ω, rotation period, T, and rotation radius, a, are $114\,\text{km\,s}^{-1}$, $136\,\text{km\,s}^{-1}$, $0.81°\text{s}^{-1}$ (or $14.1 \times 10^{-3}\,\text{rad\,s}^{-1}$), $452\,\text{s}$, and $9.8 \times 10^3\,\text{km}$, respectively. The height of the jet is evaluated as $H = 179\,\text{Mm}$.

It seems reasonable that the shortest unstable wavelength, λ_{KH}, to be equal to $10\,\text{Mm}$ (approximately half of the jet width, $\Delta\ell = 19.6\,\text{Mm}$), which implies that its position on the $k_z a$ one-dimensional space is $k_z a = 6.158$. The input parameters, necessary to find out that MHD wave mode number, whose instability range will contain the non-dimensional wavenumber of 6.158, using inequality (10.15), are accordingly (see Table 10.1) $\eta = 0.9$, $b = 1.834$, $v_{\text{Ai}} = 112.75\,\text{km\,s}^{-1}$, and $U_\phi = 136\,\text{km\,s}^{-1}$. (We note that the values of b and v_{Ai} were obtained with the help of Eq. (10.3) assuming that the magnetic field twist is $\varepsilon_1 = 0.005$.) With these

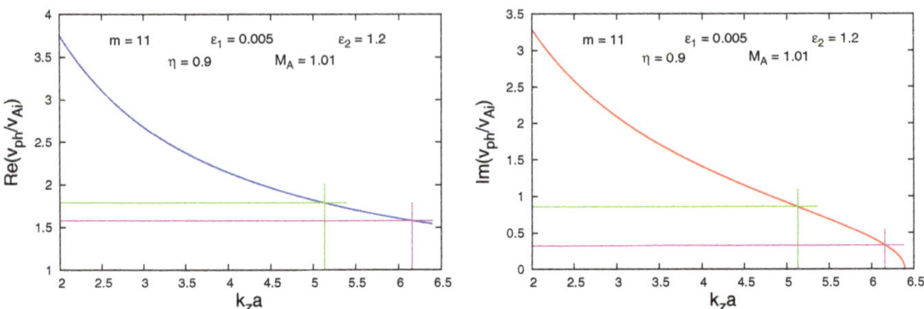

Fig. 10.3. (*Left panel*) Dispersion curve of the $m = 11$ MHD mode propagating along a twisted incompressible coronal hole jet at $\eta = 0.9$, $b = 1.834$, $M_A = 1.01$, $\varepsilon_1 = 0.005$, and $\varepsilon_2 = 1.2$. (*Right panel*) Normalized growth rate curve of the $m = 11$ MHD mode computed at the same input parameters as in the left panel. The crosses of purple and green lines yield the normalized values of the wave phase velocity and the wave growth rate at the two unstable wavelengths of 10 and 12 Mm, respectively. *Reprinted with permission.*

entry data, from inequality (10.15) one obtains that $|m| > 15$ should provide the required instability region or window. The numerical solutions to Eq. (10.10) show that this value of m is overestimated—an $m = 11$ turns out to be perfect for the case. The discrepancy between the predicted and computed value of $|m|$ is not surprising because inequality (10.15) yields only an indicative value. The input parameters for finding the solutions to the dispersion Eq. (10.10) are as follows: $m = 11$, $\eta = 0.9$, $\varepsilon_1 = 0.005$, $\varepsilon_2 = 1.2$, $b = 1.834$, and $M_A = 1.01$ (=114/112.75). The results of computations are graphically presented in Fig. 10.3. From that figure, one can obtain the normalized wave phase velocity, $\mathrm{Re}(v_{\mathrm{ph}}/v_{\mathrm{Ai}})$, and the normalized growth rate, $\mathrm{Im}(v_{\mathrm{ph}}/v_{\mathrm{Ai}})$, of the unstable $\lambda_{\mathrm{KH}} = 10\,\mathrm{Mm}$ wave, both read at the purple cross points. From the same plot, one can find the instability characteristics at another wavelength, precisely $\lambda_{\mathrm{KH}} = 12\,\mathrm{Mm}$, whose position on the $k_z a$-axis is fixed at $k_z a = 5.131$. The values of non-dimensional wave phase velocity and growth rate can be read from the green cross points. The KHI wave growth rate, γ_{KH}, growth time, $\tau_{\mathrm{KH}} = 2\pi / \gamma_{\mathrm{KH}}$, and wave velocity, v_{ph}, in absolute units, estimated from the plots in Fig. 10.3, for the two wavelengths, are

$$\gamma_{\mathrm{KH}} \cong 23.09 \times 10^{-3}\,\mathrm{s}^{-1}, \quad \tau_{\mathrm{KH}} \cong 4.5\,\mathrm{min},$$

$$v_{\mathrm{ph}} \cong 178\,\mathrm{km\,s}^{-1}, \quad \text{for } \lambda_{\mathrm{KH}} = 10\,\mathrm{Mm},$$

and

$$\gamma_{\mathrm{KH}} \cong 50.65 \times 10^{-3}\,\mathrm{s}^{-1}, \quad \tau_{\mathrm{KH}} \cong 2.1\,\mathrm{min},$$

$$v_{\mathrm{ph}} \cong 202\,\mathrm{km\,s}^{-1}, \quad \text{for } \lambda_{\mathrm{KH}} = 12\,\mathrm{Mm}.$$

Let us recall that the value of the Alfvén speed used in the normalization is $v_{\mathrm{Ai}} = 112.75\,\mathrm{km\,s^{-1}}$. We see that the two wave phase velocities are slightly super-Alfvénic and when moving along the $k_z a$-axis to the left, the normalized wave velocity becomes higher. If we fix a $k_z a$-position near the lower limit of the unstable region, $(k_z a)_{\mathrm{lhs}} = 0.344$, say, at $k_z a = 0.513$, which means $\lambda_{\mathrm{KH}} = 120\,\mathrm{Mm}$, the KHI characteristics obtained from the numerical solutions to Eq. (10.10) are $\tau_{\mathrm{KH}} = 1.4\,\mathrm{min}$ and $v_{\mathrm{ph}} = 1\,473\,\mathrm{km\,s^{-1}}$, respectively. As we have discussed in Zhelyazkov *et al.* (2018), "the KHI growth time could be estimated from the temporal evolution of the blobs in their initial stage and it was found to be about 2–4 minutes," so the instability developing times of 2.1 and 4.5 min obtained from our plots are in good agreement with the observations.

A specific property of the instability $k_z a$-ranges is that for a fixed mode number, m, their widths depend upon ε_1 and with increasing the value of ε_1, the instability window becomes narrower and at some critical ε_1 its width equals zero. In our case that happens with $\varepsilon_1^{\mathrm{cr}} = 0.653577$ at $(k_z a)_{\mathrm{lhs}} = 0.344$. In Fig. 10.4, curves of dimensionless v_{ph} and γ_{KH} have been plotted for several ε_1 values. Note that each larger value of ε_1 implies an increase in $B_{\mathrm{i}\phi}$. But that increase in $B_{\mathrm{i}\phi}$ requires an increase in B_{iz} too, in order the total pressure balance Eq. (10.3) to be satisfied under the condition that the hydrodynamic pressure term and the environment total pressure are fixed. The increase in B_{iz} (and in the full magnetic field B_{i}) implies a decrease both in the magnetic field ratio, b, and in the Alfvén Mach number, M_{A}. Thus, gradually increasing the magnetic field twist ε_1 from 0.005 to 0.653577, we get a series of dispersion and growth rate

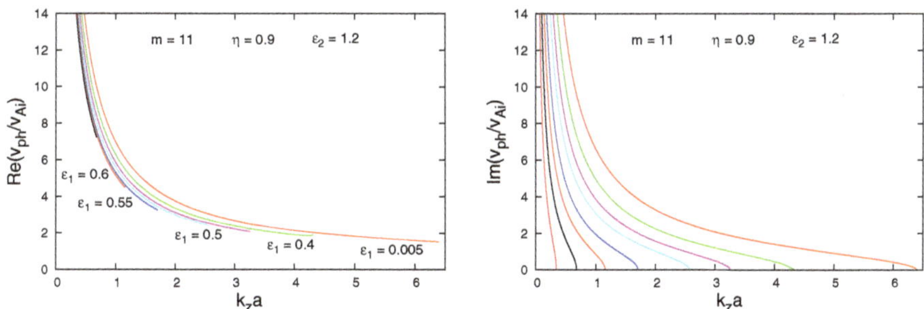

Fig. 10.4. (*Left panel*) Dispersion curves of the unstable $m = 11$ MHD mode propagating along a twisted incompressible jet in a coronal hole at $\eta = 0.9$, $\varepsilon_2 = 1.2$, and the following values of ε_1 (from right to left): 0.005, 0.4, 0.5, 0.55, 0.6, 0.625, 0.645, and 0.653577 (red curve in the right plot). Alfvén Mach numbers for these curves are respectively 1.01, 0.93, 0.88, 0.84, 0.81, 0.79, 0.77, and 0.7653. (*Right panel*) Growth rates of the unstable $m = 11$ mode for the same input parameters. The azimuthal magnetic field that corresponds to $\varepsilon_1^{\mathrm{cr}} = 0.653577$ (the instability window with zero width) and stops the KHI onset is equal to 1.4 G. *Reprinted with permission.*

curves with progressively diminishing parameters b and M_A. The red growth rate curve in the right panel of Fig. 10.4 has been obtained for $\varepsilon_1^{\mathrm{cr}} = 0.653577$ with $M_A = 0.7652$ and it visually fixes the lower limit of all other instability windows. The azimuthal magnetic field $B_{1\phi}^{\mathrm{cr}}$ that stops the KHI, computed at $B_i = 2.58\,\mathrm{G}$, is equal to $1.4\,\mathrm{G}$.

10.4.2 *Kelvin–Helmholtz instability in a blowout polar coronal hole jet*

Young and Muglach (2014a) observed a small blowout jet at the boundary of the south polar coronal hole on 2011 February 8 at around 21:00 UT. The evolution of jet observed by the AIA is displayed in Fig. 10.5. The jet activity was between 20:50 to 21:15 UT. This coronal hole is centered around $x = -400\,\mathrm{arcsec}$, $y = -400\,\mathrm{arcsec}$. The jet has very broad and faint structure and is ejected in the south direction. We could see the evolution of jet in the AIA 193 Å clearly. However, in AIA 304 Å, the whole jet is not visible. Moreover, we observe the eastern boundary of the jet in AIA 304 Å. During its evolution in 304 Å, we found the blob structures at the jet boundary. These blobs could be due to the KHI as

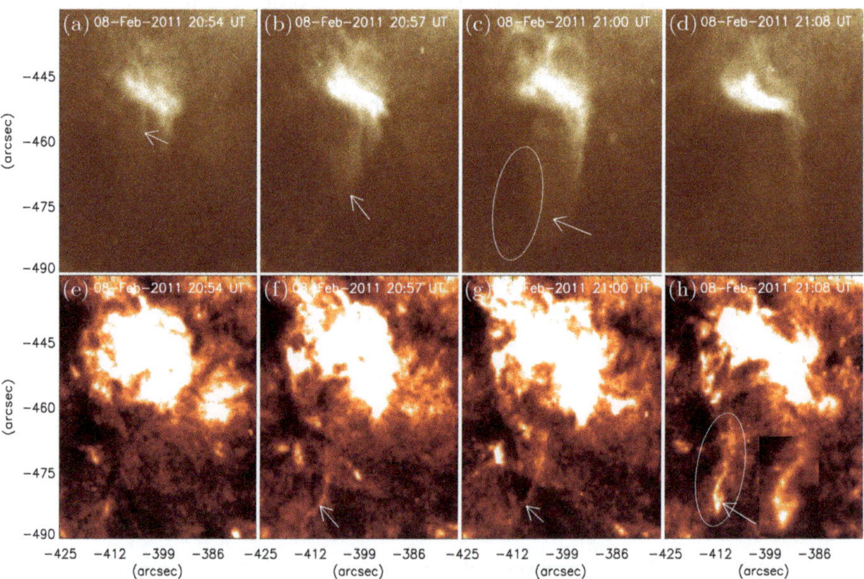

Fig. 10.5. AIA 193 top (a–d) and 304 Å bottom (e–h) images showing the evolution of the jet observed on 2011 February 8 marked by white arrows could be produced by KHI. For the better visibility of the jet, we have saturated the image at the foot-point. The ellipse in the (c–h) represents the east edge of the jets. In 304 Å at ∼21:08 UT we have observed the blob structures at the eastern boundary. The enlarged view of the blobs is shown in the inset in (h). *Reprinted with permission.*

reported in previous observations (see, for example, Zhelyazkov *et al.*, 2018). At the jet initiation/base site, we observed the coronal hole bright points. These bright points are the results of coronal low-laying loops reconnection (Madjarska, 2019).

According to Young and Muglach (2014a) estimations, the jet is extended for $H = 30\,\mathrm{Mm}$ with a width of $\Delta\ell = 15\,\mathrm{Mm}$. The jet duration is 25 min and the bright point is not significantly disrupted by the jet occurrence. The jet n is $n_\mathrm{i} = 1.7 \times 10^8\,\mathrm{cm}^{-3}$, while that of the surrounding coronal plasma we assume to be $n_\mathrm{e} = 1.5 \times 10^8\,\mathrm{cm}^{-3}$. The jet temperature is $T_\mathrm{i} = 1.7\,\mathrm{MK}$ and the environment one is $T_\mathrm{e} = 2.0\,\mathrm{MK}$. The jet axial velocity is $U_z = 250\,\mathrm{km\,s}^{-1}$ and the rotational one is $U_\phi = 90\,\mathrm{km\,s}^{-1}$. Assuming a magnetic field twist $\varepsilon_1 = 0.025$ and $B_\mathrm{e} = 3\,\mathrm{G}$, from Eq. (10.3), we obtain $\eta = 0.882$, $v_\mathrm{Ai} = 494.7\,\mathrm{km\,s}^{-1}$ (Alfvén speed in the environment is $v_\mathrm{Ae} = 534.0\,\mathrm{km\,s}^{-1}$), and $b = 1.014$. We note that while in the derivation of Eq. (10.11) we have neglected the thermal pressures, here, in using Eq. (10.3), we kept them. If we anticipate that the shortest unstable wavelength is equal to $7.5\,\mathrm{Mm}$ (with $k_z a = 2\pi$), the mode number m whose instability range would accommodate the aforementioned wavelength, according to inequality (10.15) must be at least $|m| = 71$. The numerics show that the suitable m is $|m| = 65$. Thus, the input parameters for obtaining the numerical solutions to Eq. (10.11) are: $m = 65$, $\eta = 0.882$, $\varepsilon_1 = 0.025$, $\varepsilon_2 = 0.36$ (=90/250), $b = 1.014$, and $M_\mathrm{A} = 0.505$ ($\cong$250.0/494.7). The results are illustrated in Fig. 10.6. Along with $\lambda_\mathrm{KH} = 7.5\,\mathrm{Mm}$ (purple lines), we have calculated the KHI characteristics also for $\lambda_\mathrm{KH} = 15\,\mathrm{Mm}$ (at $k_z a = \pi$) (green lines), and

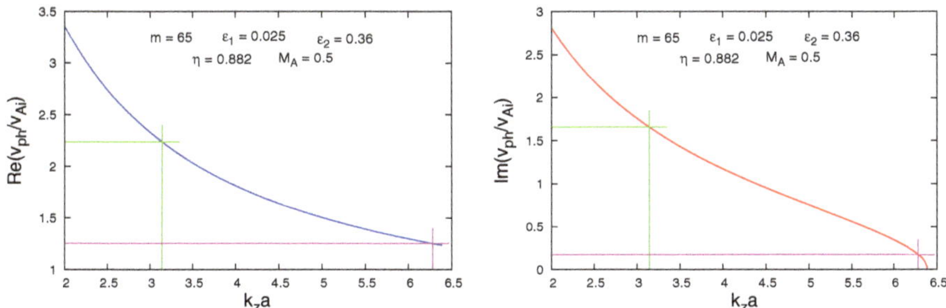

Fig. 10.6. (*Left panel*) Dispersion curve of the $m = 65$ MHD mode propagating along a twisted cool coronal hole jet at $\eta = 0.882$, $b = 1.014$, $M_\mathrm{A} = 0.5$, $\varepsilon_1 = 0.025$, and $\varepsilon_2 = 0.36$. (*Right panel*) Normalized growth rate curve of the $m = 65$ MHD mode computed at the same input parameters as in the left panel. The crosses of purple and green lines yield the normalized values of the wave phase velocity and the wave growth rate at the two unstable wavelengths of 7.5 and 15 Mm, respectively. *Reprinted with permission.*

they are

$$\gamma_{\mathrm{KH}} = 72.0 \times 10^{-3}\,\mathrm{s}^{-1}, \quad \tau_{\mathrm{KH}} \cong 1.5\,\mathrm{min},$$

$$v_{\mathrm{ph}} \cong 618\,\mathrm{km\,s}^{-1}, \quad \text{for } \lambda_{\mathrm{KH}} = 7.5\,\mathrm{Mm},$$

and

$$\gamma_{\mathrm{KH}} \cong 342.75 \times 10^{-3}\,\mathrm{s}^{-1}, \quad \tau_{\mathrm{KH}} \cong 0.3\,\mathrm{min},$$

$$v_{\mathrm{ph}} \cong 1106\,\mathrm{km\,s}^{-1}, \quad \text{for } \lambda_{\mathrm{KH}} = 15\,\mathrm{Mm}.$$

It is seen from the left panel that the unstable $m = 65$ MHD waves are generally super-Alfvénic. Since the instability developing times of the $m = 65$ mode are relatively short, that is, much less than the jet lifetime of 25 min, we can conclude that the KHI in this jet is relatively fast.

With the increase in the parameter ε_1, the instability region, as seen from the right panel of Fig. 10.7, becomes narrower and at the lower limit $(k_z a)_{\mathrm{lhs}} = \pi/2$ with $\varepsilon_1^{\mathrm{cr}} = 0.10682$ and $M_{\mathrm{A}} = 0.5026$ its width is equal to zero. In other words, there is no longer instability. Therefore, the critical azimuthal magnetic field that suppresses the KHI is $B_{1\phi}^{\mathrm{cr}} \cong 0.3$ G—obviously a relatively small value.

10.4.3 *Kelvin–Helmholtz instability in a jet emerging from a filament eruption*

Filippov *et al.* (2015) observationally studied three jets events originated from the active region NOAA 11715 (located on the west limb) on 2013 April 10–11. These

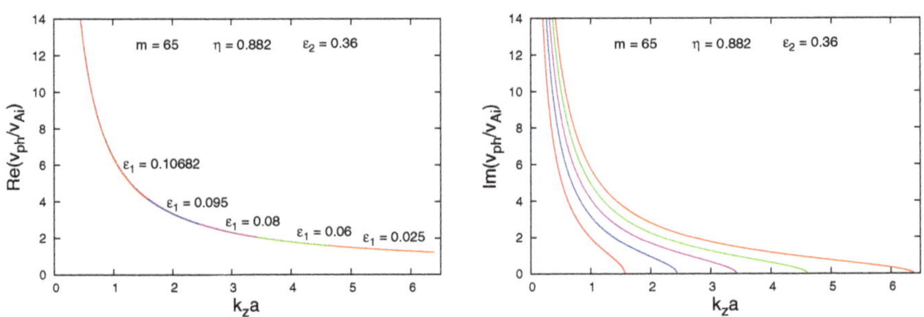

Fig. 10.7. (*Left panel*) Dispersion curves of the unstable $m = 65$ MHD mode propagating along a twisted cool jet in a coronal hole at $\eta = 0.882$, $\varepsilon_2 = 0.36$, and the following values of ε_1 (from right to left): $0.025, 0.06, 0.08, 0.095,$ and 0.10682 (red curve in the right plot). Alfvén Mach numbers for these curves are respectively $0.505, 0.505, 0.504, 0.503,$ and 0.5026. (*Right panel*) Growth rates of the unstable $m = 65$ mode for the same input parameters. The azimuthal magnetic field that corresponds to $\varepsilon_1^{\mathrm{cr}} = 0.10682$ (the instability window with zero width) and stops the KHI onset is equal to 0.3 G. *Reprinted with permission.*

authors claim that the jets originated from the emergence of a filament having a null-point (inverted Y) topology. We have considered the second event described in that paper for a detailed study. The jet electron number density, n_i, and electron temperature, T_i, both listed in Table 10.1, have been calculated by us using the techniques elaborated by Aschwanden *et al.* (2013). This technique requires the data from the six 94, 131, 171, 193, 211, and 335 Å AIA/*SDO* EUV channels. In addition to the electron number densities and electron temperatures in the jet and surrounding plasma, we have also estimated the jet width as $\Delta \ell \approx 30\,\mathrm{Mm}$, its height as $H = 180\,\mathrm{Mm}$, and have found the jet lifetime to be 30 min. The two important parameters, axial and azimuthal velocities, according to the observations, are $U_z = 100$ and $U_\phi = 180\,\mathrm{km\,s^{-1}}$, respectively. The time evolution of the jet in AIA 304 Å is shown in Fig. 10.8 and we have observed vortex type structures in the eastern side of the jet, which are indicated by arrows. These structures implicitly indicate for occurrence of KHI.

With the typical n and T_e (see Table 10.1), rotating velocity $U_\phi = 180\,\mathrm{km\,s^{-1}}$, assumed $B_\mathrm{e} = 6\,\mathrm{G}$, and $\varepsilon_1 = 0.1$, Eq. (10.3) yields $\eta = 0.864$, $b = 4.36$, and $v_{\mathrm{Ai}} = 44.00\,\mathrm{km\,s^{-1}}$ (for comparison, the Alfvén speed in the environment is $v_{\mathrm{Ae}} = 206.3\,\mathrm{km\,s^{-1}}$). We note that the choice of ε_1 was made taking into account the fact that the inclination of the treads of the jet in the event on 2013 April 10, detected by *SDO*/AIA, yields a relationship between $B_{\mathrm{i}\phi}$ and $B_{\mathrm{i}z}$, which was evaluated as $\varepsilon_1 \approx 0.1$. If we assume that the shortest unstable wavelength is $\lambda_{\mathrm{KH}} = 12\,\mathrm{Mm}$, which is located at $k_z a = 2.5\pi$ on the $k_z a$-axis, from inequality (10.15) we find that an MHD wave with $|m| = 12$ would provide an instability

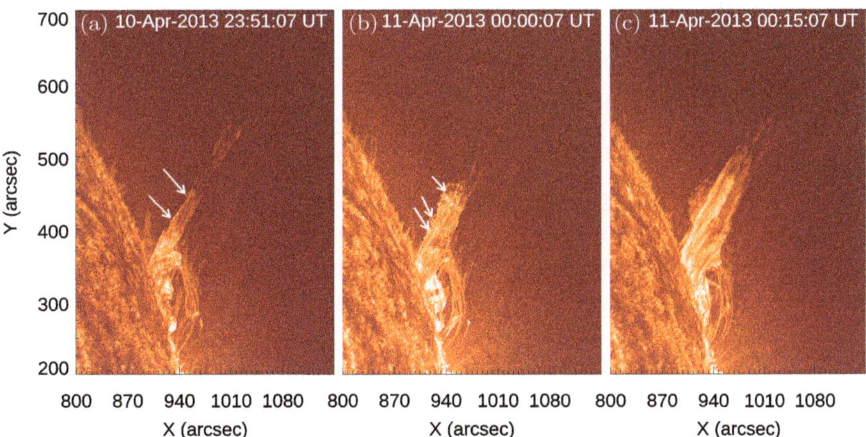

Fig. 10.8. Evolution of the jet associated with a filament eruption observed on 2013 April 10–11 in AIA 304 Å. The structure shown by arrows can be due to the KHI. *Reprinted with permission.*

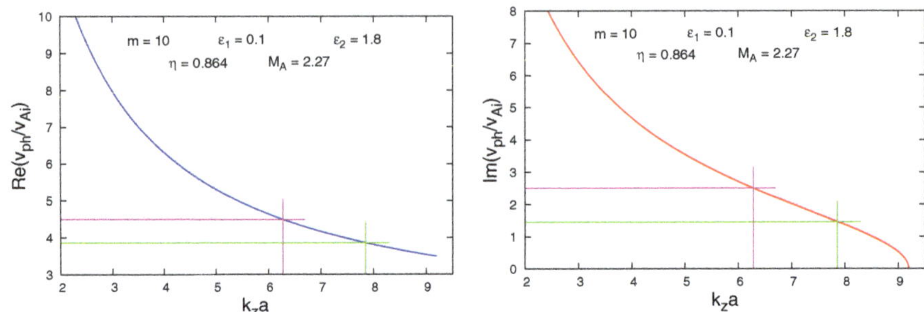

Fig. 10.9. (*Left panel*) Dispersion curves of the unstable $m = 10$ MHD mode propagating along a twisted incompressible emerging from a filament eruption jet at $\eta = 0.864$, $\varepsilon_1 = 0.1$, $\varepsilon_2 = 1.8$, and $M_A = 2.27$. (*Right panel*) Growth rate curves of the unstable $m = 10$ MHD mode propagating along a twisted incompressible emerging from a filament eruption jet at the same input parameters. *Reprinted with permission.*

region, accommodating the non-dimensional $k_z a = 2.5\pi$. It turns out that a suitable mode number is $m = 10$. The wave dispersion and growth rate diagrams are shown in Fig. 10.9. In that instability range one can also find the instability characteristics at $k_z a = 2\pi$, which corresponds to $\lambda_{KH} = 15\,\mathrm{Mm}$. The input parameters for finding the solutions to Eq. (10.10) are: $m = 10$, $\eta = 0.864$, $\varepsilon_1 = 0.1$, $\varepsilon_2 = 1.8$, $b = 4.36$, and $M_A = 2.27$ ($=100/44$). The KHI parameters at the aforementioned wavelengths are as follows:

$$\gamma_{KH} \cong 33.51 \times 10^{-3}\,\mathrm{s}^{-1}, \quad \tau_{KH} \cong 3.1\,\mathrm{min},$$

$$v_{ph} \cong 170\,\mathrm{km\,s}^{-1}, \quad \text{for } \lambda_{KH} = 12\,\mathrm{Mm},$$

and

$$\gamma_{KH} \cong 46.06 \times 10^{-3}\,\mathrm{s}^{-1}, \quad \tau_{KH} \cong 2.3\,\mathrm{min},$$

$$v_{ph} \cong 198\,\mathrm{km\,s}^{-1}, \quad \text{for } \lambda_{KH} = 15\,\mathrm{Mm}.$$

The KHI developing or growth times seem reasonable and the wave phase velocities are super-Alfvénic ones.

It is intriguing to see how the width of the instability range will shorten as the magnetic field twist ε_1 is increased. Our numerical computations indicate that for a noticeable contraction of the instability window one should change the magnitude of ε_1 with relatively large steps. The results of such computations are illustrated in Fig. 10.10. It is necessary to underline that at values of ε_1 close to 1, (i) β_i becomes <1 and the jet has to be treated as a cool medium, which implies a new wave

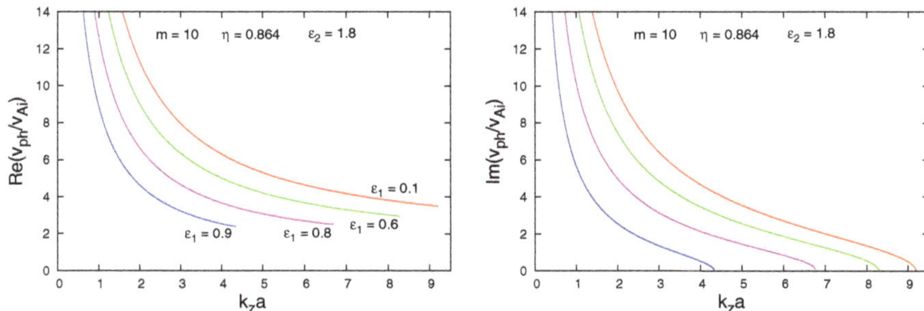

Fig. 10.10. (*Left panel*) Dispersion curves of the unstable $m = 10$ MHD mode propagating along a twisted incompressible emerging from a filament eruption jet at $\eta = 0.864$, $\varepsilon_2 = 1.8$, and the following values of ε_1 (from right to left): 0.1, 0.6, 0.8, and 0.9. Alfvén Mach numbers for these curves are respectively 2.27, 1.83, 1.37, and 0.9958. (*Right panel*) Growth rate curves of the unstable $m = 10$ MHD mode propagating along a twisted incompressible emerging from a filament eruption jet at $\eta = 0.864$, $\varepsilon_2 = 1.8$, and the following values of ε_1 (from right to left): 0.1 (orange), 0.6 (green), 0.8 (purple), and 0.9 (blue). *Reprinted with permission.*

dispersion relation and probably a higher wave mode number, m; (ii) one cannot use $\varepsilon_1 > 1$, because in that case the instability is of another kind, namely kink instability (Lundquist, 1951; Hood and Priest, 1979; Zaqarashvili *et al.*, 2014a). At this "pathological" case, one cannot reach the lower limit of the instability range, $(k_z a)_{\text{lhs}} = 0.524$, and consequently we are unable to evaluate that azimuthal magnetic field, B_ϕ^{cr}, which will stop the KHI onset!

10.4.4 *Kelvin–Helmholtz instability in a spinning macrospicule*

As we have mentioned in Sec. 1, Pike and Mason (1998) did a statistical study of the dynamics of solar transition region features, like macrospicules. These features were observed on the solar disk and also on the solar limb by using data from the *Coronal Diagnostic Spectrometer* (CDS) on board *SOHO*. In addition, in their article, Pike and Mason (1998) discussed the unique CDS observations of a macrospicule first reported by Pike and Harrison (1997) along with their own (Pike and Mason) observations from the *Normal Incidence Spectrometer* (NIS). This spectrometer covers the wavelength range from 307 to 379 Å and that from 513 to 633 Å using a microchannel plate and CCD combination detector. The details of macrospicule events observed near the limb are given in Table I in Pike and Mason (1998), while those of macrospicule events observed on the disk are presented in Table II. The main finding in the study of Pike and Mason (1998) was the rotation in these features based on the red and blue shifted emission on either side of the macrospicule axes. According to the authors, the detected rotation assuredly plays an important role in the dynamics of the transition region. Using the basic observational parameters obtained by Pike and Mason (1998), Zhelyazkov and Chandra (2019) examined the conditions for KHI rising in the macrospicule. Let us discuss that study as follows.

Our (Zhelyazkov and Chandra, 2019) choice for modeling namely the macrospicule observed on 1997 March 8 at 00:02 UT (see Table II in Pike and Mason, 1998) was made taking into account the fact that this macrospicule possesses the basic characteristics of the observed over the years tornado-like jets—the axial velocity of the jet was $U_z = 75\,\mathrm{km\,s^{-1}}$, while its rotating speed, we evaluate to be $U_\phi = 40\,\mathrm{km\,s^{-1}}$. For the other characteristics of the macrospicule such as lifetime, maximum width, average flow velocity, and maximum length or height, we used some average values obtained from a huge number of observations and specified in Kiss *et al.* (2017) as $16.75 \pm 4.5\,\mathrm{min}$, $6.1 \pm 4\,\mathrm{Mm}$, $73.14 \pm 25.92\,\mathrm{km\,s^{-1}}$, and $28.05 \pm 7.67\,\mathrm{Mm}$, respectively. For our study here, we take the macrospicule width to be $\Delta\ell = 6\,\mathrm{Mm}$, its height $H = 28\,\mathrm{Mm}$, and lifetime of the order on $15\,\mathrm{min}$. The basic macrospicule physical parameters (see Table 10.1) with $U_\phi = 40\,\mathrm{km\,s^{-1}}$, $\varepsilon_1 = 0.005$, and $B_e = 5\,\mathrm{G}$ yield [using Eq. (10.3)] $\eta = 0.1$, Alfvén speed $v_{\mathrm{Ai}} = 60.6\,\mathrm{km\,s^{-1}}$ (while in the surrounding plasma we have $v_{\mathrm{Ae}} \cong 345\,\mathrm{km\,s^{-1}}$), and $b = 1.798$. The excited MHD mode, whose instability window would contain a $\lambda_{\mathrm{KH}} = 3\,\mathrm{Mm}$, is $|m| = 52$. Performing the numerical computations with the input parameters: $m = 52$, $\eta = 0.1$, $\varepsilon_1 = 0.005$, $\varepsilon_2 = 0.53$ $(=40/75)$, $b = 1.798$, and $M_A = 1.24$ $(=75/60.6)$, we get graphics pictured in Fig. 10.11. From this plot, we can calculate the instability characteristics of the $m = 52$ MHD mode for two unstable wavelengths, equal to 3 and $5\,\mathrm{Mm}$ $(k_z a = 6\pi/5)$, respectively. The KHI wave growth rate, γ_{KH}, growth

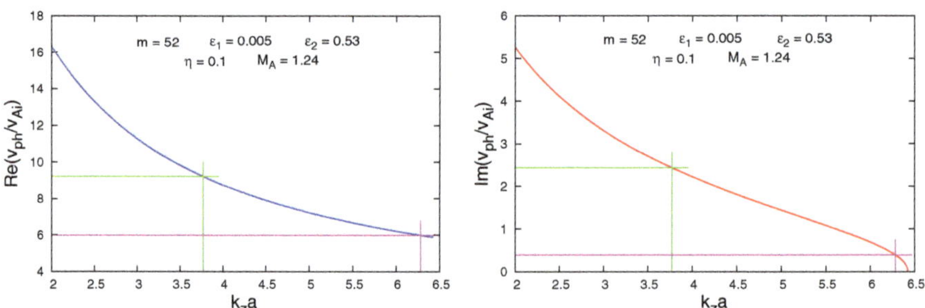

Fig. 10.11. (*Left panel*) Dispersion curve of the $m = 52$ MHD mode propagating along a twisted incompressible macrospicule at $\eta = 0.1$, $b = 1.798$, $M_A = 1.24$, $\varepsilon_1 = 0.005$, and $\varepsilon_2 = 0.53$. (*Right panel*) Normalized growth rate curve of the $m = 52$ MHD mode computed with the same input parameters as in the left panel. *Reprinted with permission.*

time, $\tau_{\mathrm{KH}} = 2\pi/\gamma_{\mathrm{KH}}$, as well as wave velocity, v_{ph}, calculated from the graphics in Fig. 10.11, for the aforementioned wavelengths are as follows:

$$\gamma_{\mathrm{KH}} \cong 48.38 \times 10^{-3}\,\mathrm{s}^{-1}, \quad \tau_{\mathrm{KH}} \cong 2.2\,\mathrm{min},$$

$$v_{\mathrm{ph}} \cong 361\,\mathrm{km\,s}^{-1}, \quad \text{for } \lambda_{\mathrm{KH}} = 3\,\mathrm{Mm},$$

and

$$\gamma_{\mathrm{KH}} \cong 184.8 \times 10^{-3}\,\mathrm{s}^{-1}, \quad \tau_{\mathrm{KH}} \cong 0.57\,\mathrm{min},$$

$$v_{\mathrm{ph}} \cong 556\,\mathrm{km\,s}^{-1}, \quad \text{for } \lambda_{\mathrm{KH}} = 5\,\mathrm{Mm}.$$

As seen, at both wavelengths phase velocities are super-Alfvénic. The two growth times of 2.2 and ≈ 0.6 min are reasonable taking into account that the macrospicule lifetime is about 15 min, that is, the KHI at the selected wavelengths is rather fast. A specific property of the instability $k_z a$ ranges is that for a fixed m, their widths depend upon the magnetic field twist parameter ε_1. A discussion on this dependence is provided in Zhelyazkov and Chandra (2018) and here we quote it, namely "with increasing the value of ε_1, the instability window becomes narrower and at some critical magnetic field twist its width equals zero. This circumstance implies that for $\varepsilon_1 \geqslant \varepsilon_1^{\mathrm{cr}}$ there is no instability, or, in other words, there exists a critical azimuthal magnetic field $B_\phi^{\mathrm{cr}} = \varepsilon_1^{\mathrm{cr}} B_{\mathrm{i}z}$ that suppresses the instability onset." In Fig. 10.12, a series of dispersion and dimensionless wave phase velocity growth rates for various increasing magnetic field twist parameter values has been plotted. Note that each larger ε_1 implies an increase in B_ϕ. The red dispersion curve in the

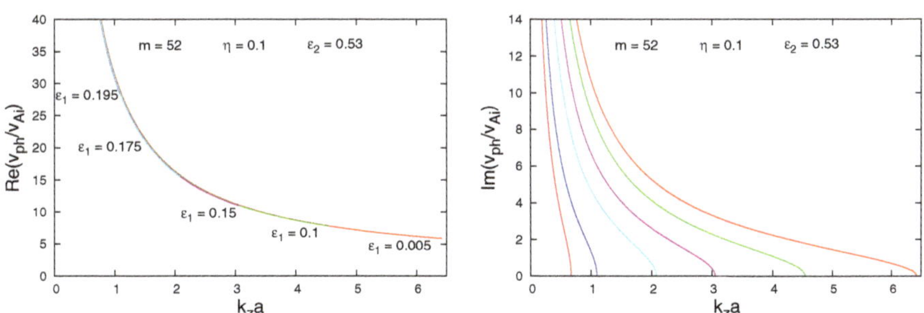

Fig. 10.12. (*Left panel*) Dispersion curves of the unstable $m = 52$ MHD mode propagating along a twisted incompressible macrospicule at $\eta = 0.1$, $\varepsilon_2 = 0.53$, and the following values of ε_1 (from right to left): $0.005, 0.1, 0.15, 0.175, 0.195$, and 0.202085 (red curve in the right plot). Alfvén Mach numbers for these curves are respectively $1.24, 1.23, 1.22, 1.22, 1.21$, and 1.2119. (*Right panel*) Growth rates of the unstable $m = 52$ mode for the same input parameters. The azimuthal magnetic field that corresponds to $\varepsilon_1 = 0.202085$ (the instability window with zero width) and stops the KHI onset is equal to 0.57 G. *Reprinted with permission.*

right panel of Fig. 10.12 has been obtained for $\varepsilon_1^{\mathrm{cr}} = 0.202085$ with $M_{\mathrm{A}} = 1.2119$, and it visually defines the left-hand-side limit of all other instability ranges. The azimuthal magnetic field B_ϕ^{cr} that stops the KHI is equal to $0.57\,\mathrm{G}$.

The instability $k_z a$-range of the $m = 52$ MHD mode pictured in Fig. 10.11 allows us to investigate how the width of the macrospicule will affect the KHI characteristics for a fixed instability wavelength. Such an appropriate wavelength is $\lambda_{\mathrm{KH}} = 4\,\mathrm{Mm}$. We calculate (and plot) the instability growth rate, γ_{KH}, the instability development or growth time, τ_{KH}, and the wave phase velocity of the $m = 52$ mode. Our choice for macrospicule widths is: 8, 6, and $4\,\mathrm{Mm}$, respectively. Note that the unstable $4\,\mathrm{Mm}$ wavelength has three different positions on the $k_z a$-axis, notably $k_z a = 2\pi$ for $\Delta\ell = 8\,\mathrm{Mm}$, $k_z a = 1.5\pi$ for $\Delta\ell = 6\,\mathrm{Mm}$, and $k_z a = \pi$ for $\Delta\ell = 4\,\mathrm{Mm}$ (see Fig. 10.13). The basic KHI characteristics of the $m = 52$ MHD mode at the wavelength $\gamma_{\mathrm{KH}} = 4\,\mathrm{Mm}$ at the aforementioned three different macrospicule widths are presented in Table 10.2. The most striking result concerns the instability growth or developing time: it is only approximately half a minute

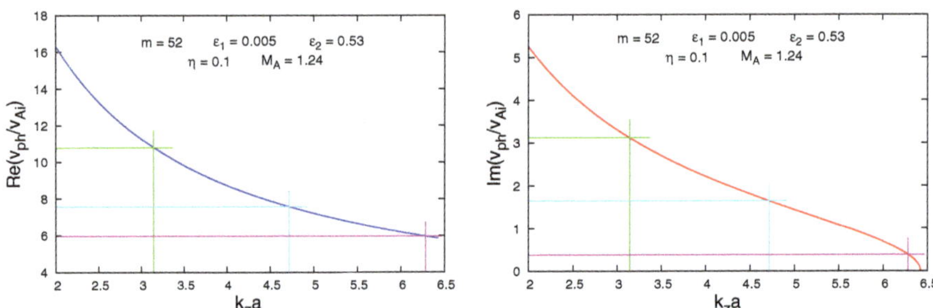

Fig. 10.13. (*Left panel*) Dispersion curve of the $m = 52$ MHD mode propagating along a twisted incompressible macrospicule at $\eta = 0.1$, $b = 1.798$, $M_{\mathrm{A}} = 1.24$, $\varepsilon_1 = 0.005$, and $\varepsilon_2 = 0.53$. The $k_z a$-positions marked with purple, cyan, and green vertical lines correspond to $\lambda_{\mathrm{KH}} = 4\,\mathrm{Mm}$ for three different macrospicule widths, equal to 8, 6, and $4\,\mathrm{Mm}$, respectively. (*Right panel*) Normalized growth rate curve of the $m = 52$ MHD mode propagating along a twisted incompressible macrospicule with the same input parameters as in the left panel. *Reprinted with permission.*

Table 10.2.　Kelvin–Helmholtz instability characteristics of the $m = 52$ MHD mode at $\lambda_{\mathrm{KH}} = 4$ Mm for three different widths of the macrospicule correspondingly equal to 8, 6, and $4\,\mathrm{Mm}$.

$\Delta\ell$ (Mm)	γ_{KH} ($\times 10^{-3}\,\mathrm{s}^{-1}$)	τ_{KH} (min)	v_{ph} ($\mathrm{km\,s}^{-1}$)	$\varepsilon_1^{\mathrm{cr}}$	B_ϕ^{cr} (G)
8	36.280	2.90	361	0.198430	0.55
6	156.48	0.67	459	0.202085	0.57
4	246.70	0.35	654	0.205465	0.58

at $\Delta\ell = 4\,\mathrm{Mm}$ and approximately seven times longer (2.9 min) when the jet width is 8 Mm. It is not surprising, bearing in mind the shape of the dispersion curve of the $m = 52$ MHD mode, that the wave phase velocities of the unstable mode quickly increase from 361 km s^{-1} at the widest macrospicule to 654 km s^{-1} at the narrowest one. In that table, we also give the critical magnetic field twist parameter, $\varepsilon_1^{\mathrm{cr}}$, at which the size of the instability range of a given jet becomes equal to zero. Those three limiting wave growth rate curves are plotted in three different colors in Fig. 10.14; from left to right the cyan curve corresponds to $\Delta\ell = 4\,\mathrm{Mm}$, the black one to 6 Mm, and the red curve to the macrospicule width of 8 Mm. It is interesting to observe that the critical azimuthal magnetic field components, B_ϕ^{cr}, that would stop the KHI appearance have very close magnitudes, roughly 0.6 G. The plots of the dimensionless dispersion curves for the last three values of ε_1 in Fig. 10.14, similar to those seen in the left panel of Fig. 10.12, are practically not usable—they possess very high values (in the range of 80–140) corresponding to extremely high wave phase velocities.

Another interesting observation is that the KHI growth or developing times of the $m = 52$ MHD mode, evaluated at wavelengths equal to the half-width of the

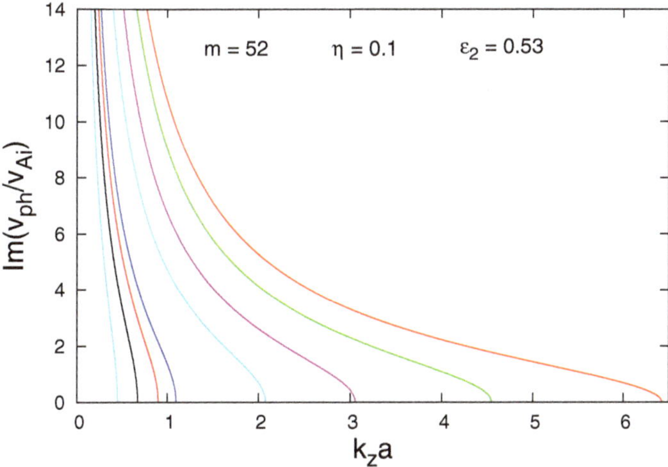

Fig. 10.14. Growth rate curves of the $m = 52$ MHD mode propagating along a twisted incompressible macrospicule at $\eta = 0.1$, $\varepsilon_2 = 0.53$, and eight different values of ε_1 equal to 0.005 (orange curve), 0.1 (green curve), 0.15 (purple curve), 0.175 (cyan curve), 0.195 (blue curve), 0.19843 (red curve), 0.202085 (black curve), and 0.205465 (cyan curve), respectively. The growth rate curves computed with the last three values of the magnetic field twist parameter ε_1 correspond to the three different macrospicule widths, equal to 8, 6, and 4 Mm, respectively. The Alfvén Mach numbers used at the computation of these three curves are accordingly equal to 1.2128, 1.2119, and 1.211. *Reprinted with permission.*

Table 10.3. Kelvin–Helmholtz instability growth times of the $m = 52$ MHD mode at wavelengths equal to the half-width of the jet and at $(\Delta\ell/2 + 2)\,\mathrm{Mm}$ for three different widths of the macrospicule correspondingly equal to 8, 6, and 4 Mm.

$\Delta\ell$ (Mm)	$\tau_{\mathrm{KH}}(\Delta\ell/2)$ (min)	$\tau_{\mathrm{KH}}\big[(\Delta\ell/2 + 2)\,\mathrm{Mm}\big]$ (min)
8	2.9	0.80
6	2.2	0.57
4	1.4	0.35

macrospicule and, say at $(\Delta\ell/2 + 2)\,\mathrm{Mm}$, for the three aforementioned widths are practically on the same order. This observation is illustrated in Table 10.3. At the thinnest jet both growth times are the shortest while at the thickest jet they are relatively longer. An observational evaluation of the macrospicule width will naturally help in finding the appropriate conditions for KHI onset of the corresponding high MHD mode.

It is interesting to see how the change in background magnetic field, B_{e}, will modify the picture. We set $B_{\mathrm{e}} = 4.8\,\mathrm{G}$ (a magnetic field at which the total balance Eq. (10.3) is satisfied)—with the same input parameters for electron number densities and electron temperatures, we have new values for Alfvén speeds, plasma betas, and magnetic fields ratio, notably $v_{\mathrm{Ai}} = 52.36\,\mathrm{km\ s^{-1}}$, $v_{\mathrm{Ae}} = 330.9\,\mathrm{km\ s^{-1}}$, $\beta_{\mathrm{i}} = 3.01$, $\beta_{\mathrm{e}} = 0.15$, and $b = 1.998$. With this new internal Alfvén speed, v_{Ai}, the Alfvén Mach number has a slightly higher magnitude, $M_{\mathrm{A}} = 1.43$. It turns out that the appropriate MHD mode number that provides the required width of the instability window at $\Delta\ell = 6\,\mathrm{Mm}$ is $m = 48$. Performing the calculations with $m = 48$, $\eta = 0.1$, $\varepsilon_1 = 0.005$, $\varepsilon_2 = 0.53$, $b = 1.998$, and $M_{\mathrm{A}} = 1.43$, we obtain dispersion and growth rate curves that are very similar to those pictured in Fig. 10.11. We do not plot these curves, but compare the instability characteristics for both mode numbers, 48 and 52, respectively, at the instability wavelength of 3 and 5 Mm. The results of this comparison are shown in Table 10.4. One observes that at both unstable wavelengths the corresponding phase velocities are super-Alfvénic. Moreover, the two growth times of 2.2 and $\sim$0.6 min seem reasonable bearing in mind the fact that the macrospicule lifetime is about 15 minutes, which implies that the KHI at the selected wavelengths is rather fast. The $B_{\mathrm{i}\phi}^{\mathrm{cr}}$ that suppresses the KHI onset, equals 0.57 G and was calculated with $\varepsilon_1^{\mathrm{cr}} = 0.202085$ and $M_{\mathrm{A}} = 1.2119$. Our study (Zhelyazkov and Chandra, 2019) shows that a decrease in the background magnetic field to $B_{\mathrm{e}} = 4.8\,\mathrm{G}$ would require the excitation of MHD wave with mode number $m = 48$, at which the KHI characteristics at the wavelengths of 3 and 5 Mm are very close to those obtained with $m = 52$.

Table 10.4. Kelvin–Helmholtz instability characteristics of the $m = 48$ and 52 MHD modes at $\lambda_{\rm KH} = 3$ and $5\,{\rm Mm}$ for the macrospicule width equal to $6\,{\rm Mm}$.

Mode number m	$\gamma_{\rm KH}$ $(\times 10^{-3}\,{\rm s}^{-1})$	$\tau_{\rm KH}$ (min)	$v_{\rm ph}$ $({\rm km\,s}^{-1})$	$\varepsilon_1^{\rm cr}$	$B_\phi^{\rm cr}$ (G)
		$\lambda_{\rm KH} = 3\,{\rm Mm}$			
48	49.23	2.1	338.0	0.234940	0.58
52	48.38	2.2	361.0	0.202085	0.57
		$\lambda_{\rm KH} = 5\,{\rm Mm}$			
48	172.96	0.61	518.0	0.234940	0.48
52	184.85	0.57	556.5	0.202085	0.58

10.5 Summary

The successful modeling of Kelvin–Helmholtz instability in rotating solar jets requires:

(i) Specification of magnetic and velocity fields topologies, which define some important input parameters like the density contrast $\eta = \rho_{\rm e}/\rho_{\rm i}$ of the jet–environment system, the ratio of axial magnetic fields $b = B_{\rm e}/B_{\rm iz}$, as well as the twists of the internal magnetic field $B_{\rm i}$ ($\varepsilon_1 = B_{\rm i\phi}(a)/B_{\rm iz}$) and jet velocity U ($\varepsilon_2 = U_\phi(a)/U_z$), where a is the radius of the magnetic flux tube which model the spinning jet. Recall that one assumes that the plasma densities are homogeneous.

(ii) No less important is the determination of the physical parameters of the jet and its environment: electron temperatures $T_{\rm e,i}$, electron number densities $n_{\rm e,i}$, jet's height H and width $\Delta\ell$, the background or jet's magnetic field, and plasma betas of both media. It is necessary to check whether the selected physical parameters and the chosen magnetic field satisfy the total pressure balance Eq. (10.3). The values of plasma betas will tell us what kind of wave dispersion relations will be used in searching the conditions under which the exited/detected high MHD modes ($m \geqslant 2$) become unstable against KH instability.

(iii) Before starting the numerical task for finding the solutions to the wave dispersion relation, it is recommended to use the instability condition (10.12), which allows to roughly determine the expected unstable $k_z a$ range [i.e., to evaluate $(k_z a)_{\rm lhs}$ using Eq. (10.14) and $(k_z a)_{\rm rhs}$ via Eq. (10.13)], as well as the MHD wave mode number m given by inequality (10.15). The magnitude of m can lie in a wide range of values depending on the choice of input parameters in (10.15). For solving given dispersion equation, it is necessary to specify the values of η, b, ε_1, ε_2, and the Alfvén Mach number $M_A = U_z/v_{Ai}$, where $v_{Ai} = B_{iz}/\sqrt{\mu\rho_i}$. Having obtained the solutions (in complex variables) to the wave dispersion relation, from the graphical plots of the normalized real and imaginary parts of the wave phase velocity as functions of the non-dimensional wavenumber $k_z a$, one can compute the KH instability parameters: growth rate γ_{KH}, growth time $\tau_{KH} = 2\pi/\gamma_{KH}$, and the wave phase velocity v_{ph} at a fixed wavelength $\lambda_{KH} = \pi \Delta\ell/k_z a$.

(iv) It is well established that the width of the instability range depends upon the value of ε_1 (the value of ε_2 is fixed by $U_\phi(a)$ and U_z). Each increase in ε_1 yields to a narrowing of the instability range and at some ε_1^{cr} that range becomes equal to zero. That critical magnetic field twist parameter allows to evaluate the critical azimuthal magnetic field $B_{i\phi}^{cr} = \varepsilon_1^{cr} B_{iz}$. It is very important to note that each change in ε_1 is associated with an increase of the axial magnetic field and consequently of the Alfvén speed v_{Ai} and Alfvén Mach number M_A. For small values of ε_1 in the range of 0.001 to 0.4, the changes in v_{Ai} and M_A are negligibly small, but for bigger values they are not. That is why it is strongly recommended for each value of ε_1, to compute the appropriate values of v_{Ai} and M_A from the total pressure balance Eq. (10.3). The latter plays an important role in such studies.

Chapter 11

Kelvin–Helmholtz Instability in Coronal Mass Ejections

11.1 Coronal mass ejections and magnetic flux ropes

Alongside the small-scale solar eruptions, which we already studied in the previous six chapters, there also exist large-scale solar mass eruptions called coronal mass ejections (CMEs), which are probably the most important sources of adverse space weather effects (see, e.g., Howard and DeForest, 2014; Aparna and Tripathi, 2016; Zhang *et al.*, 2007; Temmer *et al.*, 2010). They are often associated with dramatic changes of coronal magnetic fields (see Zhang and Low, 2005; Liu *et al.*, 2009a; Su and van Ballegooijen, 2012). Coronal mass ejections are huge clouds of magnetized plasma (about 10^{15-16} g) that erupt from the solar corona into interplanetary space. CMEs were observationally discovered 40 years ago (Gosling *et al.*, 1974)—they were observed with the white light coronagraph experiment on board *Skylab* during the first 118 days of the mission. CMEs propagate in the heliosphere with velocities ranging from 20 to 3200 km s^{-1} with an average speed of 489 km s^{-1}, according to *SOHO*/LASCO coronagraph measurements between 1996 and 2003. The CME on July 23, 2012, captured by *STEREO*-Ahead's Cor2 (coronagraph), clocked in between 2900 and 3540 kilometers per second and it is the fastest one ever seen by *STEREO*. Depending on the velocity magnitude, according to the CME SCORE Scale (Evans *et al.*, 2013), one can distinguish five categories of CMEs, notably: S-type ($<$500 km s^{-1}), C-type (Common) (500–999 km s^{-1}), O-type (Occasional) (1000–1999 km s^{-1}), R-type (Rare) (2000–2999 km s^{-1}), and ER-type (Extremely Rare) ($>$3000 km s^{-1}). The physical mechanisms that initiate and drive solar eruptions were discussed in many papers over past four decades (see, e.g., Chandra *et al.*, 2014; Schmieder *et al.*, 2013; Chandra *et al.*, 2011; Aulanier *et al.*, 2010; Forbes *et al.*, 2006; Török and

199

Kliem, 2005, and references therein). Recently, according to Aulanier (2014) "no more than two distinct physical mechanisms can actually initiate and drive prominence eruptions: the *magnetic breakout* and the *torus instability*. In this view, all other processes (including flux emergence, flux cancellation, flare reconnection, and long-range couplings) should be considered as various ways that lead to, or that strengthen, one of the aforementioned driving mechanisms." Studies using the data sets from (among others) the *SOHO*, *TRACE*, *Wind*, *ACE*, *STEREO*, *SDO*, and *IRIS* spacecrafts, along with ground-based instruments, have improved our knowledge of the origins and development of CMEs at the Sun (see, e.g., Webb and Howard, 2012; Joshi *et al.*, 2013a,b; Landi and Miralles, 2014).

It is generally accepted that CMEs are the results of eruptions of magnetic flux ropes (MFRs) (see Fig. 11.1). However, there is heated debate on whether MFRs exist prior to the eruptions or if they are formed during the eruptions. Several coronal signatures, for example, filaments, coronal cavities, sigmoid structures, and hot channels (or hot blobs), are proposed as MFRs and observed before the eruption, which support the pre-existing MFR scenario (see, e.g., Cheng *et al.*, 2011; Patsourakos *et al.*, 2013; Vourlidas, 2014; Green and Kliem, 2014; Cheng *et al.*, 2014; Chen *et al.*, 2014; Howard and DeForest, 2014; Vemareddy and Zhang, 2014; Zhang *et al.*, 2016). According to the second scenario, a new flux rope forms as a result of the reconnection of the magnetic lines of an arcade (a group of arches of field lines) during the eruption itself. Observational support for this mechanism was recently reported by Song *et al.* (2014). The authors present an intriguing observation of a solar eruptive event that occurred on 2013 November 21 with the

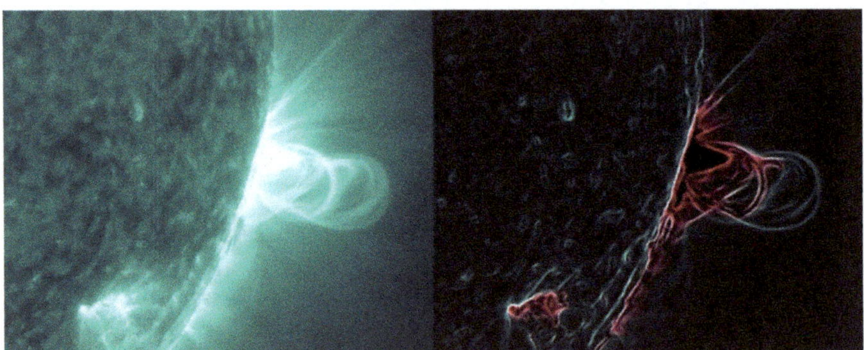

Fig. 11.1. The image on the left shows a series of magnetic loops on the Sun, as captured by NASA's *SDO* on July 18, 2012. The image on the right has been processed to highlight the edges of each loop and make the structure more clear. A series of loops such as this is known as a flux rope, and these lie at the heart of eruptions on the Sun known as CMEs. This is the first time scientists were able to discern the timing of a flux rope's formation. *Credit to: NASA*/SDO.

AIA/*SDO*, which shows the formation process of the MFR during the eruption in detail. After all, the majority of observations support the view that, in at least some CMEs, flux rope formation occurs before launch. Thus, we are convinced to accept the expanded definition of the CME suggested in Vourlidas *et al.* (2013): "A CME is the eruption of a coherent magnetic, twist-carrying coronal structure with angular width of at least $40°$ and able to reach beyond $10\,R_{\text{Sun}}$ which occurs on a timescale of a few minutes to several hours."

11.2 Kelvin–Helmholtz instability in coronal mass ejections

The first observations of the temporally and spatially resolved evolution of the magnetic KH instability in the solar corona based on unprecedented high-resolution imaging observations of vortices developing at the surface of a fast coronal mass ejecta (less than 150 Mm above the solar surface in the inner corona) taken with the Atmospheric Imaging Assembly (AIA) on board the *SDO* were reported by Foullon *et al.* (2011). The CME event they studied occurred on 2010 November 3, following a C4.9 *GOES* class flare from active region NOAA 11121. The active region was located at S21E88 on that day on the solar disk. According to *GOES* observations, the flare was initiated at 12:07 UT, peaked at 12:15 UT, and ended around 12:30 UT. The flare or eruption was clearly observed by *SDO*/AIA at different wavelengths with high spatial (0.6 arcsec) and temporal resolution (12 s) as well as by the Reuven Ramaty High-Energy Solar Spectroscopic Imager (RHESSI, Lin *et al.*, 2002) in X-ray 3–6 keV, 6–12 keV, 12–25 keV, and 25–50 keV energy bands. The temporal evolution of the flare in X-rays observed by GOES and RHESSI is presented in Fig. 11.2.

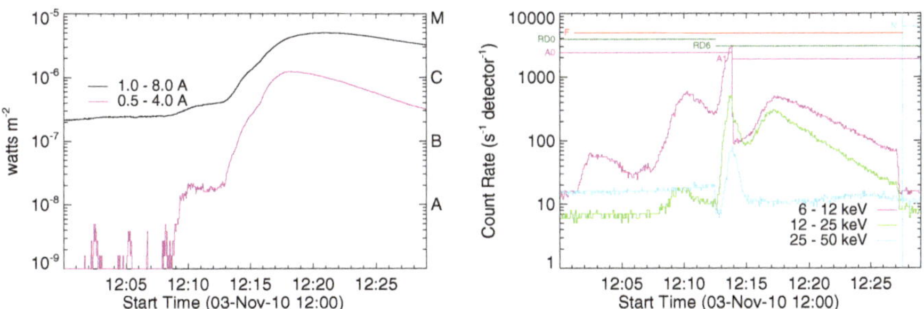

Fig. 11.2. (*Left panel*) *SDO*/AIA 131 Å image overlaid by RHESSI X-ray contours (contour level: 40%, 70%, and 95% of the peak intensity). The rectangular white box shows where the KH instability is visible. (*Right panel*) The associated difference image of the LASCO CME overlaid on the *SDO*/AIA image. *Reprinted with permission.*

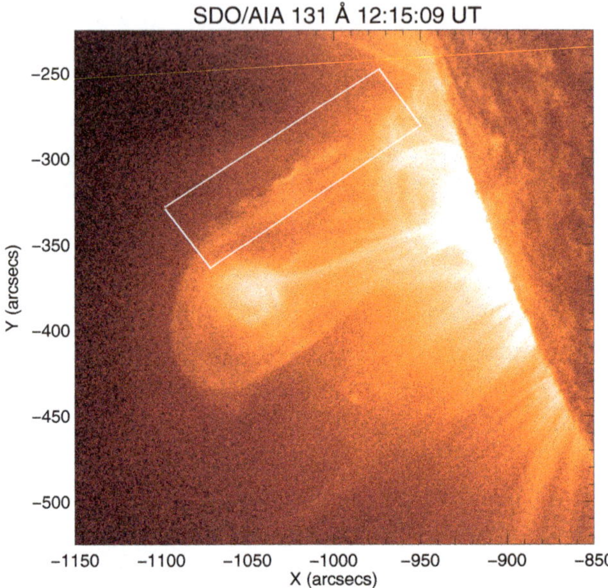

Fig. 11.3. *SDO*/AIA image showing a fast coronal mass ejecta erupting from the Sun, with KH vortices detected on its northern flank closed by rectangle. *Credit to: NASA*/SDO.

The SDO/AIA observations on 2010 November 3 at different wavelengths show evidence of a KH instability around 12:15 UT at 131 Å (see also Foullon *et al.*, 2011, 2013). Figure 11.3 shows the *SDO*/AIA 131 Å at 12:15 UT, where we can see the KH instability features (see the white rectangular box). The image is overlaid by the RHESSI 6–12 keV contours. The RHESSI image is constructed from seven collimators (3F to 9F) using the CLEAN algorithm, which yields a spatial resolution of ∼7 arcsec (Hurford *et al.*, 2002).

LASCO observed a CME at 12:36 UT in the C2 field of view with a speed of 241 km s^{-1} and an angular width of 66° associated with the flare.

The study by Foullon *et al.* (2013) of the dynamics and origin of the CME on 2010 November 3 by means of the Solar TErrestrial RElations Observatory Behind (STEREO-B) located eastward of *SDO* by 82° of heliolongitude, and used in conjunction with *SDO*, gives some indication of the magnetic field topology and flow pattern. At the time of the event, the Extreme Ultraviolet Imager (EUVI) from STEREO's Sun–Earth Connection Coronal and Heliospheric Investigation (SECCHI) instrument suite achieved the highest temporal resolution in the 195 Å bandpass: EUVI's images of the active region on the disk were taken every 5 minutes in this bandpass. The authors applied differential emission measure (DEM) techniques on the edge of the ejecta to determine the basic plasma parameters—they obtained an electron temperature of $11.6 \pm 3.8\,\text{MK}$ and an electron density of $n = (7.1 \pm 1.6) \times 10^8\,\text{cm}^{-3}$, together with a layer width $\Delta\ell = 4.1 \pm 0.7\,\text{Mm}$. Density estimates of the ejecta environment (quiet corona), according to Aschwanden and Acton (2001), vary from $(2\,\text{to}\,1) \times 10^8\,\text{cm}^{-3}$ between 0.05 and $0.15\,R_{\text{Sun}}$ (40–100 Mm), at heights where Foullon *et al.* (2013) started to see developing KH waves. The final estimate based on a maximum height of 250 Mm and the highest DEM value on the northern flank of the ejecta yields an electron density of $(7.1 \pm 0.8) \times 10^8\,\text{cm}^{-3}$. The adopted electron temperature in the ambient corona is $T = 4.5 \pm 1.5\,\text{MK}$. The other important parameters derived from using the pressure balance equation assuming a benchmark value for the magnetic field B in the environment of $10\,\text{G}$ are summarized in Table 2. The main features of the imaged KH instability presented in Table 3 include (in their notation) the speed of the 131 Å CME leading edge, $V_{\text{LE}} = 687\,\text{km s}^{-1}$, the flow shear on the 131 Å CME flank, $V_1 - V_2 = 680 \pm 92\,\text{km s}^{-1}$, the KH group velocity, $v_g = 429 \pm 8\,\text{km s}^{-1}$, the KH wavelength, $\lambda = 18.5 \pm 0.5\,\text{Mm}$, and an exponential linear growth rate, $\gamma_{\text{KH}} = 0.033 \pm 0.012\,\text{s}^{-1}$.

Since the origin of erupted material is a flux rope, the most appropriate model of the CME under consideration is a twisted magnetic flux tube of radius a $(=\Delta\ell/2)$ and density ρ_i embedded in a uniform field environment with density ρ_e (see any figure of twisted internal magnetic field from previous few chapters). The magnetic field inside the tube is helicoidal, $\boldsymbol{B}_i = (0, B_{i\phi}(r), B_{iz}(r))$, while on the outside it is uniform and directed along the z axis, $\boldsymbol{B}_e = (0, 0, B_e)$. The tube moves along its axis with a velocity of $\boldsymbol{v}_0$ with regard to the surrounding medium. The jump of the tangential velocity at the tube boundary then triggers the magnetic KH instability when the jump exceeds a critical value.

Before discussing the dispersion MHD wave equations, we specify what kind of plasma each medium is (the moving tube and its environment). The CME parameters listed in Table 2 in Foullon *et al.* (2013) show that the plasma beta inside the flux tube might be equal to 1.5 ± 1.01, while that of the coronal plasma is 0.21 ± 0.05. Hence, we can consider the ejecta as an incompressible medium and treat its environment as a cool plasma ($\beta_e = 0$). In this case, the appropriate MHD wave dispersion relation is Eq. (9.3) from Chapter 9, notably

$$\frac{\left[(\omega - \boldsymbol{k} \cdot \boldsymbol{v}_0)^2 - \omega_{\text{Ai}}^2\right] F_m(\kappa_i a) - 2mA\omega_{\text{Ai}}\sqrt{\mu\rho_i}}{\left[(\omega - \boldsymbol{k} \cdot \boldsymbol{v}_0)^2 - \omega_{\text{Ai}}^2\right]^2 - 4\omega_{\text{Ai}}^2/\mu\rho_i}$$

$$= \frac{P_m(\kappa_e a)}{\dfrac{\rho_e}{\rho_i}(\omega^2 - \omega_{\text{Ae}}^2) + A^2 P_m(\kappa_e a)/\mu\rho_i}, \tag{11.1}$$

where

$$F_m(\kappa_i a) = \frac{\kappa_i a I'_m(\kappa_i a)}{I_m(\kappa_i a)} \quad \text{and} \quad P_m(\kappa_e a) = \frac{\kappa_e a K'_m(\kappa_e a)}{K_m(\kappa_e a)}.$$

The wave attenuation coefficient inside the tube is given by

$$\kappa_i = k_z \left(1 - \frac{4A^2 \omega_{Ai}^2}{\mu \rho_i \left[(\omega - \boldsymbol{k} \cdot \boldsymbol{v}_0)^2 - \omega_{Ai}^2 \right]^2} \right)^{1/2}, \tag{11.2}$$

where

$$\omega_{Ai} = \frac{1}{\sqrt{\mu \rho_i}} (m A + k_z B_{iz}),$$

while in the environment it has the form

$$\kappa_e a = k_z a \left[1 - (\omega/\omega_{Ae})^2 \right]^{1/2}, \tag{11.3}$$

in which $\omega_{Ae} = k_z B_e / \sqrt{\mu \rho_e}$.

11.3　Numerical solutions and wave dispersion diagrams

We focus our study first on the propagation of the kink mode, $m = 1$. It is obvious that Eq. (11.1) can only be solved numerically. After performing the standard procedure of normalization of all variables in this equation, we have to specify the input parameters for its solving. Our choice for the density contrast is $\eta = 0.88$, which corresponds to electron densities $n_i = 8.7 \times 10^8$ cm^{-3} and $n_e = 7.67 \times 10^8$ cm^{-3}. With $\beta_i = 1.5$ and $\beta_e = 0$, the ratio of axial magnetic fields is $b = 1.58$. If we fix the Alfvén speed in the environment to be $v_{Ae} \cong 787$ km s^{-1} (i.e., the value corresponding to $n_e = 7.67 \times 10^8$ cm^{-3} and $B_e = 10$ G), the total pressure balance equation at $\eta = 0.88$ requires a sound speed inside the jet of $c_{si} \cong 523$ km s^{-1} and an Alfvén speed $v_{Ai} \cong 467$ km s^{-1} (more exactly, 467.44 km s^{-1}), which corresponds to a magnetic field in the flux tube $B_{iz} = 6.32$ G. With magnetic field twist parameter $\varepsilon_1 = 0.025$ (less than 1.0 to avoid the emergence of the kink instability) and M_A varied from zero (no flow) to reasonable values during computations, the solutions (in complex variables) to Eq. (11.1) yield the graphics pictured in Fig. 11.4. Note that the red curve in the left panel is in fact a marginal dispersion curve corresponding to the critical $M_A^{cr} = 2.95$ [the instability criterion, e.g., Eq. (5.5) from Chapter 5, yields a threshold Alfvén Mach number of 2.73]. The critical jet speed, $M_A^{cr} v_{Ai} \cong 1380$ km s^{-1}, which in principle is

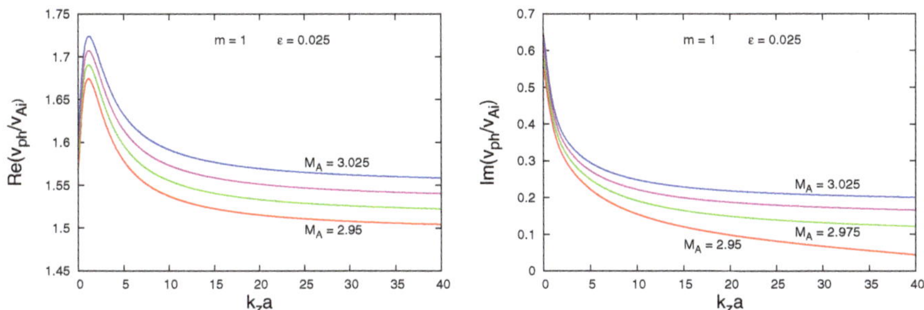

Fig. 11.4. (*Left panel*) Dispersion curves of an unstable kink ($m = 1$) mode propagating on a moving twisted magnetic flux tube with twist parameter $\varepsilon = 0.025$ at $\eta = 0.88$, $b = 1.58$ and Alfvén Mach numbers equal to 2.95, 2.975, 3, and 3.025. (*Right panel*) Growth rates of the unstable kink mode. Note that the marginal red curve reaches values lower than 0.001 beyond $k_z a = 53$. *Reprinted with permission.*

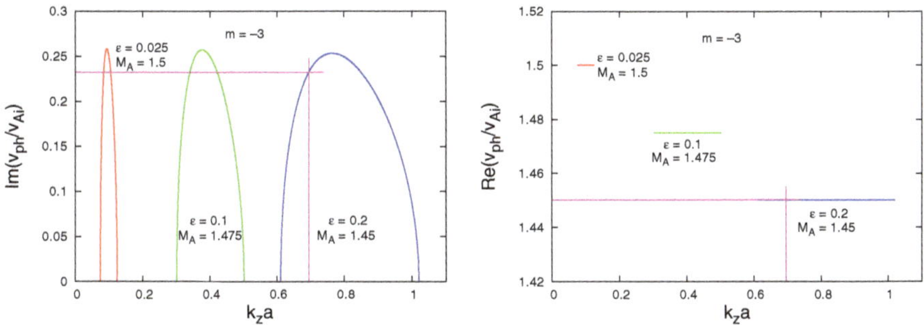

Fig. 11.5. (*Left panel*) Growth rates of an unstable $m = -3$ MHD mode in three instability windows. Adequate numerical values consistent with the observational data are obtained at $k_z a = 0.696245$, located in the third instability window, for which value of the dimensionless wavenumber the normalized mode growth rate is equal to 0.232117. (*Right panel*) Dispersion curves of unstable $m = -3$ MHD mode for three values of the twist parameter $\varepsilon = 0.025$, 0.1, and 0.2. The normalized phase velocity at $k_z a = 0.696245$ is equal to 1.45. *Reprinted with permission.*

accessible for CMEs, is, in fact, more than twice as high as the speed registered by Foullon *et al.* (2013), who reported a threshold speed of 680 km s^{-1} for observing a KH instability. Hence, the detected KH instability cannot be associated with the kink mode.

However, the situation distinctly changes for the negative $m = -3$ MHD mode number. Figure 11.5 shows the appearance of three instability windows on the $k_z a$-axis. The width of each instability window depends upon the value of the twist parameter ε—the narrowest window corresponds to $\varepsilon = 0.025$, and the widest to $\varepsilon = 0.2$. The phase velocities of unstable $m = -3$ MHD modes coincide with the magnetic flux tube speeds (the right panel of Fig. 11.5 shows that the normalized wave velocity on a given dispersion curve is equal to its label M_A). Therefore, unstable perturbations are frozen in the flow, and consequently, they are vortices rather than waves. This is firmly based on physics because the KH instability in hydrodynamics deals with unstable vortices. All critical Alfvén Mach numbers yield acceptable threshold speeds of the ejecta that ensure the occurrence of KH instability—these speeds are equal to $701\,\mathrm{km\,s^{-1}}$, $689\,\mathrm{km\,s^{-1}}$, and $678\,\mathrm{km\,s^{-1}}$ and agree very well with the speed of $680\,\mathrm{km\,s^{-1}}$ found by Foullon *et al.* (2013). The observationally detected KH instability wavelength $\lambda_{KH} = 18.5\,\mathrm{Mm}$ and ejecta width $\Delta\ell = 4.1\,\mathrm{Mm}$ define the corresponding instability dimensionless wave number, $k_z a = \pi\,\Delta\ell/\lambda$, to be equal to 0.696245. Figure 11.5 shows that $k_z a = 0.696245$ lies in the third instability window and accordingly determines a value of the dimensionless growth rate $\mathrm{Im}(v_{\mathrm{ph}}/v_{\mathrm{Ai}}) = 0.232117$ (see the left panel of Fig. 11.5), which implies a computed wave growth rate $\gamma_{KH} = 0.037\,\mathrm{s^{-1}}$, being in good agreement with the deduced from observations $\gamma_{KH} = 0.033\,\mathrm{s^{-1}}$. We also note that the wave phase velocity estimated from Fig. 11.5 (right panel) of $678\,\mathrm{km\,s^{-1}}$ is rather close to the speed of the $131\,\mathrm{\AA}$ CME leading edge, which is equal to $687\,\mathrm{km\,s^{-1}}$. The position of a given instability window, at fixed input parameters η and b, is determined chiefly by the magnetic field twist in the moving flux tube. This circumstance allows us by slightly shifting the third instability window to the right, to tune the vertical purple line (see the left panel of Fig. 11.5) to cross the growth rate curve at a value that would yield $\gamma_{KH} = 0.033\,\mathrm{s^{-1}}$. The necessary shift of only 0.023625 can be achieved by taking the magnetic field twist parameter, ε, to be equal to 0.20739—in that case, the normalized $\mathrm{Im}(v_{\mathrm{ph}}/v_{\mathrm{Ai}}) = 0.207999$ gives the registered KH instability growth rate of $0.033\,\mathrm{s^{-1}}$. (That very small instability window shift does not change the critical ejecta speed by much.) In this way, we demonstrate the flexibility of our model, which allows deriving the numerical KH instability characteristics in very good agreement with observational data.

The observed quasi-periodic vortex-like structures at the northern boundary of a filament during an eruption on 2011 February 24 by Möstl *et al.* (2013) can be modeled in a similar way. In their case, the environment electron density n_e is taken to be $1 \times 10^9\,\mathrm{cm^{-3}}$, plasma pressure p_e is $0.09\,\mathrm{Pa}$, and the magnetic field B_e is $10\,\mathrm{G}$. In the filament, they assumed that the density is at least 10 times higher than the corona value (i.e., $n_i = 1 \times 10^{10}\,\mathrm{cm^{-3}}$) and that $T_i \leqslant 1 \times 10^5\,\mathrm{K}$. From these data, one obtains that the temperature of coronal plasma is $T_e = 3.26 \times 10^6\,\mathrm{K}$ and, accordingly, $c_{se} \cong 212\,\mathrm{km\,s^{-1}}$. Alfvén speed in the environment is $v_{Ae} \cong 689\,\mathrm{km\,s^{-1}}$. With a density contrast $\eta = 0.1$, the total pressure (sum of thermal and magnetic pressures) balance equation yields the

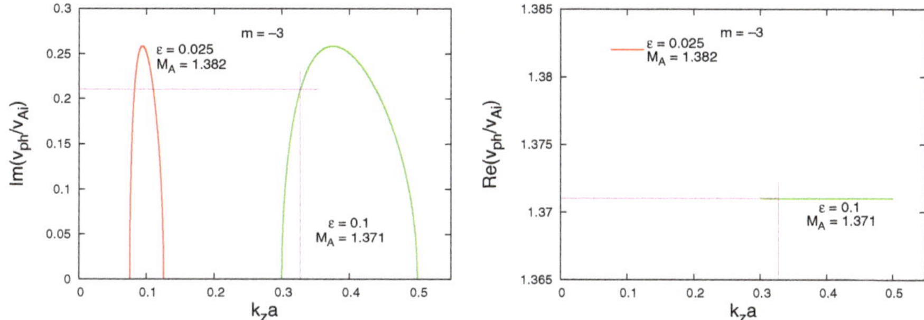

Fig. 11.6. (*Left panel*) Growth rates of an unstable $m = -3$ MHD mode in two instability windows. The best coincidence with the observational data one obtains at $k_z a = 0.32725$, for which value the normalized mode growth rate is equal to 0.2096. (*Right panel*) Dispersion curves of unstable $m = -3$ MHD mode for two values of the twist parameter $\varepsilon = 0.025$ and 0.1. The normalized phase velocity at $k_z a = 0.32725$ is equal to 1.371. *Reprinted with permission.*

Alfvén velocity inside the ejecta, $v_{Ai} = 226.4 \,\mathrm{km\,s^{-1}}$. The magnetic field there is similar to that in the corona, $B_{iz} \cong 10.4\,\mathrm{G}$, that implies a value of parameter $b = B_e / B_{iz}$ close to 1 (actually it is equal to 1.0385). Considering both media as incompressible magnetized plasmas, we can use dispersion equation (11.1) with taking the wave attenuation coefficient in the environment, κ_e, to be simply equal to k_z, that is, to use Eq. (13) of Zhelyazkov and Zaqarashvili (2012). As in the case of Foullon *et al.* (2013), the kink mode ($m = 1$) may become unstable at rather high speed of the moving flux tube, notably $\cong 1138\,\mathrm{km\,s^{-1}}$—a speed that is ~ 3.7 times higher than the detected critical velocity of $310\,\mathrm{km\,s^{-1}}$. The $m = -3$ MHD harmonic (see the left panel in Fig. 11.6) becomes unstable at threshold Alfvén Mach numbers equal to 1.382 (at $\varepsilon = 0.025$) and 1.371 (at $\varepsilon = 0.1$). According to Möstl *et al.* (2013), the half-width of the filament, $\Delta\ell/2 = a$, lies between $0.5\,\mathrm{Mm}$ and $1.0\,\mathrm{Mm}$. If one takes $a = 0.75\,\mathrm{Mm}$, the registered KH wavelength of $14.4\,\mathrm{Mm}$ corresponds to the dimensionless wavenumber $k_z a \cong 0.327$. Figure 11.6 shows that this normalized wavenumber falls in the second instability window for a twist of 0.1. For $a = 1.0\,\mathrm{Mm}$, the corresponding $k_z a = 0.436$, which also alights in the same window. However, for the smallest $a = 0.5\,\mathrm{Mm}$, the dimensionless wavenumber is equal to 0.218 and one needs to perform calculations for a smaller value of the magnetic field twist parameter ε—a value of 0.08 would yield an instability window accommodating that $k_z a = 0.218$. The normalized wave phase velocity growth rate that corresponds to $k_z a \cong 0.327$ is equal (see the left panel in Fig. 11.6) to 0.2096 which yields a linear growth rate $\gamma_{KH} \cong 0.021\,\mathrm{s^{-1}}$. For the other two dimensionless wavenumbers, the corresponding growth rates are of the same order. The computed critical filament speed in the second instability window is equal to $310.4\,\mathrm{km\,s^{-1}}$, that is, very close to the deduced from observations velocity of $310\,\mathrm{km\,s^{-1}}$. The value of the wave phase velocity, as in the case of Foullon *et al.* (2013), coincides with the critical speed of corresponding instability window.

Although similar in nature, the two ejecta (see Foullon *et al.*, 2013; Möstl *et al.*, 2013) have distinctive characteristics: the density contrast in Foullon *et al.* (2013) is small ($\eta = 0.88$) while in Möstl *et al.* (2013) it is relatively high ($\eta = 0.1$). The first ejecta is much hotter that its environment, while for the second one we have just the opposite situation. It is necessary to stress that each CME is a unique event and its successful modeling requires a full set of observational data for the plasma densities, magnetic fields, and temperatures of both media along with the detected ejecta width and speed. Concerning values of the flux tube speeds at which the instability starts, they can vary from a few kilometers per second, 6–$10\,\mathrm{km\,s^{-1}}$, as observed by Ofman and Thompson (2011), through $310 \pm 20\,\mathrm{km\,s^{-1}}$ of Möstl *et al.* (2013), $680 \pm 92\,\mathrm{km\,s^{-1}}$, deduced from Foullon *et al.* (2013), to $770\,\mathrm{km\,s^{-1}}$ from the 2.5D simulation of the event of 3 November 2010 performed by Nykyri and Foullon (2013).

Finally, we would like to comment on whether there is a required field orientations $\phi = \widehat{(\boldsymbol{k}, \boldsymbol{B})}$. From analyzing a flat (semi-infinite) geometry in their study, Foullon *et al.* (2013) concluded that it is most probable to observe a wave propagation quasi-perpendicular to the magnetic field $\boldsymbol{B}_{\mathrm{e}}$. But this restriction on the ejecta magnetic field tilt angle cancels out in our numerical investigation because the inequality $|V_{\mathrm{i}} - V_{\mathrm{e}}| \geqslant \sqrt{2}v_{\mathrm{Ai}} = 661\,\mathrm{km\,s^{-1}}$, required for wave parallel propagation, is satisfied (Zhelyazkov *et al.*, 2015c). Thus, the adopted magnetic flux rope nature of a CME and its (ejecta) consideration as a moving twisted magnetic flux tube allow us to explain the emerging instability as a manifestation of the KH instability of a suitable MHD mode.

$$\begin{array}{c}\textbf{Chapter 12}\end{array}$$

Summary and Outlook

The main goal of this book is to demonstrate how one can explore the Kelvin–Helmholtz instability of MHD modes propagating along various solar jets using relatively simple but reliable models of the plasma flows. Let us first emphasize that we are modeling and learning a particular event. This means that it is extremely important to get the fullest possible information about the physical parameters of the jet and its environment. Today's spacecrafts and missions like *Solar and Heliospheric Observatory (SOHO)*, *Ulysses*, *Hinode*, *Solar Dynamics Observatory (SDO)*, *Transition Region and Coronal Explorer (TRACE)*, *Reuven Ramaty High-Energy Solar Spectroscopic Imager (RHESSI)*, *Interface Region Imaging Spectrograph (IRIS)*, *Parker Solar Probe* along with the Earth-based telescopes Big Bear Solar Observatory, New Vacuum Solar Telescope, Swedish 1-m Solar Telescope, THÉMIS Solar Telescope and other provide the basic jets' parameters: electron number density and electron temperature, the magnetic field and its twist (if it exists). We model a particular jet as a vertical cylindrical flux tube with radius a, moving with axial velocity $\boldsymbol{v}_0$. In the case when the jet is rotating, it is more convenient that the spinning velocity to be noted by $\boldsymbol{U}$, possessing both axial U_z and azimuthal U_ϕ component. For simplicity, we assume that the jet plasma density ρ_i is homogeneous. The same assumption holds for electron temperature T_i, too. Concerning the environment's parameters, ρ_e and T_e, we usually make an assumption about their values. The latter can be greater or less than those of the plasma flow. We define the density contrast between the jet and its surrounding magnetoplasma as $\eta = \rho_\mathrm{e}/\rho_\mathrm{i}$. Note that aforementioned values cannot be arbitrary—they should satisfy the total pressure balance equation

$$\frac{\mathrm{d}}{\mathrm{d}r}\left(p + \frac{B^2}{2\mu}\right) = 0, \tag{12.1}$$

where p is thermal (plasma) pressure, $B^2/2\mu$ is the magnetic pressure, and μ is the permeability of the free space. At the jet–environment interface, $r = a$, this equation yields

$$p_{\rm i} + \frac{B_{\rm i}^2}{2\mu} = p_{\rm e} + \frac{B_{\rm e}^2}{2\mu}. \tag{12.2}$$

In the case when from observations instead of electron temperatures and magnetic fields we get the sound $v_{\rm s}$ and Alfvén speeds $v_{\rm A}$, the total pressure balance equation has the form

$$\frac{\rho_{\rm e}}{\rho_{\rm i}} = \frac{v_{\rm si}^2 + \frac{5}{6}v_{\rm Ai}^2}{v_{\rm se}^2 + \frac{5}{6}v_{\rm Ae}^2}. \tag{12.3}$$

When the internal magnetic field is twisted recall that the magnetic field and plasma pressure satisfy the equilibrium condition in the radial direction

$$\frac{\rm d}{{\rm d}r}\left(p_{\rm i} + \frac{B_{\rm i}^2}{2\mu}\right) = -\frac{B_{\rm i\phi}^2}{\mu r}, \tag{12.4}$$

where $B_{\rm i}(r) = (B_{\rm i\phi}^2 + B_{\rm iz}^2)^{1/2} = |\boldsymbol{B}_{\rm i}|$ denotes the strength of the equilibrium magnetic field. Further on, we consider the special case of an equilibrium with uniform twist with $\boldsymbol{B}_{\rm i} = (0, B_{\rm i\phi}(r), B_{\rm iz}(r))$, for which $B_{\rm i\phi}(r)/r B_{\rm iz}(r)$ is a constant. Then the magnetic field in the jet–environment configuration is

$$\boldsymbol{B}(r) = \begin{cases} (0, Ar, B_{\rm iz}) & \text{for } r \leqslant a, \\ (0, 0, B_{\rm e}) & \text{for } r > a, \end{cases}$$

where A, $B_{\rm iz}$, and $B_{\rm e}$ are constant. Then, the equilibrium condition (12.4) gives the equilibrium plasma pressure $p_{\rm i}(r)$ as

$$p_{\rm i}(r) = p_0 - \frac{A^2 r^2}{\mu},$$

where p_0 is the plasma pressure at the center of the tube. The uniform twist of the internal magnetic field is characterized by the quantity $\varepsilon_1 = B_{\rm i\phi}(r)/B_{\rm iz}$, evaluated at $r = a$, that is, $\varepsilon_1 = Aa/B_{\rm iz}$.

When along with the magnetic field twist we have a spinning jet, we assume an uniform rotation velocity whose azimuthal component is linear function of the radial position r and evaluated at the interface $r = a$ is equal to $U_\phi = \Omega a$, where

Ω is constant. Here, Ω is the jet angular speed, deduced from the observations. Similarly to the magnetic field twist, we characterizes the jet velocity twist by the quantity $\varepsilon_2 = U\phi(a)/U_z = \Omega a/U_z$. Note that in equilibrium, the rigidly rotating plasma column must satisfy the following force-balance equation:

$$\frac{\mathrm{d}}{\mathrm{d}r}\left(p_\mathrm{i} + \frac{B_\mathrm{i}^2}{2\mu}\right) = \frac{\rho_\mathrm{i} U_\phi^2}{r} - \frac{B_{\mathrm{i}\phi}^2}{\mu r}, \tag{12.5}$$

where $p_\mathrm{i} + B_\mathrm{i}^2/2\mu$ with $B_\mathrm{i}^2 = B_{\mathrm{i}\phi}^2(r) + B_{\mathrm{i}z}^2$ is the total (thermal plus magnetic) pressure. This equation, after being integrated in radial direction, yields the following total pressure balance equation:

$$p_\mathrm{i}(0) + \frac{1}{2}\rho_\mathrm{i} U_\phi^2(a) + \frac{B_{\mathrm{i}z}^2}{2\mu}(1 - \varepsilon_1^2) = p_\mathrm{e} + \frac{B_\mathrm{e}^2}{2\mu}, \tag{12.6}$$

where $p_\mathrm{i}(0)$ is the thermal pressure at the magnetic tube axis, and p_e denotes the thermal pressure in the environment. This total pressure balance equation includes all previous cases and can be used for any uniformly twisted and rotating solar jets.

Having at hand all basic physical parameters of the jet–environment system satisfying Eq. (12.6), we have to make the following important steps:

1. To determine the plasma betas of both media, recall that the plasma beta is the ratio of thermal to magnetic pressure:

$$\beta = \frac{p}{B^2/2\mu}.$$

A more convenient form of the plasma beta is $(6/5)c_\mathrm{s}^2/v_\mathrm{A}^2$, where $c_\mathrm{s} = (\gamma k_\mathrm{B} T/m_\mathrm{ion})^{1/2}$ is the sound speed (in which $\gamma = 5/3$, k_B is the Boltzmann's constant, T the electron temperature, and m_ion the ion or proton mass), and $v_\mathrm{A} = B/(\mu n_\mathrm{ion} m_\mathrm{ion})^{1/2}$ is the Alfvén speed, in which expression B is the full magnetic field $=(B_\phi^2 + B_z^2)^{1/2}$, and n_ion is the ion or proton number density.

Plasma beta in solar atmosphere varies with the height. At photospheric levels it is larger than 1, while in the chromosphere, the transition region and the solar wind β are the order of one (at times it is less than, but close to 1). In the solar corona, the plasma beta is much smaller than 1. In the case when the magnetic flux tube modeling the jet is a strait vertical cylinder without twist of the magnetic field, the numerical solutions to the wave dispersion relation governing the propagation of MHD mode can be found without any complications. However, in the presence of magnetic field twist we are forced to simplify the jet–environment configuration. If the plasma β of given medium is larger than or close to 1, we can treat it as an incompressible magnetoplasma. Otherwise, we have to consider it as a cool plasma. Hence, practically we have four different cases: (i) both media (the jet and its environment) are incompressible plasmas, (ii) the jet is an incompressible medium and its surrounding plasma is a cool medium, (iii) the reverse case: the jet is a cool magnetized medium and its environment is an incompressible plasma, and finally (iv) both media are cool plasmas. For each of these particular cases we have to specify the wave dispersion relations who govern the propagation of various MHD modes and see at which conditions, say Alfvén Mach numbers, given mode becomes unstable against the Kelvin–Helmholtz instability.

2. For the simplest case of untwisted cylindrical magnetic flux tube of radius a moving with velocity $\boldsymbol{v}_0 = (0, 0, \pm v_0)$, the wave dispersion relation has the form

$$\frac{\rho_e}{\rho_i}(\omega^2 - k_z^2 v_{Ae}^2)\kappa_i \frac{I_m'(\kappa_i a)}{I_m(\kappa_i a)} - [(\omega - \boldsymbol{k} \cdot \boldsymbol{v}_0)^2 - k_z^2 v_{Ai}^2]\kappa_e \frac{K_m'(\kappa_e a)}{K_m(\kappa_e a)} = 0, \quad (12.7)$$

where the squares of the wave attenuation coefficients κ_i and κ_e are given by

$$\kappa^2 = -\frac{\left[(\omega - \boldsymbol{k} \cdot \boldsymbol{v}_0)^2 - k_z^2 c_s^2\right]\left[(\omega - \boldsymbol{k} \cdot \boldsymbol{v}_0)^2 - k_z^2 v_A^2\right]}{(c_s^2 + v_A^2)\left[(\omega - \boldsymbol{k} \cdot \boldsymbol{v}_0)^2 - k_z^2 c_T^2\right]}.$$

Here, the so-called tube velocity, c_T, is expressed via the sound and Alfvén speeds:

$$c_T = \frac{c_s v_A}{(c_s^2 + v_A^2)^{1/2}}.$$

For the azimuthal mode number $m = 0$, Eq. (12.7) describes the propagation of so-called *sausage* waves, while with $m = 1$ it governs the propagation of the *kink* waves. It is worth mentioning that when both media are considered as incompressible plasmas, the both attenuation coefficients are equal to k_z and the wave dispersion relation (12.7) becomes a quadratic equation whose solutions for the real and imaginary parts of the complex normalized wave phase velocity are given by

$$v_{ph} \equiv \frac{\omega}{k_z} = \frac{-M_A B \pm \sqrt{D}}{\eta A - B} v_{Ai}, \quad (12.8)$$

where $M_A = v_0/v_{Ai}$ is the Alfvén Mach number,

$$A = I_m'(k_z a)/I_m(k_z a), \quad B = K_m'(k_z a)/K_m(k_z a),$$

and the discriminant D is

$$D = M_A^2 B^2 - (\eta A - B)\left[(1 - M_A^2)B - Ab^2\right].$$

Obviously, if $D \geqslant 0$, then

$$\mathrm{Re}(v_{\mathrm{ph}}) = \frac{-M_{\mathrm{A}}B \pm \sqrt{D}}{\eta A - B} v_{\mathrm{Ai}}, \quad \mathrm{Im}(v_{\mathrm{ph}}) = 0,$$

else

$$\mathrm{Re}(v_{\mathrm{ph}}) = \frac{-M_{\mathrm{A}}B}{\eta A - B} v_{\mathrm{Ai}}, \quad \mathrm{Im}(v_{\mathrm{ph}}) = \frac{\sqrt{-D}}{\eta A - B} v_{\mathrm{Ai}}.$$

We point out that our choice of the sign of $\sqrt{-D}$ in the expression for $\mathrm{Im}(v_{\mathrm{ph}})$ is plus although, in principle, it might also be minus—in that case, owing to arising instability the wave's energy is transferred to the jet. This simple form of the roots to the wave dispersion relation is very useful for finding that Alfvén Mach number M_{A}, for which the MHD mode becomes unstable.

In the case of twisted moving magnetic flux tube, the wave dispersion relation has the form

$$\frac{(\Omega^2 - \omega_{\mathrm{Ai}}^2) F_m(\kappa_{\mathrm{i}}a) - 2m A \omega_{\mathrm{Ai}} / \sqrt{\mu \rho_{\mathrm{i}}}}{(\Omega^2 - \omega_{\mathrm{Ai}}^2)^2 - 4A^2 \omega_{\mathrm{Ai}}^2 / \mu \rho_{\mathrm{i}}} = \frac{P_m(k_z a)}{\frac{\rho_{\mathrm{e}}}{\rho_{\mathrm{i}}}(\omega^2 - \omega_{\mathrm{Ae}}^2) + A^2 P_m(k_z a) / \mu \rho_{\mathrm{i}}}, \tag{12.9}$$

where $\Omega = \omega - \boldsymbol{k} \cdot \boldsymbol{v}_0$ is the Doppler-shifted wave frequency in the moving flux tube, and

$$F_m(\kappa_{\mathrm{i}}a) = \frac{\kappa_{\mathrm{i}}a I_m'(\kappa_{\mathrm{i}}a)}{I_m(\kappa_{\mathrm{i}}a)} \quad \text{and} \quad P_m(k_z a) = \frac{k_z a K_m'(k_z a)}{K_m(k_z a)}.$$

Note that the wave attenuation coefficient inside the tube, κ_{i}, is given by

$$\kappa_{\mathrm{i}} = k_z \left[1 - 4A^2 \omega_{\mathrm{Ai}}^2 / \mu \rho_{\mathrm{i}}(\Omega^2 - \omega_{\mathrm{Ai}}^2)^2\right]^{1/2}, \tag{12.10}$$

in which the Alfvén angular wave frequency, ω_{Ai}, is

$$\omega_{\mathrm{Ai}} = \frac{m A + k_z B_{\mathrm{iz}}}{\sqrt{\mu \rho_{\mathrm{i}}}}, \tag{12.11}$$

and the prime sign means a differentiation by the Bessel function argument. If the environment is a cool medium, the argument $k_z a$ of the modified Bessel function K_m should be replaced by

$$\kappa_{\mathrm{e}}a = k_z a \left[1 - (\omega/\omega_{\mathrm{Ae}})^2\right]^{1/2},$$

where $\omega_{\mathrm{Ae}} = k_z B_{\mathrm{e}} / \sqrt{\mu \rho_{\mathrm{e}}}$.

If the jet velocity is rotational, that is $U = (0, U_\phi(r), U_z)$ with $U_\phi(r) = \Omega a$, the wave dispersion relation of the MHD modes running on the spinning jet is

$$\frac{(\sigma^2 - \omega_{\mathrm{Ai}}^2) F_m(\kappa_i a) - 2m(\sigma\Omega + A\omega_{\mathrm{Ai}}/\sqrt{\mu\rho_i})}{\rho_i(\sigma^2 - \omega_{\mathrm{Ai}}^2)^2 - 4\rho_i(\sigma\Omega + A\omega_{\mathrm{Ai}}/\sqrt{\mu\rho_i})^2}$$
$$= \frac{P_m(\kappa_e a)}{\rho_e(\omega^2 - \omega_{\mathrm{Ae}}^2) - (\rho_i\Omega^2 - A^2/\mu)P_m(\kappa_e a)}, \tag{12.12}$$

where

$$F_m(\kappa_i a) = \frac{\kappa_i a I'_m(\kappa_i a)}{I_m(\kappa_i a)} \quad \text{and} \quad P_m(\kappa_e a) = \frac{\kappa_e a K'_m(\kappa_e a)}{K_m(\kappa_e a)}.$$

In above expressions, the prime means differentiation of the Bessel functions with respect to their arguments,

$$\kappa_i^2 = k_z^2\left[1 - 4\left(\frac{\sigma\Omega + A\omega_{\mathrm{Ai}}/\sqrt{\mu\rho_i}}{\sigma^2 - \omega_{\mathrm{Ai}}^2}\right)^2\right] \quad \text{and} \quad \kappa_e^2 = k_z^2[1 - (\omega/\omega_{\mathrm{Ae}})^2],$$

are the squared wave amplitude attenuation coefficients in the jet and its environment, in which

$$\omega_{\mathrm{Ai}} = \left(\frac{m}{r}B_{i\phi} + k_z B_{iz}\right)\Big/\sqrt{\mu\rho_i} \quad \text{and} \quad \omega_{\mathrm{Ae}} = k_z B_e/\sqrt{\mu\rho_e}$$

are the local Alfvén frequencies in both media, and

$$\sigma = \omega - \frac{m}{r}U_\phi - k_z U_z$$

is the Doppler-shifted angular wave frequency in the jet. We note that in the case of incompressible coronal plasma $\kappa_e = k_z$, because at an incompressible environment the argument of the modified Bessel function of second kind, K_m, and its derivative, K'_m, is $k_z a$.

In the case when both media are cool plasmas, the wave dispersion relation of the propagating MHD modes is read as

$$\frac{(\sigma^2 - \omega_{\mathrm{Ai}}^2) F_m(\kappa_i^c a) - 2m(\sigma\Omega + A\omega_{\mathrm{Ai}}/\sqrt{\mu\rho_i})}{\rho_i(\sigma^2 - \omega_{\mathrm{Ai}}^2)^2 - 4\rho_i(\sigma\Omega + A\omega_{\mathrm{Ai}}/\sqrt{\mu\rho_i})^2}$$
$$= \frac{P_m(\kappa_e^c a)}{\rho_e(\omega^2 - \omega_{\mathrm{Ae}}^2) - (\rho_i\Omega^2 - A^2/\mu)P_m(\kappa_e^c a)}, \tag{12.13}$$

where

$$F_m(\kappa_i^c a) = \frac{\kappa_i^c a I'_m(\kappa_i^c a)}{I_m(\kappa_i^c a)} \quad \text{and} \quad P_m(\kappa_e^c a) = \frac{\kappa_e^c a K'_m(\kappa_e^c a)}{K_m(\kappa_e^c a)}.$$

Here, the wave attenuation coefficient in the internal medium has the form

$$\kappa_i^c = k_z \left\{ 1 - 4 \left(\frac{\sigma\Omega + A\omega_{Ai}/\sqrt{\mu\rho_i}}{\sigma^2 - \omega_{Ai}^2} \right)^2 \right\}^{1/2} \left(1 - \frac{\sigma^2}{\omega_{Ai}^2} \right)^{1/2},$$

while that in the environment, with $\Omega = 0$ and $A = 0$, is given by

$$\kappa_e^c = k_z \left(1 - \frac{\omega^2}{\omega_{Ae}^2} \right)^{1/2}.$$

Note that (i) both dispersion relations, (12.12) and (12.13), have similar forms—the difference is in the expressions for the wave attenuation coefficient inside the jet, namely $\kappa_i^c = \kappa_i \left[1 - \sigma^2/\omega_{Ai}^2 \right]^{1/2}$; and (ii) the wave attenuation coefficients in the environments are the same, that is, $\kappa_e^c = \kappa_e \equiv k_z \left[1 - \omega^2/\omega_{Ae}^2 \right]^{1/2}$.

In the most complicated jet–environment configuration when both magnetic fields are twisted and both media are treated as incompressible plasmas, the wave dispersion relation has the form:

$$\frac{(\Omega^2 - \omega_{Ai}^2) F_m(\kappa_i a) - 2m A\omega_{Ai}/\sqrt{\mu\rho_i}}{(\Omega^2 - \omega_{Ai}^2)^2 - 4A^2\omega_{Ai}^2/\mu\rho_i} = \frac{a^2(\omega^2 - \omega_{Ae}^2) Q_\nu(\kappa_e a) - G}{L - H\left[a^2(\omega^2 - \omega_{Ae}^2) Q_\nu(\kappa_e a) - G \right]},$$

$$(12.14)$$

where the squared order of the modified Bessel function of second kind is defined as

$$n^2 = m^2 - \frac{4m^2 B_\phi^2}{\mu\rho_e a^2(\omega^2 - \omega_{Ae}^2)} + \frac{8m B_\phi \omega_{Ae}}{\sqrt{\mu\rho_e} a(\omega^2 - \omega_{Ae}^2)},$$

and

$$\kappa_i = k_z \left[1 - 4A^2\omega_{Ai}^2/\mu\rho_i(\Omega^2 - \omega_{Ai}^2)^2 \right]^{1/2} \quad \text{and}$$

$$\kappa_e = k_z \left(1 - \frac{4 B_{e\phi}^2 \omega^2}{\mu\rho_e(\omega^2 - \omega_{Ae}^2)^2 a^2} \right)^{1/2},$$

$$Q_\nu(\kappa_e a) = \frac{\kappa_e a K_\nu'(\kappa_e a)}{K_\nu(\kappa_e a)}, \quad L = a^2\rho_e(\omega^2 - \omega_{Ae}^2)^2 - \frac{4 B_{e\phi}^2 \omega^2}{\mu},$$

$$H = \frac{B_{e\phi}^2}{\mu a^2} - \frac{A^2}{\mu}, \quad G = 2a^2(\omega^2 - \omega_{Ae}^2) + \frac{2m a B_{e\phi} \omega_{Ae}}{\sqrt{\mu\rho_e}}.$$

Here, ω_{Ai} and ω_{Ai} are the local Alfvén angular frequencies given by Eq. (12.11) and

$$\omega_{\text{Ae}} = \frac{m B_{\text{e}\phi} + k_z a B_{\text{ez}}}{\sqrt{\mu \rho_{\text{e}}} a},$$

respectively, and $\Omega = \omega - \boldsymbol{k} \cdot \boldsymbol{v}_0$ is the Doppler-shifted angular frequency.

When the surrounding plasma is considered as cool medium, the wave dispersion relation has the same form, however, the complex order of the modified Bessel function of second kind is given by $\nu_{\text{c}} = \sqrt{2 + n_{\text{c}}^2}$, where

$$n_{\text{c}}^2 = m^2 - \frac{4 m^2 B_\phi^2}{\mu \rho_{\text{e}} a^2 (\omega^2 - \omega_{\text{Ae}}^2)} + \frac{4 m B_\phi \omega_{\text{Ae}}}{\sqrt{\mu \rho_{\text{e}}} a (\omega^2 - \omega_{\text{Ae}}^2)}.$$

Here, the label "c" stamps for *cool*. Note that in the cool environment the wave attenuation coefficient is given by

$$\kappa_{\text{ec}} = k_z \left(1 - \frac{4 B_\phi^2 \omega^2}{\mu \rho_{\text{e}} (\omega^2 - \omega_{\text{Ae}}^2)^2 a^2} \right)^{1/2} \left(\frac{\omega^2}{\omega_{\text{Ae}}^2} - 1 \right)^{1/2}. \tag{12.15}$$

Thus, we can use Eq. (12.14), in which we have simply to replace κ_{e} by κ_{ec}.

3. As is seen from the above listed wave dispersion relations of MHD modes traveling along solar jets, all of them (except the one for untwisted incompressible jet–environment system) are transcendental equations in complex variables. The finding the solutions to those equations is not a trivial task. It is more convenient to normalize all velocities with respect to internal Alfvén speed, $v_{\text{Ai}} = B_{\text{iz}}/\sqrt{\rho_{\text{i}}\mu}$, and the wavelengths to the tube radius a. In this way, we have to find the real and imaginary parts of the normalized complex wave phase velocity, $v_{\text{ph}}/v_{\text{Ai}}$, as functions of the normalized real wavenumber $k_z a$. Having derived $\text{Re}(v_{\text{ph}}/v_{\text{Ai}})$ and $\text{Im}(v_{\text{ph}}/v_{\text{Ai}})$, we can easily obtain in absolute units the conditions under which the KH instability appears, notably the critical flow speed, $v_0^{\text{cr}} = \text{Re}(v_{\text{ph}}/v_{\text{Ai}}) v_{\text{Ai}}$ and the wave growth rate $\gamma_{\text{KH}} = \text{Im}(v_{\text{ph}}/v_{\text{Ai}}) v_{\text{Ai}}$ occurring at a wavelength $\lambda_{\text{KH}} = \pi \Delta \ell / k_z a$ ($\Delta \ell$ is the jet's width). Note that the $\text{Re}(v_{\text{ph}}/v_{\text{Ai}})$ is in fact the Alfvén–Mach number, M_{A}, and the critical one, M_{A}^{cr}, defines the marginally dispersion curve in the $v_{\text{ph}}/v_{\text{Ai}}$–$k_z a$-plane: for each $M_{\text{A}} < M_{\text{A}}^{\text{cr}}$ the jet–environment system is stable against the KH instability, otherwise the system is unstable. The critical growth rate, γ_{KH}, determines how quickly the wave amplitude increases in time and becoming bigger enough makes the linear approach inapplicable. In that stage of the instability development, the nonlinearity plays an important role. In particular, it can decrease the wave growth rate.

Before starting the numerical task of finding that $M_{\mathrm{A}}^{\mathrm{cr}}$, at which the instability occurs, one can use some criterion, which would give an approximate value. Such a criterion is

$$|m|M_{\mathrm{A}}^2 > (1 + 1/\eta)(|m|b^2 + 1), \qquad (12.16)$$

where $b = B_{\mathrm{e}}/B_{\mathrm{i}z}$.

In most cases, likewise KH instability in rotating jets or in CMEs, the instability emerges in a finite range (window) on the one-dimensional $k_z a$-space, that is, that range possesses a left-hand side value, $(k_z a)_{\mathrm{lhs}}$, and right-hand side one, $(k_z a)_{\mathrm{rhs}}$. We have to emphasize that in the case of spinning jets the instability criterion has a more complicated form

$$\frac{a^2\Omega^2}{v_{\mathrm{Ai}}^2} > \frac{1+\eta}{1+|m|\eta}\frac{(k_z a)^2}{|m|-1}(1+b^2), \qquad (12.17)$$

applicable for an untwisted magnetic flux tube moving with negligibly small longitudinal velocity U_z. Recall that $\Omega = U_\phi(a)/a$. This inequality says that each MHD wave with mode number $m \geqslant 2$ propagating in a rotating jet can become unstable. Numerical calculations show that this instability condition can be used also in the cases of slightly twisted spinning jets, provided that the magnetic field twist parameter, ε_1, is a number lying in the range of 0.001–0.005. Assuming that the deduced from observations jet axial velocity, U_z, is the threshold speed for the KHI occurrence, then, for fixed values of m, η, $U_\phi = \Omega a$, v_{Ai}, and b, the inequality (12.17) can be rearranged to define the upper limit of the instability range on the $k_z a$-axis

$$(k_z a)_{\mathrm{rhs}} < \left\{ \left(\frac{U_\phi}{v_{\mathrm{Ai}}}\right)^2 \frac{1+|m|\eta}{1+\eta} \frac{|m|-1}{1+b^2} \right\}^{1/2}. \qquad (12.18)$$

According to the above inequality, the KHI can occur for non-dimensional wavenumbers $k_z a$ less than $(k_z a)_{\mathrm{rhs}}$. On the other hand, one can talk for instability if the unstable wavelength, $\lambda_{\mathrm{KH}} = \pi \, \Delta\ell/k_z a$, is shorter than the height of the jet, H, which means that the lower limit of the instability region is given by

$$(k_z a)_{\mathrm{lhs}} > \frac{\pi \, \Delta\ell}{H}. \qquad (12.19)$$

Hence, the instability range in the $k_z a$-space is $(k_z a)_{\mathrm{lhs}} < k_z a < (k_z a)_{\mathrm{rhs}}$. Note that the lower limit, $(k_z a)_{\mathrm{lhs}}$, is fixed by the width and height of the jet, while the upper limit, $(k_z a)_{\mathrm{rhs}}$, depends on several jet–environment parameters. At fixed U_ϕ, v_{Ai}, η, and b, the $(k_z a)_{\mathrm{rhs}}$, from the instability criterion (12.17), one can find (roughly) that mode number magnitude $|m|$, for which the expected instability window will be obtained, namely it has to satisfy the inequality

$$\eta|m|^2 + (1 - \eta)|m| - 1 - \frac{(k_z a)^2(1 + \eta)(1 + b^2)}{(U_\phi/v_{\mathrm{Ai}})^2} > 0. \qquad (12.20)$$

After these preliminary evaluations, one can begin the numerical task of finding the complex roots of the wave dispersion relation. Among the available numerical methods, at least three ones, notably the well-known Newton–Raphson algorithm, or the secant method, even better the Muller's method, can be used. It is necessary to underline that the numerical procedure does not go smoothly—sometimes one has to do many takes before finding the true root. It is recommended also to check that each found root satisfies the equation which is numerically solved.

4. When the KH instability window at fixed mode number m has a finite width on the $k_z a$-axis, a specific property is that the width depends upon ε_1 and with increasing the value of ε_1, the instability window becomes narrower and at some critical ε_1 its width equals zero. Note that each larger value of ε_1 implies an increase in $B_{\mathrm{i}\phi}$. But that increase in $B_{\mathrm{i}\phi}$ requires an increase in $B_{\mathrm{i}z}$ too, in order the total pressure balance equation (12.6) to be satisfied under the condition that the thermal pressure, the hydrodynamic pressure term and the environment total pressure are fixed. The increase in $B_{\mathrm{i}z}$ (and in the full magnetic field B_{i}) implies a decrease both in the magnetic field ratio, b, and in the Alfvén Mach number, M_{A}. It is obvious that at some $\varepsilon_1^{\mathrm{cr}}$ the window width becomes equal to zero, that is, there is no longer instability at all. In other words, there exists a critical azimuthal magnetic field $B_{\mathrm{i}\phi} = \varepsilon_1^{\mathrm{cr}} B_{\mathrm{i}z}$. Thus, in searching the condition under which KH instability stops, by gradually increasing the magnetic field twist ε_1 from a very small value, say 0.005, to $\varepsilon_1^{\mathrm{cr}}$, we get a series of dispersion and growth rate curves with progressively diminishing parameters b and M_{A}. Be careful, however, that one has to avoid values of ε_1 close or larger than 1. In that case our approach is inapplicable because the arising instability is of another type, notably it is the kink instability that emerges.

In this book, we have shown how one can obtain the conditions under which the MHD modes propagating along a solar jet can become unstable against KH instability on using our approach. We are aware that our method (applicable to ideal plasma media), based on homogeneous plasma densities and magnetic fields as well as linear dependencies of the azimuthal components of $\boldsymbol{B}_{\mathrm{i}}$ and $\boldsymbol{v}_0$, is relatively simple but very flexible. The method itself consists of two parts. In the first part, considering the appropriate magnetic fields topology and physical parameters of studied jet–environment configuration, one derives *in closed form* the wave dispersion relation of propagating along the flow various MHD modes. In the second, equally important part, the wave dispersion relation is numerically solved.

The obtained true real and imaginary roots of that equation permit graphically to represent the dependencies of normalized real and imaginary parts of the wave velocity on the non-dimensional wavenumber. From such a pair of graphics one can easily obtain the critical flow velocity, v_0^{cr}, and the wave growth rate, γ_{KH}, of the unstable mode m at the observed or assumed wavelength, λ_{KH}. In addition, it is possible to evaluate the KHI developing time, $\tau_{\mathrm{KH}} = 2\pi/\gamma_{\mathrm{KH}}$. All these three instability parameters are generally in a good agreement with their values extracted from the observations.

Our method can be refined in three directions, namely (i) to consider the radial dependence of the jet density, $\rho_{\mathrm{i}}(r)$. This assumption immediately requires the exploration of continuity spectra and the resonant absorption in the layers, in which the local wave frequency (usually of the magnetosonic mode) coincides with the Alfvén frequency. Such a consideration can modify sometimes the KHI parameters and the energy budget (i.e., the plasma heating) significantly. (ii) At large enough wave amplitudes, the nonlinearity in the basic equations begins to play an important role, more specifically can lead to saturation and subsequently diminishing of the wave growth rate, which should influence the final results. (iii) A global modeling of KHI in the jet–surrounding plasma system implies, except of radial density profiles, also random radial profiles of the azimuthal components of the magnetic field and the flow velocity. If in the previous two cases one can make some progress in deriving the governing wave equations in closed forms, then in this case one has to start the KHI modeling from the very beginning only numerically. This requires the development of complex numerical packages which is a challenge for the plasma physics scholars of all generations. The newest results on MHD waves and instabilities, their modeling, and especially the numerical methods the reader can find in Goedbloed *et al.* (2019).

Bibliography

Acton, F. S. (1990). *Numerical Methods That (Usually) Work* (Mathematical Association of America, Washington, D.C.), Chapter 14.

Adams, M., Sterling, A. C., Moore, R. L., and Gary, G. A. (2014). A small-scale eruption leading to a blowout macrospicule jet in an on-disk coronal hol, *Astrophys. J.* **783**, 11 (13pp).

Ajabshirizadeh, A., Ebadi, H., Vekalati, R. E., and Molaverdikhani, K. (2015). The possibility of Kelvin–Helmholtz instability in solar spicules, *Astrophys. Space Sci.* **357**, 33 (7pp).

Alexander, D. and Fletcher, L. (1999). High-resolution observations of plasma jets in the solar corona, *Solar Phys.* **190**, pp. 167–184.

Alfvén, H. (1942). Existence of electromagnetic-hydrodynamic waves, *Nature* **150**, pp. 405–406.

Altschuler, M. D. and Newkirk, Jr., G. (1969). Magnetic fields and the structure of the solar corona. I: Methods of calculating coronal fields, *Solar Phys.* **9**, pp. 131–149.

Andries, J. and Goossens, M. (2001). Kelvin–Helmholtz instabilities and resonant flow instabilities for a coronal plume model with plasma pressure, *Astron. Astrophys.* **368**, pp. 1083–1094.

Antiochos, S. K. (1990). Heating of the corona by magnetic singularities, *Mem. Soc. Astron. Ital.* **61**, pp. 369–382.

Antiochos, S. K. (1998). The magnetic topology of solar eruptions, *Astrophys. J.* **502**, pp. L181–L184.

Antiochos, S. K., DeVore, C. R., and Klimchuk, J. A. (1999). A model for solar coronal mass ejections, *Astrophys. J.* **510**, pp. 485–493.

Antolin, P., Schmit, D., Pereira, T. M. D., De Pontieu, B., and De Moortel, I. (2018). Transverse wave induced Kelvin–Helmholtz rolls in spicules, *Astrophys. J.* **856**, 44 (20pp).

Aparna, V. and Tripathi, D. (2016). A hot flux rope observed by *SDO*/AIA, *Astrophys. J.* **819**, 71 (9pp).

Aschwanden, M. J., Fletcher, L., Schrijver, C. J., and Alexander, D. (1999). Coronal loop oscillations observed with the *Transition Region and Coronal Explorer*, *Astrophys. J.* **520**, pp. 880–894.

Aschwanden, M. J. and Acton, L. W. (2001). Temperature tomography of the soft x-ray corona: measurements of electron densities, temperatures, and differential emission measure distributions above the limb, *Astrophys. J.* **550**, pp. 475–492.

Aschwanden, M. (2004). *Physics of the Solar Corona: An Introduction* (Springer, Berlin), Chapter 1.

Aschwanden, M. (2005). *Physics of the Solar Corona: An Introduction with Problems and Solutions* (Springer, Berlin Heidelberg), Chapters 7 and 8.

Aschwanden, M. J., Boerner, P., Schrijver, C. J., and Malanushenko, A. (2013). Automated temperature and emission measure analysis of coronal loops and active regions observed with the *Atmospheric Imaging Assembly* on the *Solar Dynamics Observatory* (SDO/AIA), *Solar Phys.* **283**, pp. 5–30.

Athay, R. G. (2000). *Solar Phys.* **197**, pp. 31–42.

Athay, R. G. and Holzer, T. E. (1982), *Astrophys. J.* **255**, pp. 743–752.

Aulanier, G., Török, T., Démoulin, P., DeLuca, and E. E. (2010). Formation of torus-unstable flux ropes and electric currents in erupting sigmoids, *Astrophys. J.* **708**, pp. 314–333.

Aulanier, G., Janvier, M., and Schmieder, B. (2012). The standard flare model in three dimensions. I. Strong-to-weak shear transition in post-flare loops, *Astron. Astrophys.* **543**, A110 (14pp).

Aulanier, G., Démoulin, P., Schrijver, C. J., Janvier, M., Pariat, E., and Schmieder, B. (2013). The standard flare model in three dimensions. II. Upper limit on solar flare energy, *Astron. Astrophys.* **549**, A66 (7pp).

Aulanier, G. (2014). The physical mechanisms that initiate and drive solar eruptions. In: Schmieder, B., Malherbe, J.-M., Wu, S. T. (Eds.), *Nature of Prominences and their Role in Space Weather*, *Proc. IAU Symposium* No. 300 (Cambridge University Press, New York), pp. 184–196.

Beckers, J. M. (1963). Study of the undisturbed chromosphere from hα-disk filtergrams, with particular reference to the identification of spicules, *Astrophys. J.* **138**, pp. 648–663.

Beckers, J. M. (1968). Solar spicules, *Solar Phys.* **3**, pp. 367–433.

Beckers, J. M. (1972). Solar spicules, *Ann. Rev. Astron. Astrophys.* **10**, pp. 73–100.

Bellot Rubio, L. R. and Beck, C. (2005). Magnetic flux cancellation in the moat of sunspots: results from simultaneous vector spectropolarimetry in the visible and the infrared, *Astrophys. J.* **626**, pp. L125–L128.

Bennett, K., Roberts, B., and Narain, U. (1999). Waves in twisted magnetic flux tubes, *Solar Phys.* **185**, pp. 41–59.

Bennett, S. M. and Erdélyi, R. (2015). On the statistics of macrospicules, *Astrophys. J.* **808** 135 (9pp).

Bodo, G., Rosner, R., Ferrari, A., and Knobloch, E. (1989). On the stability of magnetized rotating jets: the axisymmetric case, *Astrophys. J.* **341**, pp. 631–649.

Bodo, G., Rosner, R., Ferrari, A., and Knobloch, E. (1996). On the Stability of magnetized rotating jets: the nonaxisymmetric modes, *Astrophys. J.* **470**, pp. 797–805.

Bodo, G., Mamatsashvili, G., Rossi, P., and Mignone, A. (2016). Linear stability analysis of magnetized jets: the rotating case, *Mon. Not. R. Astron. Soc.* **462**, pp. 3031–3052.

Bogdanova, M., Zhelyazkov, I., Joshi, R., and Chandra, R. (2018). Solar jet on 2014 April 16 modeled by Kelvin–Helmholtz instability, *New Astron.* **63**, pp. 75–87.

Bohlin, J. D., Vogel, S. N., Purcell, J. D., *et al.* (1975). A newly observed solar feature: macrospicules in He II 304 Å, *Astrophys. J.* **197**, pp. L133–L135.

Bondenson, A., Iacono, R., and Bhattacharjee, A. (1987). Local magnetohydrodynamic instabilities of cylindrical plasma with sheared equilibrium flows, *Phys. Fluids* **30**, pp. 2167–2180.

Bong, S.-C., Cho, K.-S., and Yurchyshyn, V. (2014). Kinematics of solar chromospheric surges of AR 10930, *J. Korean Astron. Soc.* **47**, pp. 311–317.

Bostancı, Z. F. and Al Erdoğan, N. (2010). Cloud modeling of a quiet solar region in Hα, *Mem. Soc. Astron. Ital.* **81**, pp. 769–772.

Bostancı, Z. F. (2011). Cloud modeling of a network region in Hα, *Astron. Nachr.* **332**, pp. 815–820.

Bostancı, Z. F., Gültekin, A. and Al, N. (2014). Oscillatory behaviour of chromospheric fine structures in a network and a semi-active region, *Mon. Not. R. Astron. Soc.* **443**, pp. 1267–1273.

Brady, C. S. and Arber, T. D. (2016). Simulations of Alfvén and kink wave driving of the solar chromosphere: efficient heating and spicule launching, *Astrophys. J.* **829**, 80 (7pp).

Bratsolis, E., Dialetis, D., and Alissandrakis, C. E. (1993). A new determination of the mean lifetime of bright and dark chromospheric mottles, *Astron. Astrophys.* **274**, pp. 940–943.

Bray, R. J. (1973). High-resolution photography of the solar chromosphere. X: Physical parameters of Hα mottles, *Solar Phys.* **29**, pp. 317–325.

Brooks, D. H., Kurokawa, H., and Berger, T. E. (2007). An Hα surge provoked by moving magnetic features near an emerging flux region, *Astrophys. J.* **656**, pp. 1197–1207.

Brueckner, G. E., Howard, R. A., Koomen, M. J., *et al.* (1995). The large angle spectroscopic coronagraph (LASCO), *Solar Phys.* **162**, pp. 357–402.

Cally, P. S. (1986). Leaky and non-leaky oscillations in magnetic flux tubes, *Solar Phys.* **103**, pp. 277–298.

Cally, P. S. and Khomenko, E. (2015). Fast-to-Alfvén mode conversion mediated by the hall current. I. Cold plasma model, *Astrophys. J.* **814**, 106 (11pp).

Canfield, R. C., Peardon, K. P., Leka, K. D., *et al.* (1996). Hα surges and X-ray jets in AR 7260, *Astrophys. J.* **464**, pp. 1016–1129.

Carmichael, H. (1964). A process for flares. In: Wilmot Hess, N. (Ed.), *The Physics of Solar Flares, Proc. of the AAS-NASA Symposium, NASA SP-50* (National Aeronautics and Space Administration, Washington, D.C.), p. 451.

Cauzzi, G., Reardon, K., Rutten, R. J., Tritschler, A., and Uitenbroek, H. (2009). The solar chromosphere at high resolution with IBIS. IV. Dual-line evidence of heating in chromospheric network, *Astron. Astrophys.* **503**, pp. 577–587.

Cavus, H. and Kazkapan, D. (2013). Magnetic Kelvin–Helmholtz instability in the solar atmosphere, *New Astron.* **25**, pp. 89–94.

Chae, J., Wang, H., Lee, C.-Y., Goode, P. R., and U. Schühle, (1998). Photospheric magnetic field changes associated with transition region explosive events, *Astrophys. J.* **497**, pp. L109–L112.

Chae, J., Wang, H., Qiu, J., *et al.* (2001). The formation of a prominence in active region NOAA 8668. I. SOHO/MDI observations of magnetic field evolution, *Astrophys. J.* **560**, pp. 476–489.

Chae, J., Moon, Y.-J., and Park, S.-Y. (2003). Observational tests of chromospheric magnetic reconnection, *J. Korean Astron. Soc.* **36**, pp. S13–S20.

Chandra, R., Schmieder, B., Mandrini, C. H., Démoulin, P., Pariat E., Török, T., and Uddin, W. (2011). Homologous Flares and Magnetic Field Topology in Active Region NOAA 10501 on 20 November 2003, *Solar Phys.* **269**, pp 83–104.

Chandra, R., Gupta, G. R., Mulay, S., and Tripathi, D. (2014). Sunspot waves and triggering of homologous active region jets, *Mon. Not. R. Astron. Soc.* **446**, pp. 3741–3748.

Chandra, R., Mandrini, C. H., Schmieder, B., Joshi, B., Cristiani, G. D., Cremades, H., Pariat, E., Nuevo, F. A., Srivastava, A. K., and Uddin, W. (2017). Blowout jets and impulsive eruptive flares in a bald-patch topology, *Astron. Astrophys.*, **598**, A41 (15pp).

Chandrasekhar, S. (1961). *Hydrodynamic and Hydromagnetic Stability* (Clarendon Press, Oxford), Chapter 11.

Chandrashekhar, K., Bemporad, A., Banerjee, D., Gupta, G. R., and Teriaca, L. (2014a). Characteristics of polar coronal hole jets, *Astron. Astrophys.*, **561**, A104 (9pp).

Chandrashekhar, K., Morton, R., Banerjee, D., and Gupta, G. R. (2014b). The dynamical behaviour of a jet in an on-disk coronal hole observed with AIA/SDO, *Astron. Astrophys.* **562**, A98 (10pp).

Chen, H. D., Jiang, Y. C., and Ma, S. L. (2008). Observations of Hα surges and ultraviolet jets above satellite sunspots, *Astron. Astrophys.* **478**, pp 907–913.

Chen, H.-D., Zhang, J., and Ma, S.-L. (2012). The kinematics of an untwisting solar jet in a polar coronal hole observed by *SDO/*AIA, *Res. Astron. Astrophys.* **12**, pp. 573–583.

Chen, B., Bastian, T. S., and Gary, D. E. (2014). Direct evidence of an eruptive, filament-hosing magnetic flux rope leading to a fast solar coronal mass ejection, *Astrophys. J.* **794**, 149 (14pp).

Chen, N.-H. and Innes, D. E. (2016). Undercover EUV solar jets observed by the interface region imaging spectrograph, *Astrophys. J.* **833**, 22 (10pp).

Chen, P. F., Innes, D. E., and Solanki, S. K. (2008). SOHO/SUMER observations of prominence oscillation before eruption, *Astron. Astrophys.* **484**, pp. 487–493.

Cheng, X., Zhang, J., Liu, Y., and Ding, M. D. (2011). Observing flux rope formation during the impulsive phase of a solar eruption, *Astrophys. J.* **732**, L25 (6pp).

Cheng, X., Ding, M. D., Guo, Y., Zhang, J., Vourlidas, A., Liu, Y. D., Olmedo, O., Sun, J. Q., and Li, C. (2014). Tracking the Evolution of a Coherent Magnetic Flux Rope Continuously from the Inner to the Outer Corona, *Astrophys. J.* **780**, 28 (9pp).

Cheremnykh, O., Fedun, V., Ladikov-Roev, Yu., and Verth, G. (2018b). On the stability of incompressible MHD modes in magnetic cylinder with twisted magnetic field and flow, *Astrophys. J.* **866**, 86 (12pp).

Cheung, M. C. M., Pontieu De, B., Tarbell, T. D., *et al.* (2015). Homologous helical jets: observations by *IRIS*, *SDO*, and *HINODE* and magnetic modeling with data-driven simulations, *Astrophys. J.* **801**, 83 (13pp).

Chifor, C., Isobe, H., Mason, H. E., *et al.* (2008a). Magnetic flux cancellation associated with a recurring solar jet observed with *Hinode*, *RHESSI*, and *STEREO*/EUVI, *Astron. Astrophys.* **491**, pp. 279–288.

Chifor, C., Young, P. R., Isobe, H., Mason, H. E., Tripathi, D., Hara, H., and Yokoyama, T. (2008b). An active region jet observed with Hinode, *Astron. Astrophys.* **481**, pp. L57–L60.

Cirtain, J. W., Golub, L., Lundquist, L., *et al.* (2007). Evidence for Alfvén waves in solar X-ray jets, *Science* **318**, pp. 1580–1582.

Culhane, L., Harra, L. K., Baker, D., *et al.* (2007). Hinode EUV study of jets in the Sun's south polar corona, *Publ. Astron. Soc. Japan* **59**, pp. S751–S756.

Curdt, W. and Tian, H. (2011). Spectroscopic evidence for helicity in explosive events, *Astron. Astrophys.* **532**, L9 (4pp).

De Moortel, I. and Nakariakov, V. M. (2012). Magnetohydrodynamic waves and coronal seismology: an overview of recent results, *Phil. Trans. R. Soc.* A **370**, pp. 3193–3216.

De Pontieu, B., Erdélyi, R., and James, S. P. (2004). Solar chromospheric spicules from the leakage of photospheric oscillations and flows, *Nature* **430**, pp. 536–539.

De Pontieu, B. and Erdélyi, R. (2006). The nature of moss and lower atmospheric seismology, *Phil. Trans. Roy. Soc.* A **364**, pp. 383–394.

de Pontieu, B., McIntosh, S., Hansteen, V. H., *et al.* (2007). A tale of two spicules: The impact of spicules on the magnetic chromosphere, *Public. Astron. Soc. Japan* **59**, pp. S655–S662.

De Pontieu, B., McIntosh, S. W., Carlsson, M., *et al.* (2011). The origins of hot plasma in the solar corona, *Science* **331**, pp. 55–58.

De Pontieu, B., Carlsson, M., Rouppe van der Voort, L. H. M., Rutten, R. J., Hansteen, V. H., and Watanabe, H. (2012). Ubiquitous torsional motions in type II spicules, *Astrophys. J.* **752**, L12 (6pp).

De Pontieu, B., L. Rouppe van der Voort, L., McIntosh, S. W., *et al.* (2014). On the prevalence of small-scale twist in the solar chromosphere and transition region, *Science* **346**, 1255732.

De Pontieu, B., Tilte, A. M., Lemen, J. R. *et al.* (2014a). The interface region imaging spectrograph (IRIS), *Solar Phys.* **289**, pp. 2733–2779.

De Pontieu, B., De Moortel, I., Martínez-Sykora, J., and McIntosh, S. W. (2017). Observations and numerical models of solar coronal heating associated with spicules, *Astrophys. J.* **845**, L18 (7pp).

Díaz, A. J., Oliver, R., Ballester, J. L., and Soler, R. (2011). Twisted magnetic tubes with field aligned flow. I. Linear twist and uniform longitudinal field, *Astron. Astrophys.* **533**, A95 (12pp).

Domingo, V., Fleck, B., and Poland, A. I. (1995). The SOHO mission: An overview, *Solar Phys.* **162**, pp. 1–37.

Doschek, G. A., Warren, H. P., Dennis, B. R., Reep, J. W., and Caspi, A. (2015). Flare footnote regions and a surge observed by *Hinode*/EIS, *RHESSI*, and *SDO*/AIA, *Astrophys. J.* **813**, 32 (17pp).

Dungey, J. W. and Loughhead, R. E. (1954). Twisted magnetic fields in conducting fluids, *Aust. J. Phys.* **7**, pp. 5–13.

Ebadi, H. (2016) Kelvin–Helmholtz instability in solar spicules, *Iranian J. Phys. Res.* **16**, pp. 41–45.

Ebadi, H., Zaqarashvili, T. V., and Zhelyazkov, I. (2012). Observation of standing kink waves in solar spicules, *Astrophys. Space Sci.* **337**, pp. 33–37.

Ebadi, H. and Ghiassi, M. (2014). Observation of kink waves and their reconnection-like origin in solar spicules, *Astrophys. Space Sci.* **353**, pp. 31–36.

Edwin, P. M. and Roberts, B. (1983). Wave propagation in a magnetic cylinder, *Solar Phys.* **88**, pp. 179–191.

Erdélyi, R. and Fedun, V. (2007). Are there Alfvén waves in the solar atmosphere? *Science* **318**, pp. 1572–1574.

Erdélyi, R. and Fedun, V. (2010). Magneto-acoustic waves in compressible magnetically twisted flux tubes, *Solar Phys.* **263**, pp. 63–85.

Evans, R. M., Pulkkinen, A. A., Zheng, Y., Mays, M. L., Taktakishvili, A., Kuznetsova, M. M., and Hesse, M. (2013). The SCORE scale: A coronal mass ejection typification system based on speed, *Space Weather* **11**, pp. 333–334.

Fan, Y. and Giibson, S. E. (2004). Numerical simulations of three-dimensional coronal magnetic fields resulting from the emergence of twisted magnetic flux tubes, *Astrophys. J.* **609**, pp. 1123–1133.

Fang, F., Fan, Y., and McIntosh, S. (2014). Rotating solar jets in simulations of flux emergence with thermal conduction, *Astrophys. J.* **789**, L19 (6pp).

Filippov, B., Golub, L., and Koutchmy, S. (2009). X-ray jet dynamics in a polar coronal hole region, *Solar Phys.* **254**, pp. 259–269.

Filippov, B., Srivastava, A. K., Dwivedi, B. N., Masson, S., Aulanier, G., Joshi, N. C., and Uddin, W. (2015). Formation of a rotating jet during the filament eruption on 2013 April 10–11, *Mon. Not. R. Astron. Soc.* **451**, pp. 1117–1129.

Forbes, T. G. (2000). A review on the genesis of coronal mass ejections, *J. Geophys. Res.* **105**, pp. 23153–23166.

Forbes, T. G. and Isenberg, P. A. (1991). A catastrophe mechanism for coronal mass ejections, *Astrophys. J.* **373**, pp. 294–307.

Forbes, T. G. (2001). The nature of Petschek-type reconnection, *Earth Planets Space* **53**, pp. 423–429.

Forbes, T. G., Linker, J. A., Chen, J., *et al.* (2006). CME theory and models, *Space Sci. Rev.* **123**, pp. 251–302.

Foullon, C., Verwichte, E., Nakariakov, V. M., Nykyri, K., and Farrugia, C. J. (2011). Magnetic Kelvin–Helmholtz instability at the sun, *Astrophys. J.* **729**, L8 (4pp).

Foullon, C., Verwichte, E., Nykyri, K., Aschwanden, M. J., and Hannah, I. G. (2013). Kelvin–Helmholtz instability of the CME reconnection outflow layer in the low corona, *Astrophys. J.* **767**, 170 (18pp).

Foukal, P. V. (2004). *Solar Astrophysics*, Second, revised edition (Wiley-VCH, Weinheim), Chapter 9.

Furno, I., Intrator, T. P., Lapenta, G., *at al.* (2007). Effects of boundary conditions and flow on the kink instability in a cylindrical plasma column, *Phys. Plasmas* **14**, 022103 (11pp).

Gary, G. A. (2001). Plasma beta above a solar active region: Rethinking the paradigm, *Solar Phys.* **203**, pp. 71–86.

Gent, F. A., Fedun, V., Mumford, S. J., and Erdélyi, R. (2013). Magnetohydrostatic equilibrium – I. Three-dimensional open magnetic flux tube in the stratified solar atmosphere, *Mon. Not. R. Astron. Soc.* **435**, pp. 689–697.

Georgakilas, A. A., Koutchmy, S., and Alissandrakis, C. E. (1999). Polar surges and macrospicules: simultaneous Hα and He II 304 Åobservations, *Astron. Astrophys.* **341**, pp. 610–616.

Goedbloed, J. P. (1971). Stabilization of magnetohydrodynamic instabilities by force-free magnetic fields: I. Plane plasma layer, *Physica* **53**, pp. 412–444.

Goedbloed, H. J. P. and Poedts, S. (2004). *Principles of Magnetohydrodynamics: With Applications to Laboratory and Astrophysical Plasmas* (Cambridge University Press, Cambridge), Chapters 9–11.

Goedbloed, J. P., Keppens, R., and Poedts, S. (2010). *Advanced Magnetohydrodynamics: With Applications to Laboratory and Astrophysical Plasmas* (Cambridge University Press, Cambridge), Chapters 12–14.

Goedbloed, H., Keppens, R., and Poedts, S. (2019). *Magnetohydrodynamics of Laboratory and Astrophysical Plasmas* (Cambridge University Press, Cambridge).

Gontikakis, C., Archontis, V., and Tsinganos, K. (2009). Observations and 3D MHD simulations of a solar active region jet, *Astron. Astrophys.* **506**, pp. L45–L48.

Goossens, M., Hollweg, J. V., and Sakurai, T. (1992). Resonant behaviour of MHD waves on magnetic flux tubes. III. Effect of equilibrium flow, *Solar Phys.* **138**, pp. 233–255.

Goossens, M., Andries, J., Soler, R., Van Doorsselaere, T., Arregui, I., and Terradas, J. (2012). Surface Alfvén waves in solar flux tubes, *Astrophys. J.* **753**, 111 (12pp).

Gosain, S., Schmieder, B., Venkatakrishnan, P., Chandra, R., and Artzner, G. (2009). 3D evolution of a filament disappearance event observed by STEREO, *Solar Phys.* **259**, 13 (pp20).

Gosling, J. T., Hildner, E., MacQueen, R. M., Munro, R. H., Poland, A. I., and Ross, C. L. (1974). Mass ejections from the Sun: A view from Skylab, *J. Geophys. Res.: Space Phys.* **79**, pp. 4581–4587.

Green, L. M. and Kliem, B. (2014). Observations of flux rope formation prior to coronal mass ejections. In: Schmieder, B., Malherbe, J.-M., Wu, S. T. (Eds.), *Nature of Prominences and their Role in Space Weather, Proc. IAU Symposium* No. 300, (Cambridge University Press, New York), pp. 209–214.

Gu, X. M., Lin, J., Li, K. J., *et al.* (1994). Kinematic characteristics of the surge on March 19, 1989, *Astron. Astrophys.* **282**, pp. 240–251.

Guo, Y., Démoulin, P., Schmieder, B., Ding, M. D., Vargas Domínguez, S., and Liu, Y. (2013). Recurrent coronal jets induced by repetitively accumulated electric currents, *Astron. Astrophys.* **555**, A19 (9pp).

Hain, K. and Lüst, R. (1958). Zur Stabilität zylindersymmetrischer Plasmakonfigurationen mit Volumenströmen, *Z. Naturforschung* **13 a**, pp. 936–940.

Hansteen, V. H, de Pontieu, Rouppe van der Voort, L., van Noort, M., and Carlsson, M. (2006). Dynamic fibrils are driven by magnetoacoustic shocks, *Astrophys. J.* **647**, pp. L73–L76.

He, J.-S., Tu, C.-Y., Marsch, E., Guo, L.-J., Yao, S., and Tian, H. (2009). Upward propagating high-frequency Alfvén waves as identified from dynamic wave-like spicules observed by SOT on *Hinode, Astron. Astrophys.* **497**, pp. 525–535.

Heggland, L., De Pontieu, B., and Hansteen, V. H. (2009). Observational signatures of simulated reconnection events in the solar chromosphere and transition region, *Astrophys. J.* **703**, pp. 1–18.

von Helmholtz, H. L. F. (1868). Über discontinuierliche Flüssigkeits-Bewegungen, *Monatsberichte König. Preuss. Akad. Wiss. Berl.* **23**, pp. 215–228.

Heyvaerts, J., Priest, E., and Rust, D. M. (1997). An emerging flux model for solar flares, *Solar Phys.* **53**, pp. 255–258.

Hirayama, T. (1974). Theoretical model of flares and prominences. I: Evaporating flare model, *Solar Phys.* **34**, pp. 323–338.

Hirayama, T. (1985). Modern observations of solar prominences, *Solar Phys.* **100**, pp. 415–434.

Holzwarth, V., Schmitt, D., and Schüssler, M. (2007). Flow instabilities of magnetic flux tubes. II. Longitudinal flow, *Astron. Astrophys.* **469**, pp. 11–17.

Hong, J.-C., Jiang, Y.-C., Yang, J.-Y., Zheng, R.-S., Bi, Y., Li, H.-D., Yang, B., and Yang, D. (2013). Twist in a polar blowout jet, *Res. Astron. Astrophys.* **13**, pp. 253–258.

Hood, A. W. and Priest, E. R. (1979). Kink instability of solar coronal loops as the cause of solar flares, *Solar Phys.* **64**, pp. 303–321.

Howard, T. A., Moses, J. D., Vourlidas, A., *et al.* (2008). Sun earth connection coronal and heliospheric investigation (SECCHI), *Space Sci. Rev.* **136**, pp. 67–115.

Howard, T. A. and DeForest, C. E. (2014). the formation and launch of a coronal mass ejection flux rope: a narrative based on observations, *Astrophys. J.* **796**, 33 (15pp).

Howson, T. A., De Moortel, I., and Antolin, P. (2017). Energetics of the Kelvin–Helmholtz instability induced by transverse waves in twisted coronal loops, *Astron. Astrophys.* **607**, A77 (13pp).

Hundhausen, A. J., Sawyer, C. B., House, L., Illing, R. M. E., and Wagner, W. J. (1984). Coronal mass ejections observed during the Solar Maximum Mission: Latitude distribution and rate of occurrence, *J. Geophys. Res.* **89**, 2639–2646.

Hurford, G. J., Schmahl, E. J., Schwartz, R. A., *et al.* (2002). The RHESSI imaging concept, *Solar Phys.* **210**, pp. 61–86.

Iijima, H. and Yokoyama, T. (2017). A three-dimensional magnetohydrodynamic simulation of the formation of solar chromospheric jets with twisted magnetic field lines, *Astrophys. J.* **848**, 38 (16pp).

Innes, D. E., McIntosh, S. W., and Pietarila, A. (2010). STEREO quadrature observations of coronal dimming at the onset of mini-CMEs, *Astron. Astrophys.* **517**, L7 (4pp).

Innes, D. E., Cameron, R. H., and Solanki, S. K. (2011). EUV jets, type III radio bursts and sunspot waves investigated using SDO/AIA observations, *Astron. Astrophys.* **531**, L13 (4pp).

Innes, D. E., Bučík, R., Guo, L.-J., and Nitta, N. (2016). Observations of solar X-ray and EUV jets and their related phenomena, *Astron. Nachr.* **337**, pp. 1024–1032.

Inoue, S., Kusano, K., Büchner, J., and Skála, J. (2018). Formation and dynamics of a solar eruptive flux tube, *Nature Commun.* **9**, 174 (11pp).

Janvier, M., Aulanier, G., Pariat, E., and Démoulin, P. (2013). The standard flare model in three dimensions. III. Slip-running reconnection properties, *Astron. Astrophys.* **555**, A77 (14pp).

Jardin, S. (2010). *Computational Methods in Plasma Physics* (CRC Press, Boca Raton, FL), Ch. 8.

Ji, H., Yamada, M., Hsu, S., Kulsrud, R., Carter, T., and Zaharia, S. (1999). Magnetic reconnection with Sweet-Parker characteristics in two-dimensional laboratory plasmas, *Phys. Plasmas* **6**, pp. 1743–1750.

Jiang, F., J. Zhang, J. and S. Yang, S. (2015). Interaction between an emerging flux region and a preexisting fan-spine dome observed by IRIS and SDO, *Publ. Astron. Soc. Japan* **67**, 78 (11pp).

Joshi, N. C., Srivastava, A. K., Filippov, B., Kayshap, P., Uddin, W., and Chandra, R. (2013a). A study of a failed coronal mass ejection associated with an asymmetric filament eruption, *Astrophys. J.* **771**, 65 (13pp).

Joshi, N. C., Srivastava, A. K., Uddin, W., Chandra, R., *et al.* (2013b). A multiwavelength study of eruptive events on January 23, 2012 associated with a major solar energetic particle event, *Adv. Space Res.* **52**, pp. 1–14.

Joshi, N. C., Chandra, R., Guo, Y., Magara, T., Zhelyazkov, I., Moon, Y-J., and Uddin, W. (2017). Investigation of recurrent EUV jets from highly dynamic magnetic field region, *Astrophys. Space Sci.* **362**, 10 (14pp).

Kamio, S., Hara, H., Watanabe, T., Matsuzaki, K., Shibata, K., Culhane, L., and Warren, H. P. (2007). Velocity structure of jets in a coronal hole, *Publ. Astron. Soci. Japan* **59**, pp. S757–S762.

Kamio, S., Curdt, W., Teriaca, L., Inhester, B., and Solanki, S. K. (2010). Observations of a rotating macrospicule associated with an X-ray jet, *Astron. Astrophys.* **510**, L1 (4pp).

Karpen J. T., Antiochos, S. K., Dahlburg, R. B., and Spicer, D. S. (1993). The Kelvin–Helmholtz instability in photospheric flows: effects on coronal heating and structures, *Astrophys. J.* **403**, pp. 769–779.

Katsukawa, Y., Berger, T. E., Ichimoto, K., *et al.* (2007). Small-scale jetlike features in penumbral chromospheres, *Science* **318**, pp. 1594–1597.

Kayshap, P., Srivastava, A. K., and Murawski, K. (2013). The kinematics and plasma properties of a solar surge triggered by chromospheric activity in AR 11271, *Astrophys. J.* **763**, 24 (12pp).

Kelvin, Lord (1871). Hydrokinetic solutions and observations, *Phil. Mag.* **42**, pp. 362–377.

Kim, Y.-H., Moon, Y.-J., Park, Y.-D., Sakurai, T., Chae, J., Cho, K. S., and Bong, S.-C. (2007). Small-scale X-ray/EUV jets seen in Hinode XRT and TRACE, *Publ. Astron. Soc. Japan* **59**, pp. S763–S769.

Kiss, T. S., Gyenge, N., and Erdélyi, R. (2017). Systematic observations of microspicule properties observed by *SDO*/AIA over Half a Decade, *Astrophys. J.* **835**, 47 (10pp).

Kiss, T. S., Gyenge, N., and Erdélyi, R. (2018). Quasi-biennial oscillations in the cross-correlation of properties of macrospicules, *Adv. Space Res.* **61**, pp. 611–616.

Kliem, B. and Török, T. (2006). Torus instability, *Phys. Rev. Lett.* **96**, 255002–255006.

Kopp, R. A. and Pneuman, G. W. (1976). Magnetic reconnection in the corona and the loop prominence phenomenon, *Solar Phys.* **50**, pp. 85–98.

Kruskal, M. and Tuck, J. L. (1958). The instability of a pinched fluid with a longitudinal magnetic field, *Proc. Roy. Soc.* A **245**, pp. 222–237.

Kubo, M., Katsukawa, Y., Suematsu, Y., Kano, R., Bando, T., Narukage, N.,Ishikawa, R., Hara, H., Giono, G, Tsuneta, S., Ishikawa, S., Shimizu, T., Sakao, T., Winebarger, A., Kobayashi, K., Cirtain, J., Champey, P., Auchère, F., Bueno, J. T., Ramos, A. A., Štěpán, J., Belluzzi, L., Sainz, R. M., De Pontieu, B., Ichimoto, K., Carlsson, M., Casini, R., and Goto, M. (2016). Discovery of ubiquitous fast-propagating intensity disturbances by the Chromospheric Lyman Alpha Spectropolarimeter (CLASP), *Astrophys. J.* **832**, 141 (9pp).

Kudoh, T. and Shibata, K. (1999). Alfvén wave model of spicules and coronal heating, *Astrophys. J.* **514**, pp. 493–505.

Kukhianidze, V., Zaqarashvili, T. V., and Khutsishvili, E. (2006). Observation of kink waves in solar spicules, *Astron. Astrophys.* **449**, pp. L35–L38.

Kulsrud, R. M. (2001). Magnetic reconnection: Sweet–Parker versus Petschek, *Earth Planets Space* **53**, pp. 417–422.

Kuridze, D., Mathioudakis, M., Jess, D. B., Shelyag, S., Christian, D. J., Keenan, F. P., and Balasubramaniam, K. S. (2011). Small-scale Hα jets in the solar chromosphere, *Astron. Astrophys.* **533**, A76 (5pp).

Kuridze, D., Morton, R. J., Erdélyi, R., Dorrian, G. D., Mathioudakis, M., Jess, D. B., and Keenan, F. P. (2012). Transverse oscillations in chromospheric mottles, *Astrophys. J.* **750**, 51 (5pp).

Kuridze, D., Verth, G., Mathioudakis, M., Erdélyi, R., Jess, D. B., Morton, R. J., Christian, D. J., and Keenan, F. P. (2013). Characteristics of transverse waves in chromospheric mottles, *Astrophys. J.* **779**, 82 (8pp).

Kurokawa, H. and Kawai, G. (1993). H alpha surge activity at the first stage of magnetic flux emergence. In: Zirin, H., Ai, G., Wang, H. (Eds.), *The Magnetic and Velocity Fields of Solar Active Regions*, IAU Colloquium **141**, (ASP, San Francisco), pp. 507–512.

Kuźma, B., Murawski, K., Zaqarashvili, T. V., Konkol, P., and Mignone, A. (2017a). Numerical simulations of solar spicules: Adiabatic and non-adiabatic studies, *Astron. Astrophys.* **597**, A133 (8pp).

Kuźma, B., Murawski, K., Kayshap, P., Wóójcik, D., Srivastava, A. K., and Dwivedi, B. N. (2017b). Two-fluid numerical simulations of solar spicules, *Astrophys. J.* **849**, 78 (9pp).

Labrosse, N., Heinzel, P., Vial, J. C., Kucera, T., Parenti, S., Gunar, S., Schmieder, B., Kilper, G. (2010). Physics of solar prominences: I—Spectral diagnostics and non-LTE modelling, *Space Sci. Rev.* **151**, pp. 243–332.

Landi, E. and Miralles, M. P. (2014). Density diagnostics of coronal mass ejection cores with the *Solar Dynamics Observatory*/Atmospheric Imaging Assembly, *Astrophys. J.* **780**, L7 (5pp).

Leenaarts, J., Carlsson, M. and Rouppe van der Voort, L. (2012). The formation of the Hα line in the solar chromosphere, *Astrophys. J.* **749**, 136 (14pp).

Lemen J. R., Title A. M., Akin D. J., *et al.* (2012). The *Atmospheric Imaging Assembly* (AIA) on the *Solar Dynamics Observatory* (SDO), *Solar Phys.* **275**, pp. 17–40.

Lin, R. P., Dennis, B. R., Hurford, G. J., *et al.* (2002). The Reuven Ramaty High-Energy Solar Spectroscopic Imager (RHESSI), *Solar Phys.* **210**, pp. 3–32.

Liu, Y., Luhmann, J. G., Lin, R. P., Bale, S. D., Vourlidas, A., and Petrie, G. J. D. (2009a). Coronal mass ejections and global coronal magnetic field reconfiguration, *Astrophys. J.* **698**, L51–L55.

Liu, W., Berger, T. E., Title, A. M., and Tarbell, T. D. (2009b). An intriguing chromospheric jet observed by *Hinode*: Fine structure kinematics and evidence of unwinding twists, *Astrophys. J.* **707**, pp. L37–L41.

Liu, C., Deng, N., Liu, R., Ugarte-Urra, I., Wang, S., and Wang, H. (2011). A standard-to-blowout Jet, *Astrophys. J.* **735**, L18 (7pp).

Liu, J., Wang, Y., Shen, C., Liu, K., Pan, Z., and Wang, S. (2015). A solar coronal jet event triggers a coronal mass ejection, *Astrophys. J.* **813**, 115 (6pp).

Longcope, D. W. (1998). A model for current sheets and reconnection in X-ray bright points, *Astrophys. J.* **507**, pp. 433–442.

Luna, M., Su, Y., Schmieder, B., Chandra, R., and Kucera, T. A. (2017). Large-amplitude longitudinal oscillations triggered by the merging of two solar filaments: Observations and magnetic field analysis, *Astrophys. J.* **850**, 143 (13pp).

Lundquist, S. (1951). On the stability of magneto-hydrostatic fields, *Phys. Rev.* **83**, pp. 307–311.

Mackay, D. H., Karpen, J. T., Ballester, J. L., Schmieder, B., and Aulanier, G. (2010). Physics of solar prominences: II—Magnetic structure and dynamics, *Space Sci. Rev.* **151**, pp. 333–399.

Madjarska, M. S. (2011). Dynamics and plasma properties of an X-ray jet from SUMER, EIS, XRT, and EUVI A & B simultaneous observations, *Astron. Astrophys.* **526**, 19 (14pp).

Madjarska, M., Huang, Z., Subramanian, S., and Doyle, G. (2013). Jets from coronal holes—possible source of the slow solar wind. In *EGU General Assembly Conference Abstracts* (pp. EGU2013-2455). Volume 15 of *EGU General Assembly Conference Abstracts* **15**, p. 2455.

Madjarska, M. S. (2019). Coronal bright points, *Living. Rev. Sol. Phys.* **16**, 2 (79pp).

Mandrini, C. H., Démoulin, P., van Driel-Gesztelyi, L., Schmieder, B., Cauzzi, G., and Hofmann, A. (1996). 3D magnetic reconnection at an X-ray bright point, *Solar Phys.* **168**, pp. 115–133.

Martínez-Sykora, J., De Pontieu, B., Leenaarts, J., Pereira, T. M. D., Carlsson, M., Hansteen, V., Stern, J. V., Tian, H., McIntosh, S. W., and Rouppe van der Voort, L. (2013). A detailed comparison between the observed and synthesized properties of a simulated type II spicule, *Astrophys. J.* **771**, 66 (25pp).

Martínez-Sykora, J., De Potieu, B., Hansteen, V. H., Rouppe van der Voort, L., Carlsson, M., and Pereira, T. M. D. (2017). On the generation of solar spicules and Alfvénic waves, *Science* **356**, pp. 1269–1272.

Mathioudakis, M., Jess, D. B., and Erdélyi, R. (2013). Alfvén waves in the solar atmosphere, *Space Sci. Rev.* **175**, pp. 1–27.

Matsumoto, T. and Shibata, K. (2010). Nonlinear propagation of Alfvén waves driven by observed photospheric motions: application to the coronal heating and spicule formation, *Astrophys. J.* **710**, pp. 1857–1867.

McMath, R. R. and Mohler, O. (1948). Simultaneous observations of solar flares, surges, and high-speed dark flocculi, *The Observatory* **68**, pp. 110–111.

Mishin, V. V. and Tomozov, V. M. (2016). Kelvin–Helmholtz instability in the solar atmosphere, solar wind and geomagnetosphere, *Solar Phys.* **291**, pp. 3165–3184.

Miteva, R., Zhelyazkov, I., and Erdélyi, R. (2004). Hall-magnetohydrodynamic surface waves in solar wind flow-structures, *New J. Phys.* **6**, 14 (18pp).

Miteva, R., Ivanovski, S., and Zhelyazkov, I. (2006). Parameter study on Kelvin–Helmholtz instability in solar wind type flowing structures, *Proc. Int. Astron. Union* **2**, pp. 309–310.

Miyagoshi, T. and Yokoyama, T. (2003). Magnetohydrodynamic numerical simulations of solar X-ray jets based on the magnetic reconnection model that includes chromospheric evaporation, *Astrophys. J.* **593**, pp. L133–L136.

Moore, R. L., Sterling, A. C., Hudson, H. S., and Lemen, J. R. (2001). Onset of the magnetic explosion in solar flares and coronal mass ejections, *Astrophys. J.* **552**, pp. 833–848.

Moore, R. L., Cirtain, J. W., Sterling, A. C., and Falconer, D. A. (2010). Dichotomy of solar coronal jets: standard jets and blowout jets, *Astrophys. J.* **720**, pp. 757–770.

Moore, R. L., Sterling, A. C., Cirtain, J. W., and Falconer, D. A. (2011). Solar X-ray jets, type-ii spicules, granule-size emerging bipoles, and the genesis of the heliosphere, *Astrophys. J.* **731**, L18 (5pp).

Moore, R. L., Sterling, A. C., Falconer, D. A., and Robe, D. (2013). The Cool component and the dichotomy, lateral expansion, and axial rotation of solar X-ray jets, *Astrophys. J.* **769**, 134 (19pp).

Moore, R. L., Sterling, A. C., and Falconer, D. A. (2015). Magnetic untwisting in solar jets that go into the outer corona in polar coronal holes, *Astrophys. J.* **806**, 11 (20pp).

Moreno-Insertis, F., Galsgaard, K., and Ugarte-Urra, I. (2008). Jets in coronal holes: *Hinode* observations and three-dimensional computer modeling, *Astrophys. J.* **673**, L211–L214.

Moreno-Insertis, F. and Galsgaard, K. (2013). Plasma jets and eruptions in solar coronal holes: A three-dimenstional flux emergence experiment, *Astrophys. J.* **771**, 20 (18pp).

Morgan, H. and Druckmüller, M. (2014). Multi-scale gaussian normalization for solar image processing, *Solar Phys.* **289**, pp. 2945–2955.

Morton, R. J. (2012). Chromospheric jets around the edges of sunspots, *Astron. Astrophys.* **543**, A6 (9pp).

Moschou, S. P., Tsinganos, K., Vourlidas, A., and Archontis, V. (2013). SDO observations of solar jets, *Solar Phys.* **284**, pp. 427–438.

Möstl, U. V., Temmer, M., and Veronig, A. M. (2013). The Kelvin–Helmholtz instability at coronal mass ejection boundaries in the solar corona: Observations and 2.5D MHD simulations, *Astrophys. J.* **766**, L12 (6pp).

Muller, D. E. (1956). A method for solving algebraic equations using an automatic computer, *Math. Tables Aids Comput.* **10**, pp. 208–215.

Murawski, K., Srivastava, A. K., and Zaqarashvili, T. V. (2011). Numerical simulations of solar macrospicules, *Astron. Astrophys.* **535**, A58 (9pp).

Murawski, K., Chmielewski, P., Zaqarashvili, T. V., and Khomenko, E. (2016). Numerical simulations of magnetic Kelvin–Helmholtz instability at a twisted solar flux tube, *Mon. Not. R. Astron. Soc.* **459**, pp. 2566–2572.

Nakariakov, V. M. (2007). MHD oscillations in solar and stellar coronae: Current results and perspectives, *Adv. Space Res.* **39**, pp. 1804–1813.

Narang, N, Arbacher, R. T., Tian, H., Banerjee, D., Cranmer, S. R., DeLuca, E. E., and McKillop, S. (2016). Statistical study of network jets observed in the solar transition region: A comparison between coronal holes and quiet-sun regions, *Solar Phys.* **291**, pp. 1129–1142.

Newton, H. W. (1942). Solar activity, *Mon. Not. R. Astron. Soc.* **102**, pp. 108–109.

Nindos, A., Kundu, M. R., Raulin, J.-P., *et al.* (1998). A microwave study of coronal and chromospheric ejecta. In: Bastian, T., Gopalswamy, N., and Shibasaki, K. (Eds.) *Proceedings of Nobeyama Symposium*, NRO Report **479**: *Solar Physics with Radio Observations*, The Nobeyama Radio Observatory (Nobeyama), pp. 135–140.

Nishizuka, N., Shimizu, M., Nakamura, T., *et al.* (2008). Giant chromospheric anemone jet observed with *Hinode* and comparison with magnetohydrodynamic simulations: Evidence of propagating Alfvén waves and magnetic reconnection, *Astrophys. J.* **683**, pp. L83–L86.

Nisticò, G., Bothmer, V., Patsourakos, S., and Zimbardo, G. (2009). Characteristics of EUV coronal jets observed with STEREO/SECCHI, *Sol. Phys.* **259**, pp. 87–108.

Nisticò, G., Bothmer, V., Patsourakos, S., and Zimbardo, G. (2010). Observational features of equatorial coronal hole jets, *Ann. Geophys.* **28**, pp. 687–696.

Nykyri, K. and Foullon, C. (2013). First magnetic seismology of the CME reconnection outflow layer in the low corona with 2.5-D MHD simulations of the Kelvin–Helmholtz instability. *Geophys. Res. Lett.* **40**, pp. 4154–4159.

Ofman, L., Davila, J. M., and Steinolfson, R. S. (1994). Nonlinear studies of coronal heating by the resonant absorption of Alfvén waves, *Geophys. Res. Lett.* **21**, pp. 2259–2262.

Ofman, L. and Thompson, B. J. (2011). *SDO*/AIA Observation of Kelvin–Helmholtz Instability in the Solar Corona, *Astrophys. J.* **734**, L11 (5pp).

Okamoto, T. J. and De Pontieu, B. (2011). Propagating waves along spicules, *Astrophys. J.* **736**, L24 (6pp).

Panesar, N. K., Sterling, A. C., Moore, R. L., and Chakrapani, P. (2016). Magnetic flux cancellation as the trigger of solar quiet-region coronal jets, *Astrophys. J.* **832**, L7 (7pp).

Panesar, N. K., Sterling, A. C., Moore, R. L. (2017). Magnetic flux cancellation as the origin of solar quiet-region pre-jet minifilaments, *Astrophys. J.* **844**, 131 (13pp).

Paraschiv, A. R., Lacatus, D. A., Badescu, T., Lupu, M. G., Simon, S., Sandu, S. G., Mierla, M., and Rusu, M. V. (2010). Study of coronal jets during solar minimum based on STEREO/SECCHI observations, *Solar Phys.* **264**, pp. 365–375.

Parenti, S. (2014). Solar Prominences: observations, *Living Rev. Solar Phys.* **11**, pp. 1–88.

Pariat, E., Antiochos, S. K., and DeVore, C. R. (2009). A model for solar polar jets, *Astrophys. J.* **691**, pp. 61–74.

Pariat, E., Antiochos, S. K., and DeVore, C. R. (2010). Three-dimensional modeling of quasihomologous solar jets, *Astrophys. J.* **714**, pp. 1762–1778.

Pariat, E., Dalmasse, K., DeVore, C. R., Antiochos, S. K., and Karpen, J. T. (2015). Model for straight and helical solar jets – I. Parametric studies of the magnetic field geometry, *Solar Phys.* **573**, A130 (15pp).

Pariat, E., Dalmasse, K., DeVore, C. R., Antiochos, S. K., and Karpen, J. T. (2016). Model for straight and helical solar jets. II. Parametric study of the plasma beta. *Astron. Astrophys.* **596**, A36 (20pp).

Pariat, E., Leake, J. E., Valori, G., Linton, M. G., Zuccarello, F. P., and Dalmasse, K. (2017). Relative magnetic helicity as a diagnostic of solar eruptivity, *Astron. Astrophys.* **601**, A125 (16pp).

Parker, E. N. (1957). Sweet's Mechanism for merging magnetic fields in conducting fluids, *J. Geophys. Res.* **62**, pp. 509–520.

Pataraya, A. D. and Zaqarashvili, T. V. (1995). Generation of waves by the solar magnetic field polarity reversal, *Solar Phys.* **157**, pp. 31–44.

Patsourakos, S., Pariat, E., Vourlidas, A., Antiochos, S. K., and Wuelser, J. P. (2008). *STEREO* SECCHI stereoscopic observations constraining the initiation of polar coronal jets, *Astrophys. J.* **680**, pp. L73–L76.

Patsourakos, S., Vourlidas, A., and Stenborg, G. (2013). Direct evidence for a fast coronal mass ejection driven by the prior formation and subsequent destabilization of a magnetic flux rope, *Astrophys. J.* **764**, 125 (13pp).

Pereira, T. M. D., De Pontieu, B., and Carlsson, M. (2012). Quantifying spicules, *Astrophys. J.* **759**, 18 (16pp).

Pereira, T. M. D., De Pontieu, B., and Carlsson, M., *et al.* (2014). An *Interface Region Imaging Spectrograph* first view on solar spicules, *Astrophys. J.* **792**, L15 (6pp).

Pereira, T. M. D., Rouppe van der Voort, L., and Carlsson, M. (2016). The appearance of spicules in high resolution observations of Ca ii H and Hα, *Astrophys. J.* **824**, 65 (9pp).

Pesnell, D. W., Thompson, B. J., and Chamberlin, P. C. (2012). The *Solar Dynamics Observatory* (SDO), *Solar Phys.* **275**, pp. 3–15.

Petschek, H. E. (1964). Magnetic field annihilation. In: Hess, W. N. (Ed.) *The Physics of Solar Flares*, NASA **SP-50**, pp. 425–439.

Pietarila, A., Aznar Cuadrado, R., Hirzberger, J. and Solanki, S. K. (2011). Kink waves in an active region dynamic fibril, *Astrophys. J.* **739**, 92, (7pp).

Pike, C. D. and Harrison, R. A. (1997). Euv observations of a macrospicule: Evidence for solar wind acceleration?, *Solar Phys.* **175**, pp. 457–465.

Pike, C. D. and Mason, H. E. (1998). Rotating transition region features observed with the SOHO coronal diagnostic spectrometer, *Sol. Phys.* **182**, pp. 333–348.

Priest, E. R. and Forbes, T. G. (2002). The magnetic nature of solar flares, *Astron. Astrophys. Rev.* **10**, pp. 313–377.

Priest, E. R., Parnell, C. E., and Martin, S. F. (1994). A converging flux model of an X-ray bright point and an associated canceling magnetic feature, *Astrophys. J.* **427**, pp. 459–474.

Priest, E. (2014) *Magnetohydrodynamics of the Sun* (Cambridge University Press, Cambridge), Chs. 1, 2, 4, 7.

Pucci, S., Poletto, G., Sterling, A. C. and Romoli, M. (2013). Physical parameters of standard and blowout jets, *Astrophys. J.* **776**, 16 (12pp).

Rae, I. C. (1983). The MHD Kelvin–Helmholtz instability in the solar photosphere, *Astron. Astrophys.* **126**, pp. 209–215.

Raouafi, N.-E., Patsourakos, S., Pariat, E., Young, P. R., Sterling, A. C., Savcheva, A., Shimojo, M., Moreno-Insertis, F., DeVore, C. R., Archontis, V., Török, T., Mason, H., Curdt, W., Meyer, K., Dalmasse, K., and Matsui, Y. (2016). Solar coronal jets: observations, theory, and modeling, *Space Sci. Rev.* **201**, pp. 1–53.

Rayleigh, L. (1882). Investigation of the character of the equilibrium of an incompressible heavy fluid of variable density, *Proc. London Math. Soc.* **14**, pp. 170–177.

Roberts, B. (2019). *MHD Waves in the Solar Atmosphere* (Cambridge University Press, Cambridge), Chs. 3–12.

Roberts, B. and Webb, A. R. (1978). Vertical motions in an intense magnetic flux tube, *Solar Phys.* **56**, pp. 5–35.

Rouppe van der Voort, L. H. M., De Pontieu, B., Hansteen, V. H., Carlsson, M., and van Noort, M. (2007). Magnetoacoustic shocks as driver of quiet sun mottles, *Astrophys. J.* **660**, pp. L169–L172.

Rouppe van der Voort, L., B. De Pontieu, B., T. M. D. Pereira, T. M. D., Carlsson, M., and Hansteen, V. (2015). Heating signatures in the disk counterparts of solar spicules in interface region imaging spectrograph observations, *Astrophys. J.* **799**, L3 (6pp).

Roy, J.-René. (1973a). The dynamics of solar surges, *Solar Phys.* **32**, pp. 139–151.

Roy, J.-René. (1973b). The magnetic properties of solar surges, *Solar Phys.* **28**, pp. 95–114.

Ryutova, M. (2018). *Physics of Magnetic Flux Tubes* (Springer Nature Switzerland, Cham), Chapters. 2 and 3.

Sakurai, T., Goossens, M., and Hollweg, J. V. (1991). Resonant behaviour of MHD waves on magnetic flux tubes. I. Connection formulae at the resonant surfaces, *Solar Phys.* **133** pp. 227–245.

Savcheva, A., Cirtain, J., DeLuca, E. E., *et al.* (2007). A study of polar jet parameters based on Hinode XRT observations, *Publ, Astron. Soc. Japan* **59**, pp. S771–S778.

Schatten, K. H., Wilcox, J. M., and Ness, N. F. (1969). A model of interplanetary and coronal magnetic fields, *Solar Phys.* **6**, pp. 442–455.

Schmieder, B., Mein, P., Simnett, G. M., and Tandberg-Hanssen, E. (1988). An example of the association of X-ray and UV emission with H-alpha surges, *Astron. Astrophys.* **201**, pp. 327–338.

Schmieder, B., Shibata, K., van Driel-Gesztelyi, L., and Freeland, S. (1994). Hα surges and associated Soft X-ray loops, *Solar Phys.* **156**, pp. 245–264.

Schmieder, B., Guo, Y., Moreno-Insertis, F., *et al.* (2013). Twisting solar coronal jet launched at the boundary of an active region, *Astron. Astrophys.* **559**, A1 (11p).

Schou, J., Scherrer, P. H., Bush, R. I., and Wachter, R. (2012). Design and ground calibration of the *Helioseismic and Magnetic Imager* (HMI) instrument on the *Solar Dynamics Observatory* (SDO), *Solar Phys.* **275**, pp. 229–259.

Schrijver, C. J., DeRosa, M. L., Metcalf, T., *et al.* (2008). Nonlinear force-free field modeling of a solar active region around the time of a major flare and coronal mass ejection, *Astrophys. J.* **675**, pp. 1637–1644.

Schwabe, S. H. (1843). Sonnenbeobachtungen im Jahre 1843, *Astron. Nachr.* **21**, pp. 233–236.

Secchi, A. (1877). *Le Soleil* Vol. 1 (Gauthier-Villars, Paris).

Senteno, R., Bueno, J. T., and Ramos, A. A. (2010). On the magnetic field of off-limb spicules, *Astrophys. J.* **708**, 1579–1584.

Shafranov, V. D. (1956). Stability of a cylindrical gaseous conductor in a magnetic field, *At. Energy* **5**, pp. 38–41.

Sheeley Jr., N. R. (2005). Surface evolution of the sun's magnetic field: A historical review of the flux-transport mechanism, *Living Rev. Solar Phys.* **2**, 5 (27pp).

Shetye, J., Doyle, J. G., Scullion, E., Nelson, C. J., Kuridze, D., Henriques, V., Woeger, F., and Ray, T. (2016). High-cadence observations of spicular-type events on the Sun, *Astron. Astrophys.* **589** A3 (15pp).

Shibata, K., Ishido, Y., Acton, L. W., Strong, K. T., Hirayama, T., Uchida, Y., McAllister, A. H., Matsumoto, R., Tsuneta, S., Shimizu, T., Hara, H., Sakurai, T., Ichimoto, K., Nishino, Y., and Ogawara, Y. (1992). Observation of X-ray jets with the yohkoh Soft X-ray telescope, *Publ. Astron. Soc. Japan* **44**, pp. L173–L179.

Shibata, K., Nozawa, S., and Matsumoto, R. (1992a). Magnetic reconnection associated with emerging magnetic flux, *Publ. Astron. Soc. Japan* **44**, pp. 265–272.

Shibata, K., Nitta, N., Strong, K. T., Matsumoto, R., Yokoyama, T., Hirayama, T., Hudson, H., and Ogawara, Y. (1994). A gigantic coronal jet ejected from a compact active region in a coronal hole, *Astrophys. J.* **431**, pp. L51–L53.

Shibata, K. (2000). "Solar X-ray Jets." In: Maran, S. P. (Ed.), *Encyclopedia of Astronomy and Astrophysics* (Cambridge University Press, Cambridge), p. 2272.

Shibata, K., Nakamura, T., Matsumoto, T., *et al.* (2007). Chromospheric anemone jets as evidence of ubiquitous reconnection, *Science* **318**, pp. 1591–1594.

Shimojo, M., Hashimoto, S., Shibata, K., Hirayama, T., Hudson, H. S., and Acton, L. W. (1996). Statistical study of solar X-ray jets observed with the Yohkoh Soft X-Ray Telescope. *Publ. Astron. Soc. Japan* **48**, pp. 123–136.

Shimojo, M. and Shibata, K. (2000). Physical Parameters of Solar X-Ray Jets, *Astrophys. J.* **542**, pp. 1100–1108.

Shimojo, M., Shibata, K., Yokoyama, T., and Hori, K. (2001). One-dimensional and pseudo-two dimensional hydrodynamic simulations of solar X-ray jets, *Astrophys. J.* **550**, pp. 10510–1063.

Skogsrud, H., Rouppe van der Voort, L., De Pontieu, B., and Pereira, T. M. D. (2015). On the temporal evolution of spicules observed with *IRIS*, *SDO*, and *Hinode*, *Astrophys. J.* **806**, 170 (10pp).

Song, H. Q., Zhang, J., Chen, Y., and Cheng, X. (2014). Direct observations of magnetic flux rope formation during a solar coronal mass ejection, *Astrophys. J.* **792**, L40 (6pp).

Spruit, H. C. (1982). Propagation speeds and acoustic damping of waves in magnetic flux tubes, *Solar Phys.* **75**, pp. 3–17.

Srivastava, A. K., Zaqarashvili, T. V., Kumar, Pankaj, and Khodachenko, M. L. (2010). Observation of kink instability during small B5.0 solar flare on 04 June, 2007, *Astrophys. J.* **715**, pp. 292–299.

Srivastava, A. K. and Murawski, K. (2011). Observations of a pulse-driven cool polar jet by SDO/AIA, *Astron. Astrophys.* **534**, A62 (7pp).

Srivastava, A. K., Shetye, J., Murawski, K., *et al.* (2017). High-frequency torsional Alfvén waves as an energy source for coronal heating, *Sci. Reports* **7**, 43147 (6pp).

Sterling, A. C. (2000). Solar spicules: A review of recent models and targets for future observations—(Invited Review), *Solar Phys.* **196**, pp. 79–111.

Sterling, A. C. and Moore, R. L. (2004). Evidence for gradual external reconnection before explosive eruption of a solar filament, *Astrophys. J.* **602**, pp. 1024–1036.

Sterling, A. C., Moore, R. L., and DeForest, C. E. (2010). *Hinode* solar optical telescope observations of the source regions and evolution of "type II" spicules at the solar polar limb, *Astrophys. J.* **714**, pp. L1–L6.

Sterling, A. C., Moore, R. L., Falconer, D. A., and Adams, M. (2015). Small-scale filament eruptions as the driver of X-ray jets in solar coronal holes, *Nature* **523**, pp. 437–440.

Sterling, A. C., Moore, R. L., Falconer, D. A., Panesar, N. K., Akiyama, S., Yashiro, S., and Gopalswamy, N. (2016). Minifilament eruptions that drive coronal jets in a solar active region, *Astrophys. J.* **821**, 100 (17pp).

Sterling, A. C., Moore, R. L., Falconer, D. A., Panesar, N. K., and Martinez, F. (2017). Solar active region coronal jets. II. triggering and evolution of violent jets. *Astrophys. J.* **844**, 28 (20pp).

Stix, M. (2002). *The Sun: An Introduction*, Second edition (Springer, Berlin), Chs. 1–3.

Sturrock, P. A. (1996). Model of the high-energy phase of solar flares, *Nature* **211**, pp. 695-697.

Su, Y. and van Ballegooijen, A. (2012). Observations and magnetic field modeling of a solar polar crown prominence, *Astrophys. J.* **757**, 168 (17pp).

Švestka, Z. (1976). *Solar Flares* (Reidel, Dordrecht).

Sweet, P. A. (1958). The neutral point theory of solar flares. In: Lehnert, B. (Ed.) *Electromagnetic Phenomena in Cosmical Physics* (Cambridge University Press, Cambridge), pp. 123–134.

Tandberg-Hassen, E. (1995). Physical parameters of the prominence plasma. In: *The Nature of Solar Prominences* Astrophysics and Space Science Library **199** (Springer, Dordrecht), pp. 81–111.

Tavabi, E., Koutchmy, S., and Ajabshirizadeh, A. (2011). A statistical analysis of the SOT-Hinode observations of solar spicules and their wave-like behavior, *New Astron.*, **16**, pp. 296–305.

Taylor, G. I. (1950). The instability of liquid surfaces when accelerated in a direction perpendicular to their planes, *Proc. Roy. Soc. London. Series A, Math. Phys. Sci.* **201**, pp. 192–196.

Temmer, M., Veronig, A. M., Kontar, E. P., Krucker, S., and Vršnak, B. (2010). Combined STEREO/RHESSI study of CME acceleration and particle acceleration in solar flares, *Astrophys. J.* **712**, pp. 1410–1420.

Terra-Homem, M., Erdélyi, R., and Ballai, I. (2003). Linear and non-linear MHD wave propagation in steady-state magnetic cylinders, *Solar Phys.* **217**, pp. 199–223.

Titov, V. S., Priest, E. R., and Demoulin, P. (1993). Conditions for the appearance of "bald patches" at the solar surface, *Astron. Astrophys.* **276**, pp. 564–570.

Török, T. and Kliem, B. (2005). Confined and ejective eruptions of kink-unstable flux ropes, *Astrophys. J.* **630**, pp. L97–L100.

Török, T., Aulanier, G., Schmieder, B., Reeves, K. K., and Golub, L. (2009). Fan-spine topology formation through two-step reconnection driven by twisted flux emergence, *Astrophys. J.* **704**, pp. 485–495.

Tsiropoula, G. and Schmieder, B. (1997). Determination of physical parameters in dark mottles, *Astron. Astrophys.* **324**, pp. 1183–1189.

Tsiropoula, G., Alissandrakis, C. E. and Schmieder, B. (1993). The fine structure of chromospheric rosette, *Astron. Astrophys.* **271**, pp. 574–586.

Tsiropoula, G., Madi, C., Schmieder, B., and Preka-Papadema, P. (1999). Analysis of Hα profiles. Physical parameters of chromospheric mottles: A case study, *Astron. Astrophys. Trans.* **18**, pp. 455–461.

Tsiropoula, G. and Tziotziou, K. (2004). The role of chromospheric mottles in the mass balance and heating of the solar atmosphere, *Astron. Astrophys.* **424**, pp. 279–288.

Tsiropoula, G., Tziotziou, K., Kontogiannis, I., Madjarska, M. S., Doyle, J. G., and Suematsu, Y. (2012). Solar fine-scale structures. I. Spicules and other small-scale, jet-like events at the chromospheric level: Observations and physical parameters, *Space Sci. Rev.* **169**, pp. 181–244.

Tsuneta, S., Acton, L., Bruner, M., *et al.* (1991). The Soft X-ray Telescope for the SOLAR-A mission, *Solar Phys.* **136**, pp. 37–67.

Tziotziou, K., Tsiropoula, G., and Mein, P. (2003). On the nature of the chromospheric fine structure. I. Dynamics of dark mottles and grains, *Astron. Astrophys.* **402**, pp. 361–372.

Tziotziou, K., Tsiropoula, G., and Sütterlin, P. (2005). Flare footpoint regions and a surge observed by *Hinode*/EIS, *RHESSI*, and *SDO*/AIA, *Astron. Astrophys.* **444**, pp. 265–274.

Uddin, W., Schmieder, B., Chandra, R., Srivastava, A. K., Kumar, P., and Bisht, S. (2012). Observations of multiple surges associated with magnetic activities in AR 10484 on 2003 October 25, *Astrophys. J.* **752**, 70 (10pp).

van Ballegooijen, A. A., Asgari-Targhi, M., Cranmer, S. R., and DeLuca, E. E. (2011). Heating of the solar chromosphere and corona by Alfvén wave turbulence, *Astrophys. J.* **736**, 3 (27pp).

Vargas Dominguez, S., Kosovichev, A. G., and Yurchyshyn, V. (2014). Emergence of a small-scale magnetic flux tube and the response of the solar atmosphere, *Cent. Eur. Astrophys. Bull.* **38**, pp. 25–30.

Vasheghani Farahani, S., Van Doorsselaere, T., Verwichte, E., and Nakariakov, V. M. (2009). Propagating transverse waves in soft X-ray coronal jets, *Astron. Astrophys.* **498**, pp. L29–L32.

Vemareddy, P. and Zhang, J. (2014). Initiation and eruption process of magnetic flux rope from solar active region NOAA 11719 to earth-directed CME, *Astrophys. J.* **797**, 80 (12pp).

Verth, G., Goossens, M., and He, J.-S. (2011). Magnetoseismological determination of magnetic field and plasma density height variation in a solar spicule, *Astrophys. J.* **733**, L15, (5pp).

Vourlidas, A., Lynch, B. J., Howard, R. A., and Li, Y. (2013). How many CMEs have flux ropes? Deciphering the signatures of shocks, flux ropes, and prominences in coronagraph observations of CMEs, *Solar Phys.* **284**, pp. 179–201.

Vourlidas, A. (2014). The flux rope nature of coronal mass ejections, *Plasma Phys. Control. Fusion* **56**, 064001 (6pp).

Wang, H. and Liu, C. (2012). Circular ribbon flares and homologous jets, *Astrophys. J.* **760**, 101 (9pp).

Wang, Y.-M., and Sheeley, Jr., N. R. (1992). On potential field models of the solar corona, *Astrophys. J.* **392**, pp. 310–319.

Wang, Y.-M., and Sheeley, Jr., N. R. (2002). Coronal white-light jets near sunspot maximum, *Astrophys. J.* **575**, pp. 542–552.

Wang, Y.-M., Sheeley, Jr., N. R., Socker, D. G., *et al.* (1998). Observations of correlated white-light and extreme-ultraviolet jets from polar coronal holes, *Astrophys. J.* **508**, pp. 899–907.

Webb, D. F. and Howard, T. A. (2012). Coronal mass ejections: observations, *Living Rev. Solar Phys.* **9**, 3–83.

Wilhelm, K., Curdt, W., Marsch, E., *et al.* (1995). SUMER – Solar ultraviolet measurements of emitted radiation, *Solar Phys.* **162**, pp. 189–231.

Williams, D. R., Török, T., Demoulin, P., van Driel-Gesztelyi, L., and Kliem, B. (2005). Eruption of a kink-unstable filament in NOAA Active Region 10696. *Astrophys. J.* **628**, pp. L163–L166.

Wolf, R. (1848). Sonnenflecken-Beobachtungen, *Mitt. Nat. Forsch. Ges. Bern* **130**, p. 169.

Woodford, C. and Phillips, C. (1997). *Numerical Methods with Worked Examples* (Chapman & Hall, London), Chapter 2.

Wyper, P. F., Antiochos, S. K., and DeVore, C. R. (2017). A universal model for solar eruptions, *Nature* **544**, pp. 452–455.

Wyper, P. F., DeVore, C. R., Karpen, J. T., Antiochos, S. K., and Yeates, A. R. (2017). A model for coronal hole bright points and jets due to moving magnetic elements, *Astrophys. J.* **864**, 174 (16pp).

Yang, J.-Y., Jiang, Y.-C., Yang, D., Bi, Y., Yang, B., Zheng, R.-S., and Hong, J.-C. (2012). The surge-like eruption of a miniature filament, *Res. Astron. Astrophys.* **12**, pp. 300–312.

Yang, L.-H., Jiang, Y.-C., Yang, J.-Y., Bi, Y., Zheng, R.-S., and Hong, J.-C. (2011). Observations of EUV and soft X-ray recurring jets in an active region, *Res. Astron. Astrophys.* **11**, pp. 1229–1242.

Yokoyama, T. and Shibata, K. (1995). Magnetic reconnection as the origin of X-ray jets and Hα surges on the Sun, *Nature* **375**, pp. 42–44.

Yokoyama, T. and Shibata, K. (1996). Numerical simulation of solar coronal X-ray jets based on magnetic reconnection model, *Publ. Astron. Soc. Japan* **48**, pp. 353–376.

Yoshimura, K, Kurokawa, H., Shimojo, M., and Shine, R. (2003). Close correlation among Hα surges, magnetic flux cancellations, and UV brightenings found at the edge of an emerging flux region, *Publ. Astron. Soc. Japan* **55**, pp. 313–320.

Young, P. R., Del Zanna, G., Mason, H. E., *et al.* (2007). EUV emission lines and diagnostics observed with Hinode/EIS, *Publ. Astron. Soc. Japan* **59**, pp. S857–S864.

Young, P. R., and Muglach, K. (2014a). A coronal hole jet observed with Hinode and the Solar Dynamics Observatory, *Publ. Astron. Soc. Japan* **66**, S12 (9pp).

Young, P. R. and Muglach, K. (2014b). *Solar Dynamics Observatory* and *Hinode* observations of a blowout jet in a coronal hole, *Sol. Phys.* **289**, pp. 3313–3329.

Zachariadis, Th. G., Dara, H. C., Alissandrakis, C. E., Koutchmy, S., and Gontikakis, C. (2001). Fine structure of the quiet solar chromosphere: Limb-crossing features, *Solar Phys.* **202**, pp. 41–52.

Zank, G. P. and Matthaeus, W. H. (1993). Nearly incompressible fluids. II: Magnetohydrodynamics, turbulence, and waves, *Phys. Fluids* **5**, pp. 257–273.

Zaqarashvili, T. V., Khutsishvili, E., Kukhianidze, V., and Ramishvili, G. (2007). Doppler-shift oscillations in solar spicules, *Astron. Astrophys.* **474**, pp. 627–632.

Zaqarashvili, T. V. and Erdélyi, R. (2009). Oscillations and waves in solar spicules, *Space Sci. Rev.* **149**, pp. 355–388.

Zaqarashvili, T. V., Díaz, A. J., Oliver, R., and Ballester, J. L. (2010). Instability of twisted magnetic tubes with axial mass flows, *Astron. Astrophys.* **516**, A84 (8pp).

Zaqarashvili, T. V., Vörös, Z., and Zhelyazkov, I. (2014a). Kelvin–Helmholtz instability of twisted magnetic flux tubes in the solar wind, *Astron. Astrophys.* **561**, A62 (7pp).

Zaqarashvili, T. V., Vörös, Z., Narita, Y., and Bruno, R. (2014b). Twisted magnetic flux tubes in the solar wind, *Astrophys. J.* **783**, L19 (5pp).

Zaqarashvili, T. V., Zhelyazkov, I., and Ofman, L. (2015). Stability of rotating magnetized jets in the solar atmosphere. I. Kelvin–Helmholtz instability, *Astrophys. J.* **813**, 123 (13pp).

Zhang, M. and Low, B. C. (2005). The hydromagnetic nature of solar coronal mass ejections, *Ann. Rev. Astron. Astrophys.* **43**, pp. 103–137.

Zhang, J., Richardson, I. G., Webb, D. F., *et al.* (2007) Solar and interplanetary sources of major geomagnetic storms ($Dst \leq -100$ nT) during 1996–2005, *J. Geophys. Res.: Space Phys.* **112**, A10102 (19pp).

Zhang, J., Cheng, X., and Ding, M. (2016). Observation of an evolving magnetic flux rope before and during a solar eruption, *Nat. Commun.* **3**, 747 (5pp).

Zhang, Q. M. and Ji, H. S. (2014). A swirling flare-related EUV jet, *Astron. Astrophys.* **561**, A134 (7pp).

Zhelyazkov, I. (2008). Waves and stability of flowing solar wind structures in the framework of the Hall magnetohydrodynamics. In: Hartfuss, H.-J., Dudeck, M., J. Musielok, J., Sadowski, M. J. (Eds), *PLASMA 2007: International Conference on Research and Applications of Plasmas, Greifswald, 16–19 October 2007, AIP Conf. Proc.* **993**, pp. 281–284.

Zhelyazkov, I. (2009) MHD waves and instabilities in flowing solar flux-tube plasmas in the framework of Hall magnetohydrodynamics, *Eur. Phys. J. D* **55**, pp. 127–137.

Zhelyazkov, I. (2010) Hall-magnetohydrodynamic waves in flowing ideal incompressible solar-wind plasmas, *Plasma Phys. Control. Fusion* **52**, 065008 (16pp).

Zhelyazkov, I. (2012). Magnetohydrodynamic waves and their stability status in solar spicules, *Astron. Astrophys.* **537**, A124 (8pp).

Zhelyazkov, I. and Zaqarashvili, T. V. (2012). Kelvin–Helmholtz instability of kink waves in photospheric twisted flux tubes. *Astron. Astrophys.* **547**, A14 (11pp).

Zhelyazkov, I. (2013). Kelvin–Helmholtz instability of kink waves in photospheric, chromospheric, and X-ray solar jets. In: Zhelyazkov, I., Mishonov, T. (Eds), *SPACE PLASMA PHYSICS: Proceedings of the 4th School and Workshop on Space Plasma Physics, Albena, 19–26 August 2012, AIP Conf. Proc.* **1551**, pp. 150–164.

Zhelyazkov, I. (2015). On modeling the Kelvin–Helmholtz instability in solar atmosphere, *J. Astrophys. Astr.* **36**, pp. 233–254.

Zhelyazkov, I., Chandra, R., Srivastava, A. K., and Mishonov, T. (2015a). Kelvin–Helmholtz instability of magnetohydrodynamic waves propagating on solar surges, *Astrophys. Space Sci.* **356**, pp. 231–240.

Zhelyazkov, I., Zaqarashvili, T. V., Chandra, R., Srivastava, A. K., and Mishonov, T. (2015b). Kelvin–Helmholtz instability in solar cool surges, *Adv. Space Res.* **56**, pp. 2727–2737.

Zhelyazkov, I., Zaqarashvili, T. V., and Chandra, R. (2015c). Kelvin–Helmholtz instability in coronal mass ejecta in the lower corona, *Astron. Astrophys.* **575**, A55 (7pp).

Zhelyazkov, I., Chandra, R., and Srivastava, A. K. (2016). Kelvin–Helmholtz instability in an active region jet observed with *Hinode*, *Astrophys. Space Sci.* **361**, 51 (14pp).

Zhelyazkov, I., Chandra, R., and Srivastava, A. K. (2017). Modeling Kelvin–Helmholtz instability in soft X-ray solar jets. *Adv. Astron.* **2017**, 2626495 (18pp).

Zhelyazkov, I., Zaqarashvili, T. V., Ofman, L., and Chandra, R. (2018). Kelvin–Helmholtz instability in a twisting solar polar coronal hole jet observed by *SDO*/AIA, *Adv. Space Res.* **61**, pp. 628–638.

Zhelyazkov, I. and Chandra, R. (2018). High mode magnetohydrodynamic waves propagation in a twisted rotating jet emerging from a filament eruption, *Mon. Not. R. Astron. Soc.* **478**, pp. 5505–5518.

Zhelyazkov, I. and Chandra, R. (2019). Can high-mode magnetohydrodynamic waves propagating in a spinning macrospicule be unstable due to the Kelvin–Helmholtz Instability? *Solar Phys.* **294**, 20 (19pp).

Zuccarello, F., Guglielmino, S. L., and Romano, P. (2011). Magnetic reconnection structures in the solar atmosphere: results from multi-wavelength observations, *Mem. Soc. Astron. Ital.* **82**, pp. 149–153.

Index

9 789811 223747